SECOND EDITION

Tunable Laser Optics

SECOND EDITION

Tunable
Laser
Optics

F.J. Duarte
INTERFEROMETRIC OPTICS, ROCHESTER, NEW YORK, USA

CRC Press
Taylor & Francis Group
Boca Raton London New York

CRC Press is an imprint of the
Taylor & Francis Group, an **informa** business

CRC Press
Taylor & Francis Group
6000 Broken Sound Parkway NW, Suite 300
Boca Raton, FL 33487-2742

First issued in paperback 2017

© 2015 by Taylor & Francis Group, LLC
CRC Press is an imprint of Taylor & Francis Group, an Informa business

No claim to original U.S. Government works

ISBN-13: 978-1-4822-4529-5 (hbk)
ISBN-13: 978-1-138-89375-7 (pbk)

Visit the Taylor & Francis Web site at
http://www.taylorandfrancis.com

and the CRC Press Web site at
http://www.crcpress.com

Dedication

To my family.

Contents

List of Figures

List of Tables

Preface

Broadly tunable lasers have had, and continue to have, an enormous impact in many and diverse fields of science and technology. From a renaissance in spectroscopy, to laser guide stars, to laser cooling, the nexus is the tunable laser. *Tunable Laser Optics* was conceived from a utilitarian perspective to distill into a single, and concise, volume the necessary optics to provide the foundations necessary to work efficiently and productively in an environment employing tunable lasers. The theoretical tools presented in this book use humble, practical mathematics. Mainly derived from the application of Dirac's notation, these tools are widely applicable throughout optics: from interference, to diffraction, to refraction, to dispersion, and to reflection. Although the emphasis is on optics involving macroscopic low-divergence, narrow-linewidth lasers, some of the principles described are also applicable to the microscopic and nanoscopic domains.

The style, and selection of subject matter, in this book was determined by a desire to reduce entropy in the search for information in this wonderful and fascinating field. Albeit the physics and the optics included are based on firm and ageless principles, this second edition offers a revised, expanded, and updated version of the first edition published in 2003.

F. J. Duarte
Interferometric Optics

Author

F. J. Duarte graduated with first-class honors in physics from the School of Mathematics and Physics at Macquarie University, Sydney, Australia, where he was also awarded a PhD in physics for his research on optically pumped molecular lasers. At Macquarie, he was a student of the well-known quantum physicist J. C. Ward. Duarte's research has taken place at a number of institutions in academia, industry, and the defense establishment. He is a research physicist at Interferometric Optics, Rochester, New York, and an adjunct professor at the electrical and computer engineering department, University of New Mexico, Albuquerque, New Mexico. He is the author of the generalized multiple-prism dispersion theory and has made various unique contributions to the physics and architecture of tunable laser oscillators. He has also pioneered the use of Dirac's quantum notation in interferometry, oscillator physics, and classical optics. These contributions have found applications in the design of laser resonators, laser pulse compression, imaging, microscopy, medicine, optics communications, and the nuclear industry. He is the lead author of numerous refereed papers and several US patents. Duarte is the author and editor of *Dye Laser Principles* (Academic Press, 1990), *High-Power Dye Lasers* (Springer-Verlag, 1991), *Selected Papers on Dye Lasers* (SPIE, 1992), *Tunable Lasers Handbook* (Academic Press, 1995), *Tunable Laser Applications* (1st edition, Marcel Dekker, 1995; 2nd edition, CRC Press, 2009), and *Coherence and Ultrashort Pulsed Laser Emission* (InTech, 2010). He is also the sole author of *Tunable Laser Optics* (1st edition, Elsevier Academic Press, 2003) and *Quantum Optics for Engineers* (CRC Press, 2014). Dr. Duarte is a fellow of the Australian Institute of Physics and a fellow of the Optical Society of America. He received the Engineering Excellence Award from the Optical Society of America.

1 Introduction to Lasers

1.1 INTRODUCTION

Lasers are widely applied in academic, medical, industrial, and military research. They are also used beyond the boundary of research, in numerous engineering and manufacturing applications that continue to expand.

Optics principles and optical elements are applied to build laser resonators and to propagate laser radiation. Optical instruments are utilized to characterize laser emission and lasers have been incorporated into new optical instrumentation. This book focuses on the optics and optical principles needed to build lasers, the optical instrumentation necessary to characterize laser emission, and the laser-based optical instrumentation. The emphasis is on practical and utilitarian aspects of relevant optics including the necessary theory. Albeit this book refers explicitly to macroscopic lasers, many of the principles and ideas, described here, are applicable to microscopic lasers.

This book was written for advanced undergraduate physics students, non-optics graduate students using lasers, engineers, and scientists from other fields seeking to incorporate lasers and optics into their work.

This second edition is a revised, expanded, and improved version of its first edition (Duarte 2003). In addition to new and additional material on tunable lasers, this edition also includes some relevant topics of quantum optics as described in *Quantum Optics for Engineers* (Duarte 2014).

The organization of this book is fairly straightforward. It has three stages. It begins with an introduction to laser concepts and continues with Chapters 1 through 7 that introduce the ideas necessary to quantify the propagation of laser radiation and that are central to the design tunable laser oscillators. The second stage begins with Chapter 8 on nonlinear optics that has intra- and extracavity applications.

The attention is then focused on a survey of the emission characteristics of most well-known lasers with an added emphasis on tunable semiconductor and tunable solid-state lasers. The third stage includes Chapter 11 on interferometry that describes most well-known interferometers via Dirac's quantum notation. This chapter is followed by Chapter 12 on the principles of spectrometry and dispersion-based instrumentation. In each chapter, a number of examples are described in detail to illustrate the applications of the theory. A set of fairly straightforward problems is added at the end of each chapter to assist the reader to assess the assimilation of the subject matter.

Thus, the book begins with an introduction to some basic concepts of laser excitation mechanisms and laser resonators in Chapter 1. The focus then turns to optical principles with Dirac optics being discussed in Chapter 2 and the uncertainty principle introduced in Chapter 3. The principles of dispersive optics are described in Chapter 4, whereas classical polarization is discussed in Chapter 5. Next, propagation matrices are introduced in Chapter 6. The optical principles discussed in Chapters 1 through 6 can all be applied to the design and construction of tunable laser oscillators as described in Chapter 7. Nonlinear optics, with an emphasis on frequency conversion, is outlined in Chapter 8. A brief but fairly comprehensive survey of gas lasers, liquid lasers, solid-state lasers, and semiconductor lasers is given in Chapter 9. The attention is focused on the emission characteristics of the various lasers. At this second stage, it is hoped that the student, or reader, should have gained sufficient confidence and familiarity with the subject of laser optics as to select an appropriate gain medium and resonator architecture for its efficient use in an applied field. The optical architecture and applications of N-slit laser interferometers are considered in Chapter 10, whereas interferometric based diagnostic instrumentation is described in Chapter 11. Dispersion-based spectrometry is described in Chapter 12. The book concludes with Chapter 13 with a survey of useful physical constants and optical quantities.

It should be emphasized that the material in this book does not require mathematical tools above those available to a third-year undergraduate physics student. Also, perhaps with the exception of Chapter 7, individual chapters can be studied independently. Finally, in this second edition, some equations and figures that are deemed important are reproduced again, when needed, rather than forcing the reader to go back to the text to find that material, thus easing the lecture process.

1.1.1 HISTORICAL REMARKS

A considerable amount has been written about the history of the maser and the laser. For brief, and yet informative, historical summaries, the reader should refer to Willett (1974), Siegman (1986), Townes (1999), and Silfvast (2008). Here, remarks will be limited to mention that the first experimental visible laser was demonstrated by Maiman (1960), and this laser was an optically pumped solid-state laser. More specifically, it was a flashlamp-pumped ruby laser. This momentous development was followed shortly afterward by the introduction of the first electrically excited gas laser (Javan et al. 1961). This was the He–Ne laser emitting in the near infrared. From a practical perspective, the demonstration of these laser devices also signaled the birth of experimental laser optics since the laser resonators, or laser optical cavities, are an integral and essential part of the laser.

Two apparently unrelated publications on the laser are mentioned next. The first is the description that Dirac gave on interference in his book *The Principles of Quantum Mechanics* first published in 1930 (Dirac 1978). In his statement on interference, Dirac first refers to a source of monochromatic light, and then to *a beam of light consisting of a large number of photons*. In his discussion, it is this beam composed of a large number of *undistinguishable photons* that are divided and then recombined to undergo interference. In this regard, Dirac could have been describing a high-intensity laser beam with a very narrow linewidth (Duarte 1998). Regardless of the prophetic value of Dirac's description, this was probably the first discussion, in physical optics, including a coherent beam of light. In other words, Dirac wrote the first chapter in *Laser Optics* (Duarte 2003).

The second publication of interest is *The Feynman Lectures on Physics* authored by Feynman et al. (1965). In Chapter 9 of the volume on quantum mechanics, Feynman uses Dirac's notation to describe the quantum mechanics of stimulated emission. In Chapter 10, he applies that physics to several physical systems, including dye molecules. Notice that this was done just prior to the discovery of the dye laser by Sorokin and Lankard (1966) and Schäfer et al. (1966). In this regard, Feynman could have predicted the existence of the tunable laser (Duarte 2003). Further, he made accessible Dirac's quantum notation via his thought experiments on two-slit interference with electrons. This provided the foundations for the subject of *Dirac optics* described in Chapter 2 where the method outlined by Feynman is extended to generalized transmission gratings using photons rather than electrons.

1.2 LASERS

The word *laser* has its origin in an acronym of the words *light amplification* by *stimulated emission* of *radiation*. Although the laser is readily associated with the spatial and spectral coherence characteristics of its emission, to some the physical meaning of the concept still remains shrouded in mystery. Looking up the word *laser* in a good dictionary does not help much.

A laser is a device that transforms electrical energy, chemical energy, or incoherent optical energy, into coherent optical emission. This coherence is both spatial and spectral. Spatial coherence means a highly directional light beam, with little divergence, whereas spectral coherence means an extremely pure color of emission. An alternative way to cast this idea is to think of the laser as a device that transforms ordinary energy into an extremely well-defined form of energy in both the spatial and the spectral domain. However, this is only the manifestation of the phenomenon since the essence of this energy transformation lies in the device called laser.

Physically, the laser consists of an atomic or molecular gain medium optically aligned within an optical resonator or optical cavity as depicted in Figure 1.1. When excited by electrical energy, or optical energy, the atoms or molecules in the gain medium oscillate at optical frequencies. This oscillation is maintained and sustained by the optical resonator or optical cavity. In this regard, the laser is analogous to a mechanical or radio oscillator but oscillates at extremely high frequencies. For the green color of $\lambda = 500$ nm, the equivalent frequency is $v \approx 5.99 \times 10^{14}$ Hz. A direct comparison between a laser and a radio oscillator makes the atomic or molecular gain medium equivalent to the transistor and the elements of the optical cavity equivalent to the resistances, capacitances, and inductances. Thus, from a physical perspective, the gain medium in conjunction with the optical cavity behaves like an *optical*

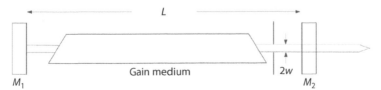

FIGURE 1.1 Basic laser resonator. It is composed of an atomic, or molecular, gain medium and two mirrors aligned along the optical axis. The length of the cavity is L, and the diameter of the beam is $2w$. The gain medium can be excited optically or electrically.

oscillator (Duarte 1990a). The spectral purity of the emission of a laser is related to how narrow its linewidth is. High-power narrow-linewidth lasers can have linewidths of $\Delta v \approx 300$ MHz, low-power narrow-linewidth lasers can have $\Delta v \approx 100$ kHz, while stabilized lasers can yield $\Delta v \approx 1$ kHz, or even much narrower linewidths. In all the instances mentioned here, the emission is in the form of a single longitudinal mode, that is, all the emission radiation is contained in a single electromagnetic mode.

In the language of the laser literature, a laser emitting narrow-linewidth radiation is referred to as a *laser oscillator* or a *master oscillator* (MO). High-power narrow-linewidth emission is attained when a MO is used to inject a *laser amplifier* or a *power amplifier* (PA). Large high-power systems include several MOPA chains, with each chain including several amplifiers. The difference between an oscillator and an amplifier is that the amplifier simply stores energy to be released up on the arrival of the narrow-linewidth oscillator signal. In some cases, the amplifiers are configured within unstable resonator cavities in what is referred to as a *forced oscillator* (FO). When that is the case, the amplifier is called a FO and the integrated configuration is referred to as a MOFO system. This subject is considered in more detail in Chapter 7.

1.2.1 LASER OPTICS

Laser optics refers to the individual optics elements that comprise laser cavities, the optics ensembles that comprise laser cavities, and the physics that results from the propagation of laser radiation. In addition, the subject of laser optics includes instrumentation employed to characterize laser radiation and instrumentation that incorporates lasers.

One of the main functions, and perhaps the principal function, of skillfully designed properly applied laser optics is to produce high-quality coherent emission for utilitarian purposes. The essence of high-quality coherent radiation is single-transverse-mode (TEM_{00}), and narrow-linewidth single-longitudinal-mode (SLM), emission.

1.2.2 LASER CATEGORIES

Lasers emitting radiation confined in TEM_{00} beams, and characterized as narrow-linewidth emission sources, are used in a variety of applications in research and development laboratories. These lasers can be categorized in two classes:

1. Narrow-linewidth tunable lasers, used in applications such as communications, laser cooling, laser isotope separation, medicine, metrology, and spectroscopy
2. Narrow-linewidth fixed-frequency lasers, used in applications such as imaging, medicine, metrology, and material processing

Lasers emitting radiation confined in TEM_{00} beams, and characterized as ultra-short-pulse emission, are mainly used in spectroscopy and the study of light-matter interactions.

1.3 EXCITATION MECHANISMS AND RATE EQUATIONS

There are various methods and approaches to describe the dynamics of excitation in the gain media of lasers. Approaches range from complete quantum mechanical treatments to rate equation descriptions (Haken 1970). A complete survey of energy-level diagrams corresponding to gain media in the gaseous, liquid, and solid states is given by Silfvast (2008). Here, a basic description of laser excitation mechanisms is given using energy levels and classical rate equations applicable to tunable molecular gain media. The link to the quantum mechanical nature of the laser is made via the cross sections of the transitions.

1.3.1 RATE EQUATIONS

Rate equations are widely applied in physics and laser physics in particular. They, for example, can be used to describe and quantify the process of molecular recombination in metal vapor lasers or to describe the dynamics of the excitation mechanism in a multiple-level gain medium. Here, the basic concept of rate equations is introduced using dye laser gain media as a vehicle, but the principles also apply to gas and solid-state gain media. Organic laser dyes are rather large molecules with molecular weights ranging from ~175 to ~830 m_u. Dye lasers using this class of gain media have been shown to span with their emission the electromagnetic spectrum from the near ultraviolet to the near infrared. Figure 1.2 illustrates the wavelength coverage provided by three classes of dye molecular species: the coumarins, the xanthenes, and the cyanines. Figure 1.3 depicts the dye molecular structure of the laser dye coumarin 545 tetramethyl (C545T), which has a molecular weight of 430.56 m_u. In a polarization-selective mirror-grating cavity, C545T provides a fairly large laser intensity dynamic range and an excellent tunability in the $501 \leq \lambda \leq 574$ nm range as shown in Figure 1.4 (Duarte et al. 2006).

An ideally simplified two-level molecular system is depicted in Figure 1.5. Here, the pump excitation intensity $I_p(t)$ populates the upper energy level N_1 from the ground state N_0. Emission from the upper state is designated as $I_l(x,t,\lambda)$ since it

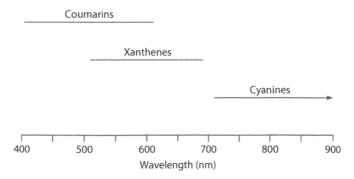

FIGURE 1.2 Approximate emission ranges for three classes of laser dyes: the coumarins, the xanthenes, and the cyanines. The rhodamines belong to the xanthenes. The cyanines reach up to 1100 nm. The wavelength range shown is covered using several laser dyes.

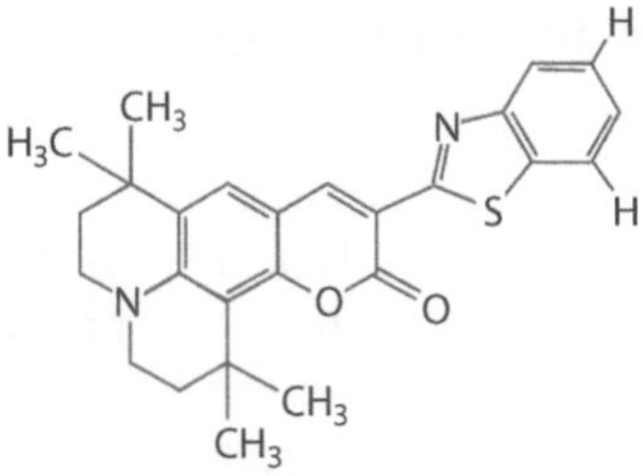

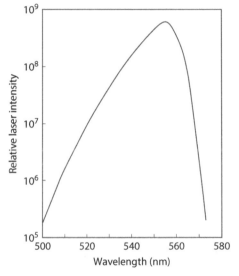

FIGURE 1.3 Molecular structure of C545T. This laser dye has a molecular weight of 430.56 m_u. (Reproduced with permission from Duarte, F.J., et al., *J. Opt. A: Pure Appl. Opt.*, 8: 172–174. Copyright 2006, Institute of Physics.)

FIGURE 1.4 Wavelength tuning range of a grating mirror laser cavity using C545T as gain medium. (Reproduced with permission from Duarte, F.J., et al., *J. Opt. A: Pure Appl. Opt.* 8: 172–174. Copyright 2006, Institute of Physics.)

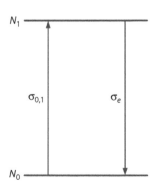

FIGURE 1.5 Simple two-level energy system including a ground level and an upper level.

is a function of position x in the gain medium, time t, and wavelength λ. The time evolution of the upper- or excited-state population can be written as

$$\frac{\partial N_1}{\partial t} = N_0 \sigma_{0,1} I_p(t) - N_1 \sigma_e I_l(x,t,\lambda) \tag{1.1}$$

which has a positive term due to excitation from the ground level and a negative component due to the emission from the upper state. Here, $\sigma_{0,1}$ is the absorption cross section and σ_e is the emission cross section. Cross sections have units of cm², time has units of seconds (s), the populations have units of molecules cm⁻³, and the intensities have units of photons cm⁻² s⁻¹.

The pump intensity $I_p(t)$ undergoes absorption due to its interaction with a molecular population N_0. A process that is described by the equation

$$c^{-1}\left[\frac{\partial I_p(t)}{\partial t}\right] = -N_0 \sigma_{0,1} I_p(t) \tag{1.2}$$

where:
  c is the speed of light

The process of emission is described by the time evolution of the intensity $I_l(x,t,\lambda)$ given by

$$c^{-1}\left[\frac{\partial I_l(x,t,\lambda)}{\partial t}\right] + \left[\frac{\partial I_l(x,t,\lambda)}{\partial x}\right] = \left(N_1 \sigma_e - N_0 \sigma_{0,1}^l\right) I_l(x,t,\lambda) \tag{1.3}$$

In the steady state, this equation reduces to

$$\left[\frac{\partial I_l(x,\lambda)}{\partial x}\right] \approx \left(N_1 \sigma_e - N_0 \sigma_{0,1}^l\right) I_l(x,\lambda) \tag{1.4}$$

which can be integrated to yield

$$I_l(x,\lambda) = I_l(0,\lambda) e^{\left(N_1 \sigma_e - N_0 \sigma_{0,1}^l\right)L} \tag{1.5}$$

Thus, if $N_1 \sigma_e > N_0 \sigma_{0,1}^l$ the intensity increases exponentially, and there is amplification that corresponds to laser-like emission. Exponential terms such as the one in Equation 1.5 are referred to as the *gain*.

1.3.2 DYNAMICS OF MULTIPLE-LEVEL SYSTEM

In this section, the rate equation approach is used to describe in some detail the excitation dynamics in a multiple-level energy system relevant to a well-known tunable molecular laser known as the dye laser. This approach applies to laser dye gain media in either the *liquid* or the *solid state*. The literature on rate equations for dye lasers is fairly extensive, which includes the works of Ganiel et al. (1975),

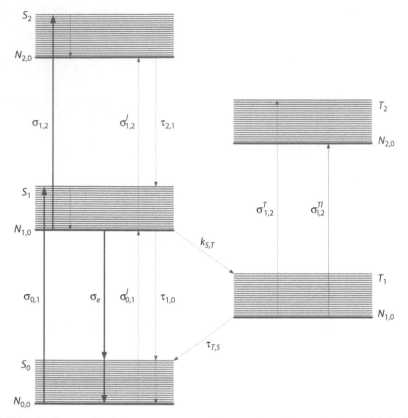

FIGURE 1.6   Energy-level diagram corresponding to a laser dye molecule, which includes three electronic levels (S_0, S_1, and S_2) and two triplet levels (T_1 and T_2). Each electronic level contains a large number of vibrational and rotational levels. Laser emission takes place due to $S_1 \rightarrow S_0$ transitions.

Teschke et al. (1976), Penzkofer and Falkenstein (1978), Dujardin and Flamant (1978), Munz and Haag (1980), Haag et al. (1983), Nair and Dasgupta (1985), Hillman (1990), Schäfer (1990), and Jensen (1991). An energy-level diagram for a laser dye molecule is depicted in Figure 1.6.

Usually, three *electronic states* S_0, S_1, and S_2 are considered in addition to two *triplet states* T_1 and T_2, which are detrimental to laser emission. Laser emission takes place due to $S_1 \rightarrow S_0$ transitions. Each electronic state contains a large number of overlapping vibrational–rotational levels. This plethora of closely lying vibrational–rotational levels is what gives origin to the broadband gain and the intrinsic tunability of dye lasers. This is because $E = h\nu$, where ν is frequency. Thus, a change in energy ΔE implies a change in frequency $\Delta \nu$, which also means a change in the wavelength domain or $\Delta \lambda$.

In reference to the energy-level diagram of Figure 1.6, and considering only vibrational manifolds at each electronic state, a set of rate equations for transverse excitations was written by Duarte (1995a):

$$N = \sum_{S=0}^{m} \sum_{v=0}^{m} N_{S,v} + \sum_{T=1}^{m} \sum_{v=0}^{m} N_{T,v} \tag{1.6}$$

$$\frac{\partial N_{1,0}}{\partial t} \approx \sum_{v=0}^{m} N_{0,v}\sigma_{0,1_{0,v}} I_p(t) + \sum_{v=0}^{m} N_{0,v}\sigma^l_{0,1_{0,v}} I_l(x,t,\lambda_v) + \frac{N_{2,0}}{\tau_{2,1}}$$

$$- N_{1,0}\left[\sum_{v=0}^{m} \sigma_{1,2_{0,v}} I_p(t) + \sum_{v=0}^{m} \sigma_{e_{0,v}} I_l(x,t,\lambda_v) \right. \tag{1.7}$$

$$\left. + \sum_{v=0}^{m} \sigma^l_{1,2_{0,v}} I_l(x,t,\lambda_v) + \left(k_{S,T} + \tau_{1,0}^{-1}\right) \right]$$

$$\frac{\partial N_{T_{1,0}}}{\partial t} \approx N_{1,0}k_{S,T} - \frac{N_{T_{1,0}}}{\tau_{T,S}} - N_{T_{1,0}}\left[\sum_{v=0}^{m} \sigma^T_{1,2_{0,v}} I_p(t) + \sum_{v=0}^{m} \sigma^{Tl}_{1,2_{0,v}} I_l(x,t,\lambda_v) \right] \tag{1.8}$$

$$c^{-1}\frac{\partial I_P(t)}{\partial t} \approx -\left(N_{0,0}\sum_{v=0}^{m} \sigma_{0,1_{0,v}} + N_{1,0}\sum_{v=0}^{m} \sigma_{1,2_{0,v}} + N_{T_{1,0}}\sum_{v=0}^{m} \sigma^T_{1,2_{0,v}} \right) I_P(t) \tag{1.9}$$

$$c^{-1}\frac{\partial I_l(x,t,\lambda)}{\partial t} + \frac{\partial I_l(x,t,\lambda)}{\partial x} \approx N_{1,0}\sum_{v=0}^{m} \sigma_{e_{0,v}} I_l(x,t,\lambda_v) - \sum_{v=0}^{m} N_{0,v}\sigma^l_{0,1_{0,v}} I_l(x,t,\lambda_v)$$

$$\tag{1.10}$$

$$- N_{1,0}\sum_{v=0}^{m} \sigma^l_{1,2_{0,v}} I_l(x,t,\lambda_v) - N_{T_{1,0}}\sum_{v=0}^{m} \sigma^{Tl}_{1,2_{0,v}} I_l(x,t,\lambda_v)$$

$$I_l(x,t,\lambda) = \sum_{v=0}^{m} I_l(x,t,\lambda_v) \tag{1.11}$$

$$I_l(x,t,\lambda) = I_l^+(x,t,\lambda) + I_l^-(x,t,\lambda) \tag{1.12}$$

In this set of equations, frequency dependence is incorporated via the summation terms and variables depending on the vibrational assignment v. Now, using the style of Duarte (2014), the equation parameters are as follows (see Figure 1.6):

1. $I_p(t)$ is the intensity of the pump laser beam. Units are photons cm^{-2} s^{-1}.
2. $I_l(x,t,\lambda)$ is the laser emission from the gain medium. Units are photons cm^{-2} s^{-1}.
3. $N_{S,v}$ refers to the population of the S electronic state at the v vibrational level. It is given as a number per unit volume (cm^{-3}).
4. $N_{T,v}$ refers to the population of the T triplet state at the v vibrational level. It is given as a number per unit volume (cm^{-3}).
5. The absorption cross sections, such as $\sigma_{0,1_{0,v}}$, are identified by a subscript $S'',S'_{v'',v'}$ that refers the electronic $S'' \rightarrow S'$ transition and the vibrational transition $v'' \rightarrow v'$. The same convention applies to the triplet levels. Units are cm^2.

6. The *emission* cross sections, $\sigma_{e_{0,v}}$, are identified by the subscript $e_{v',v''}$. Vibrational transitions denoting emission are designated $v' \to v''$. Units are cm^2.
7. Radiationless decay times, such as $\tau_{1,0}$, are identified by subscripts that denote the corresponding $S' \to S''$ transition. Units are s.
8. $k_{S,T}$ is a radiationless decay rate from the singlet to the triplet. Units are s^{-1}.

The broadband nature of the emission is a consequence of the involvement of the vibrational manifold of the ground electronic state represented by the summation terms of Equations 1.9 through 1.11. The usual approach to solve an equation system as described here is numerical.

Since the gain medium exhibits homogeneous broadening, the introduction of intracavity frequency selective optics (see Chapter 7) enables all the molecules to contribute efficiently to narrow-linewidth emission.

A simplified set of equations can be obtained by replacing the vibrational manifolds by single energy levels and by neglecting a number of mechanisms including spontaneous decay from S_2 and absorption of the pump laser by T_1. Thus, Equations 1.6 through 1.10 reduce to

$$N = N_0 + N_1 + N_T \tag{1.13}$$

$$\frac{\partial N_1}{\partial t} \approx N_0\sigma_{0,1}I_p(t) + \left(N_0\sigma_{0,1}^l - N_1\sigma_e - N_1\sigma_{1,2}^l\right)I_l(x,t,\lambda) - N_1\left(k_{S,T} + \tau_{1,0}^{-1}\right) \tag{1.14}$$

$$\frac{\partial N_T}{\partial t} = N_1 k_{S,T} - N_T\,\tau_{T,S}^{-1} - N_T\sigma_{1,2}^{Tl}I_l(x,t,\lambda) \tag{1.15}$$

$$c^{-1}\frac{\partial I_p(t)}{\partial t} = -\left(N_0\sigma_{0,1} + N_1\sigma_{1,2}\right)I_p(t) \tag{1.16}$$

$$c^{-1}\left[\frac{\partial I_l(x,t,\lambda)}{\partial t}\right] + \left[\frac{\partial I_l(x,t,\lambda)}{\partial x}\right] = \left(N_1\sigma_e - N_0\sigma_{0,1}^l - N_1\sigma_{1,2}^l - N_T\sigma_{1,2}^{Tl}\right)I_l(x,t,\lambda) \tag{1.17}$$

This set of equations is similar to the set of equations considered by Teschke et al. (1976). This type of equations has been applied to simulate numerically the behavior of the output intensity and the gain as a function of the laser–pump intensity and to optimize laser performance. Relevant cross sections and excitation rates are given in Tables 1.1 and 1.2.

For long-pulse, or continuous-wave (CW) excitation, the time derivatives approach zero and Equations 1.14 through 1.17 reduce to

$$N_0\sigma_{0,1}I_p + \left(N_0\sigma_{0,1}^l - N_1\sigma_e - N_1\sigma_{1,2}^l\right)I_l(x,\lambda) = N_1\left(k_{S,T} + \tau_{1,0}^{-1}\right) \tag{1.18}$$

$$N_1 k_{S,T} = N_T\,\tau_{T,S}^{-1} + N_T\sigma_{1,2}^{Tl}I_l(x,\lambda) \tag{1.19}$$

$$N_0\sigma_{0,1} = -N_1\sigma_{1,2} \tag{1.20}$$

TABLE 1.1

Molecular Transition Cross Sections for Rhodamine 6G

Symbol	Cross Section (cm²)	λ (nm)	Reference
$\sigma_{0,1}$	1.66×10^{-16}	510	Hargrove and Kan (1980)
$\sigma_{0,1}$	4.5×10^{-16}	530	Everett (1991)
$\sigma_{1,2}$	$\sim 0.4 \times 10^{-16}$	510	Hammond (1979)
σ_e	1.86×10^{-16}	572	Hargrove and Kan (1980)
σ_e	1.3×10^{-16}	600	Everett (1991)
$\sigma_{0,1}^l$	1.0×10^{-19}	600	Everett (1991)
$\sigma_{1,2}^l$	1.0×10^{-17}	600	Everett (1991)
$\sigma_{1,2}^T$	1.0×10^{-17}	530	Everett (1991)
$\sigma_{1,2}^{Tl}$	4.0×10^{-17}	600	Everett (1991)

TABLE 1.2

Molecular Transition Rates and Decay Times for Rhodamine 6G

Symbol	Rate (s⁻¹)	Time (s)	Reference
$k_{S,T}$	2.0×10^7		Everett (1991)
$\tau_{T,S}$		0.5×10^{-7}	Everett (1991)
$\tau_{1,0}$		3.5×10^{-9}	Everett (1991)
$\tau_{2,1}$		$\sim 1.0 \times 10^{-12}$	Hargrove and Kan (1980)

$$\frac{\partial I_l(x,\lambda)}{\partial x} = \left(N_1\sigma_e - N_0\sigma_{0,1}^l - N_1\sigma_{1,2}^l - N_T\sigma_{1,2}^{Tl} \right) I_l(x,\lambda) \qquad (1.21)$$

From these equations, some characteristic features of CW dye lasers become apparent.

1.3.2.1 Example

As indicated by Dienes and Yankelevich (1998), from Equation 1.18 just below threshold, that is, $I_l(x,\lambda) \approx 0$,

$$I_p \approx \sigma_{0,1}^{-1} \left(k_{S,T} + \tau_{10}^{-1} \right) \frac{N_1}{N_0} \qquad (1.22)$$

which means that to approach population inversion, using rhodamine 6G under visible laser excitation, pump intensities exceeding 10^{22} photons cm⁻² s⁻¹ are necessary.

1.3.2.2 Example

A problem unique to long-pulse and CW dye lasers is intersystem crossing from N_1 into T_1. Thus, researchers use *triplet-level quenchers* such as O_2 and C_8H_8 (see, e.g., Duarte 1990) to neutralize the effect of that level. Under these circumstances, from Equation 1.21, the gain factor can be written as

$$g = \left[N_1(\sigma_e - \sigma_{1,2}^t) - N_0\sigma_{0,1}^t \right] L \tag{1.23}$$

From this equation, it can be deduced that amplification can occur in the absence of triplet losses, when the ratio of the populations becomes

$$\frac{N_1}{N_0} > \frac{\sigma_{0,1}^t}{\sigma_e - \sigma_{1,2}^t} \tag{1.24}$$

From the values of the cross sections available in Table 1.1, this ratio can be approximately $(10^{-19}/10^{-16}) \approx 0.001$.

1.3.3 TRANSITION PROBABILITIES AND CROSS SECTIONS

The dynamics described with the classical approach of rate equations depends on the cross sections listed in Table 1.1, which are measured experimentally. The origin of these cross sections, however, is not classical. Its origin is quantum mechanical. Here, the quantum mechanical probability for a two-level transition is introduced and its relation to the cross section of the transition is outlined. The style adopted here follows the treatment given to this problem by Feynman et al. (1965) which uses Dirac notation. An introduction to Dirac notation is given in Chapter 2.

This approach is based on the basic Dirac principles of quantum mechanics described by

$$\langle \phi | \psi \rangle = \sum_j \langle \phi | j \rangle \langle j | \psi \rangle \tag{1.25}$$

and

$$\langle \phi | \psi \rangle = \langle \psi | \phi \rangle^* \tag{1.26}$$

For $j = 1, 2$, Equation 1.25 leads to

$$\langle \phi | \psi \rangle = \langle \phi | 2 \rangle \langle 2 | \psi \rangle + \langle \phi | 1 \rangle \langle 1 | \psi \rangle \tag{1.27}$$

which can be expressed as

$$\langle \phi | \psi \rangle = \langle \phi | 2 \rangle C_2 + \langle \phi | 1 \rangle C_1 \tag{1.28}$$

where:

$$C_1 = \langle 1 | \psi \rangle \tag{1.29}$$

and

$$C_2 = \langle 2 | \psi \rangle \tag{1.30}$$

Here, the amplitudes change as a function of time according to the Hamiltonian

$$i\hbar\frac{dC_j}{dt} = \sum_{k}^{2} H_{jk}C_k \tag{1.31}$$

Now, Feynman et al. (1965) define new amplitudes C_I and C_{II} as linear combinations of C_1 and C_2. However, since

$$\langle II | II \rangle = \langle II | 1 \rangle \langle 1 | II \rangle + \langle II | 2 \rangle \langle 2 | II \rangle \tag{1.32}$$

must equal unity, the normalization factor $2^{-1/2}$ is introduced in the definitions of the new amplitudes

$$C_{II} = \frac{1}{\sqrt{2}}(C_1 + C_2) \tag{1.33}$$

and

$$C_I = \frac{1}{\sqrt{2}}(C_1 - C_2) \tag{1.34}$$

For a molecule under the influence of an electric field, the components of the Hamiltonian are

$$H_{11} = E_0 + \mu\mathcal{E} \tag{1.35}$$

$$H_{12} = -A \tag{1.36}$$

$$H_{21} = -A \tag{1.37}$$

$$H_{22} = E_0 - \mu\mathcal{E} \tag{1.38}$$

where:

$$\mathcal{E} = \mathcal{E}_0(e^{i\omega t} + e^{-i\omega t}) \tag{1.39}$$

μ corresponds to the electric dipole moment

Expanding the Hamiltonian given in Equation 1.31, and then subtracting and adding yield

$$i\hbar\frac{dC_I}{dt} = (E_0 + A)C_I + \mu\mathcal{E}\, C_{II} \tag{1.40}$$

$$i\hbar\frac{dC_{II}}{dt} = (E_0 - A)C_{II} + \mu\mathcal{E}\, C_I \tag{1.41}$$

Assuming a small electric field, the solutions are of the form:

$$C_I = D_I e^{-iE_I/\hbar t} \tag{1.42}$$

$$C_{II} = D_{II}e^{-iE_{II}/\hbar t}$$ (1.43)

where:

$$E_I = E_0 + A$$ (1.44)

and

$$E_{II} = E_0 - A$$ (1.45)

Hence, neglecting the term $(\omega + \omega_0)$ because it oscillates too rapidly to contribute to the average value of the rate of change of D_I and D_{II}, the following expressions D_I and D_{II} are found:

$$i\hbar \frac{dD_I}{dt} = \mu\mathcal{E}_0 D_{II} e^{-i(\omega-\omega_0)t}$$ (1.46)

$$i\hbar \frac{dD_{II}}{dt} = \mu\mathcal{E}_0 D_I e^{i(\omega-\omega_0)t}$$ (1.47)

If at $t = 0$, $D_I \approx 1$, then integration of Equation 1.46 yields (Feynman et al. 1965)

$$D_{II} = \frac{\mu\mathcal{E}_0}{\hbar}\left[\frac{1-e^{i(\omega-\omega_0)T}}{\omega-\omega_0}\right]$$ (1.48)

and following multiplication with its complex conjugate

$$|D_{II}|^2 = \left(\frac{\mu\mathcal{E}_0}{\hbar}\right)^2 \left[\frac{2-2\cos(\omega-\omega_0)T}{(\omega-\omega_0)^2}\right]$$ (1.49)

which can be written as (Feynman et al. 1965)

$$|D_{II}|^2 = \left(\frac{\mu\mathcal{E}_0 T}{\hbar}\right)^2 \frac{\sin^2\left[\frac{1}{2}(\omega-\omega_0)T\right]}{\left[\frac{1}{2}(\omega-\omega_0)T\right]^2}$$ (1.50)

which is the probability of the transition $I \to II$. It can be further shown that

$$|D_I|^2 = |D_{II}|^2$$ (1.51)

which means that the probability of emission is equal to the probability of absorption. This result is central to the theory of absorption and radiation of light by atoms and molecules.

Using $\mathcal{S} = 2\varepsilon_0 c\mathcal{E}_0^2$, and replacing μ by $3^{-1/2}\mu$ (Sargent et al. 1974), the expression for the probability of the transition becomes

$$|D_{II}|^2 = \frac{2\pi}{3}\left(\frac{\mu^2 T^2}{4\pi\varepsilon_0 c\hbar^2}\right)\mathcal{S}(\omega_0)\frac{\sin^2\left[\frac{1}{2}(\omega-\omega_0)T\right]}{\left[\frac{1}{2}(\omega-\omega_0)T\right]^2}$$ (1.52)

where:
  μ is the dipole moment in units of Cm
  $(1/4\pi\varepsilon_0)$ is in units of Nm^2 C^{-2}
  $\mathcal{S}(\omega_0)$ is the intensity in units of J s^{-1} m^{-2}

Integrating Equation 1.52 with respect to the frequency ω, the dimensionless transition probability becomes

$$|D_{II}|^2 = \frac{4\pi^2}{3}\left(\frac{\mu^2}{4\pi\varepsilon_0 c\hbar^2}\right)\left[\frac{\mathcal{S}(\omega_0)}{\Delta\omega}\right]T \tag{1.53}$$

It then follows that the cross section for the transition can be written as

$$\sigma = \frac{4\pi^2}{3}\left(\frac{\mu^2}{4\pi\varepsilon_0 c\hbar}\right)\left(\frac{\omega}{\Delta\omega}\right) \tag{1.54}$$

in units of m^2 (although the more widely used unit is cm^2; see Table 1.1). For a simple atomic or molecular system, the dipole moment can be calculated from the definition (Feynman et al. 1965)

$$\mu_{mn}\xi = \langle m|H|n\rangle = H_{mn} \tag{1.55}$$

where:
  H_{mn} is the matrix element of the Hamiltonian

For diatomic molecules, the dependence of this matrix element on the Franck–Condon factor $(q_{v',v''})$ and the square of the transition moment $(|R|_e^2)$ is described by Chutjian and James (1969). Byer et al. (1972) wrote an expression for the gain of vibrational–rotational transitions of the form

$$g = \sigma NL \tag{1.56}$$

or more specifically

$$g = \frac{4\pi^2}{3}\left(\frac{1}{4\pi\varepsilon_0 c\hbar}\right)\left(\frac{\omega}{\Delta\omega}\right)|R_e|^2 q_{v',v''}\left[\frac{S_{J''}}{(2J''+1)}\right]NL \tag{1.57}$$

where:
  $S_{J''}$ is known as the line strength
  J'' identifies a specific rotational level

In practice, however, cross sections are mostly determined experimentally as in the case of those listed in Table 1.1.

Going back to Equation 1.53, and rearranging its terms, it follows that the intensity can be expressed as a function of the transition probability (Duarte 2014)

$$\mathcal{S}(\omega_0) = \frac{3}{\pi}\left(\frac{\varepsilon_0 c\hbar^2}{\mu^2}\right)\left(\frac{\Delta\omega}{T}\right)|D_{II}|^2 \tag{1.58}$$

or

$$\mathcal{G}(\omega_0) = \frac{3}{\pi} \kappa \left(\frac{\Delta\omega}{T}\right) |D_{II}|^2 \tag{1.59}$$

with

$$\kappa = \left(\frac{\varepsilon_0 c \hbar^2}{\mu^2}\right) \tag{1.60}$$

where the units for the constant κ are J s m^{-2}. Subsequently, the intensity $\mathcal{G}(\omega_0)$ has units of J s^{-1} m^{-2} or W m^{-2}.

1.3.3.1   Example

Within resonance, where $\Delta\omega \approx 2\pi/T$, Equation 1.53 can be approximated as

$$|D_{II}|^2 \approx \frac{\pi}{3} \left(\frac{\mu^2}{\varepsilon_0 c \hbar^2}\right) \frac{2\pi}{(\Delta\omega)^2} \mathcal{G}(\omega_0) \tag{1.61}$$

which can be reexpressed as

$$\mathcal{G}(\omega_0) \approx \frac{3}{2\pi^2} \kappa \ (\Delta\omega)^2 |D_{II}|^2 \tag{1.62}$$

This approximation indicates that the intensity is proportional to the square of the frequency difference multiplied by the probability of the transition, or $(\Delta\omega)^2 |D_{II}|^2$, and is given in units of J s^{-1} m^{-2} or W m^{-2}.

1.4   THE SCHRÖDINGER EQUATION AND SEMICONDUCTOR LASERS

The Schrödinger equation is central to the emission properties of semiconductor lasers that are widely tunable sources of coherent radiation in various important segments of the electromagnetic spectrum, as illustrated in Figure 1.7.

The Schrödinger equation can be derived via Schrödinger's way (Schrödinger 1926), a heuristic path, or Dirac's notation. Here, the Schrödinger equation is introduced via a heuristic path originally sketched by Haken (1981) and via Dirac's notation in an approach modeled after Feynman et al. (1965), as recently reviewed by Duarte (2014).

1.4.1   A Heuristic Introduction to the Schrödinger Equation

This approach uses a mixture of classical, relativistic, and quantum concepts. The presentation given here follows the review given by Duarte (2014), which was modeled using concepts outlined by Haken (1981). This heuristic approach begins by considering a free particle moving with classical kinetic energy

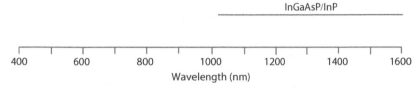

Wavelength (nm)

FIGURE 1.7 Approximate wavelength range covered by various types of semiconductors used in tunable lasers. This is only an approximate depiction and some of the ranges might not be continuous. Not shown are the InGaAs/InP semiconductors that emit between 1600 and 2100 nm or the quantum cascade lasers that emit deep into the infrared (see Chapter 9).

$$E = \frac{1}{2}mv^2 \tag{1.63}$$

and using $p = mv$, the kinetic energy equation can be restated as

$$E = \frac{p^2}{2m} \tag{1.64}$$

Next, using Planck's quantum energy (Planck 1901),

$$E = h\nu$$

in conjunction with $\lambda = c/\nu$, and $E = mc^2$, the de Broglie expression for momentum can be expressed as (de Broglie 1923)

$$p = \hbar k \tag{1.65}$$

The next step consists in incorporating the classical "wave functions of ordinary wave optics" (Dirac 1978)

$$\psi(x,t) = \psi_0 e^{-i(\omega t - kx)} \tag{1.66}$$

whose derivative with respect to time becomes

$$\frac{\partial \psi(x,t)}{\partial t} = (-i\omega)\psi_0 e^{-i(\omega t - kx)} \tag{1.67}$$

Similarly, the first and second derivatives with respect to displacement are

$$\frac{\partial \psi(x,t)}{\partial x} = (+ik)\psi_0 e^{-i(\omega t - kx)} \tag{1.68}$$

and

$$\frac{\partial^2 \psi(x,t)}{\partial x^2} = (-k^2)\psi_0 e^{-i(\omega t - kx)} \tag{1.69}$$

Multiplying the first time derivative by $(-i\hbar)$ yields

$$-i\hbar \frac{\partial \psi(x,t)}{\partial t} = (-\hbar \omega)\psi_0 e^{-i(\omega t - kx)} \tag{1.70}$$

and multiplying the second displacement derivative by $(-\hbar^2/2m)$ yields

$$-\frac{\hbar^2}{2m}\frac{\partial^2 \psi(x,t)}{\partial x^2} = \frac{\hbar^2 k^2}{2m}\psi_0 e^{-i(\omega t - kx)} \tag{1.71}$$

Recognizing that $E = \hbar^2 k^2/2m$,

$$-\frac{\hbar^2}{2m}\frac{\partial^2 \psi(x,t)}{\partial x^2} = +i\hbar \frac{\partial \psi(x,t)}{\partial t} \tag{1.72}$$

which is the basic form of Schrödinger's equation. As can be seen, this heuristic approach to Schrödinger's equation utilizes Planck's quantum energy, the classical kinetic energy of a free particle, and the classical "wave functions of ordinary wave optics." That is, Schrödinger's equation describes the motion of a free particle propagating in according to ordinary wave optics. The crucial quantum step consists in incorporating, in the development, Planck's energy equation $E = h\nu$. Once again, the central role of classical wave equation equations of the form

$$\psi(x,t) = \psi_0 e^{-i(\omega t - kx)}$$

is highlighted.

1.4.2 THE SCHRÖDINGER EQUATION VIA DIRAC'S NOTATION

As explained in Section 1.4.1, the Hamiltonian H_{ij} is related to the time-dependent amplitude C_i by (Dirac 1978)

$$i\hbar \frac{dC_i}{dt} = \sum_j H_{ij}C_i \tag{1.73}$$

For $C_i = \langle i|\psi \rangle$, this equation can be rewritten as

$$i\hbar \frac{d\langle i|\psi \rangle}{dt} = \sum_j \langle i|H|j \rangle \langle j|\psi \rangle \tag{1.74}$$

which, for $i = x$, can be written as

$$i\hbar \frac{d\langle x|\psi\rangle}{dt} = \sum_j \langle x|H|x'\rangle \langle x'|\psi\rangle \qquad (1.75)$$

Since $\langle x|\psi\rangle = \psi(x)$, this equation can be reexpressed as

$$i\hbar \frac{d\psi(x)}{dt} = \int H(x,x')\psi(x')dx' \qquad (1.76)$$

The integral on the right-hand side is given by

$$\int H(x,x')\psi(x')dx' = -\frac{\hbar^2}{2m}\frac{d^2\psi(x)}{dx^2} + V(x)\psi(x) \qquad (1.77)$$

According to Feynman, the right-hand side of the above equation simply came "from the mind of Schrödinger" (Feynman et al. 1965). A methodical description of the Schrödinger approach is given by Duarte (2014).

Next, combining Equations 1.76 and 1.77, we get the formal Schrödinger equation

$$+i\hbar \frac{d\psi(x)}{dt} = -\frac{\hbar^2}{2m}\frac{d^2\psi(x)}{dx^2} + V(x)\psi(x) \qquad (1.78)$$

In three dimensions, the wave function becomes $\psi(x,y,z)$, the potential is $V(x,y,z)$, and introducing

$$\nabla^2 = \frac{\partial^2}{\partial x^2} + \frac{\partial^2}{\partial y^2} + \frac{\partial^2}{\partial z^2} \qquad (1.79)$$

and Schrödinger's equation in three dimensions takes the succinct form

$$+i\hbar \frac{\partial\psi}{\partial t} = -\frac{\hbar^2}{2m}\nabla^2\psi + V\psi \qquad (1.80)$$

1.4.3 THE TIME-INDEPENDENT SCHRÖDINGER EQUATION

As suggested by Feynman, using a solution of the form

$$\psi(x) = \Psi(x)e^{-iEt/\hbar} \qquad (1.81)$$

Equation 1.80, in one dimension, becomes

$$\frac{\partial^2\Psi(x)}{\partial x^2} - \frac{2m}{\hbar^2}\left[V(x) - E\right]\Psi(x) = 0 \qquad (1.82)$$

or

$$\frac{\partial^2\Psi(x)}{\partial x^2} = \frac{2m}{\hbar^2}\left[V(x) - E\right]\Psi(x) \qquad (1.83)$$

which is known as the one-dimensional time-independent Schrödinger equation. This simple form of the Schrödinger equation is of enormous significance to semiconductor physics and semiconductor lasers.

1.4.4 SEMICONDUCTOR EMISSION

The description of semiconductor photon emission can be described using the time-independent Schrödinger equation

$$\frac{\partial^2 \Psi(x)}{\partial x^2} - \frac{2m}{\hbar^2}\left[V(x)-E\right]\Psi(x) = 0 \tag{1.84}$$

and using the spatial component of the wave function as solution

$$\Psi(x) = \Psi_0 e^{-ikx} \tag{1.85}$$

leads directly to

$$E = \frac{k^2\hbar^2}{2m} + V(x) \tag{1.86}$$

This energy expression is the sum of kinetic and potential energy ($E = E_K + E_P$) so that

$$E_K + E_P = \frac{k^2\hbar^2}{2m} + V(x) \tag{1.87}$$

and

$$E_K = \frac{k^2\hbar^2}{2m} \tag{1.88}$$

The kinetic energy E_K as a function of the wave number $k = 2\pi/\lambda$ is shown graphically in Figure 1.8. The graph is a positive parabola for positive values of m and is known as the *conduction* band. The negative parabola is known as the *valence* band. The separation between the two bands is known as the *band gap*, E_G. If an electron is excited and a transition from the valence band to the conduction band occurs, it is said to leave behind a vacancy or *hole*. Under certain conditions, radiation might also occur higher from the conduction band as suggested in Figure 1.9; however, such process is undermined by fast phonon relaxation.

1.4.4.1 Example

Electrons can transition from the bottom of the conduction band to the top of valence band by recombining with holes. For a material such as gallium arsenide (GaAs), the band gap at $T = 300°$K is $E_G \approx 1.43$ eV (Kittel 1971). Now, using the conversion $1\text{eV} = 1.602176565 \times 10^{-19}$ J, $h = 6.62606957 \times 10^{-34}$ Js, $c = 2.99792458 \times 10^8$ ms^{-1} (see Chapter 13), and Planck's energy equation $E = h\nu$,

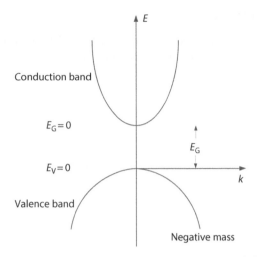

FIGURE 1.8 Conduction and valence bands according to $E_K = \pm(k^2\hbar^2/2m)$. (Reproduced from Duarte, F.J., *Quantum Optics for Engineers*, CRC Press, New York, 2014. With permission from Taylor & Francis.)

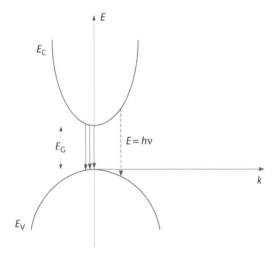

FIGURE 1.9 Emission due to recombination transitions from the bottom of the conduction band to the top of the valence band. (Reproduced from Duarte, F.J., *Quantum Optics for Engineers*, CRC Press, New York, 2014. With permission from Taylor & Francis.)

$$\lambda = \frac{ch}{E} \approx 867.02 \text{ nm}$$

In other words, the recombination emission occurs around 867 nm. Here, it should be noted that the energy band gap varies as a function of temperature so that at $T = 0°\text{K}$ it is $E_G \approx 1.52 \text{ eV}$ (Kittel 1971), which would change the wavelength to $\lambda \approx 815.68 \text{ nm}$, thus demonstrating one avenue to tune the wavelength.

1.4.5   QUANTUM WELLS

Starting with a potential well as described by Silfvast (2008), $V(x) = 0$ for $0 < x < L$ and $V(x) = \infty$ for $x = 0$ or $x = L$, illustrated in Figure 1.10, then Equation 1.84

$$\frac{\partial^2 \Psi(x)}{\partial x^2} - \frac{2m}{\hbar^2}\left[V(x) - E\right]\Psi(x) = 0$$

becomes

$$\frac{\partial^2 \Psi(x)}{\partial x^2} + \frac{2m}{\hbar^2}E\Psi(x) = 0 \tag{1.89}$$

for $0 < x < L$. This is a wave equation of the form

$$\frac{d^2 \Psi(x)}{dx^2} + k_x^2\Psi(x) = 0 \tag{1.90}$$

where:

$$k_x^2 = \frac{2m_c}{\hbar^2}E \tag{1.91}$$

The solution to Equation 1.90 is

$$\Psi(x) = \sin k_x x \tag{1.92}$$

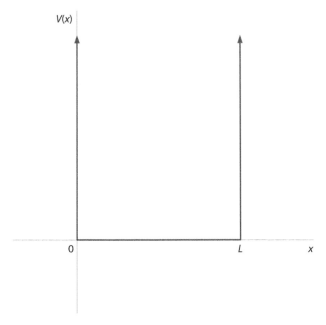

FIGURE 1.10   Potential well $V(x) = 0$ for $0 < x < L$ and $V(x) = \infty$ for $x = 0$ or $x = L$. (Reproduced from Duarte, F.J., *Quantum Optics for Engineers*, CRC Press, New York, 2014. With permission from Taylor & Francis.)

Since $\psi(x) = 0$ at $x = 0$ or $x = L$, we have

$$k_x = \frac{n\pi}{L} \tag{1.93}$$

for $n = 1, 2, 3, \dots$ Substituting Equation 1.93 into 1.91 leads to

$$E = \frac{n^2\pi^2\hbar^2}{2m_cL^2} \tag{1.94}$$

which should be labeled as E_n to account for the quantized nature of the energy, that is,

$$E_n = \frac{n^2\pi^2\hbar^2}{2m_cL^2} \tag{1.95}$$

This quantized energy E_n indicates a series of possible discrete energy levels above the lowest point of the conduction band so that the total energy above the valence band becomes (Silfvast 2008)

$$E = E_c + E_n \tag{1.96}$$

1.4.6 Quantum Cascade Lasers

Quantum cascade lasers (QCL; Faist et al. 1994) are tunable sources emitting in the infrared from a few micrometers to beyond 20 μm (see Chapters 7 and 9). These lasers operate via transitions between quantized levels *within* the conduction band of multiple-quantum well structures. The carriers involved are electrons generated in an n-doped material. A single stage includes an injector and an active region. The electron is injected into the active region at n_2 and the transition occurs down to n_1 (see Figure 1.11). Following emission, the electron continues into the next injector region. Practical devices include a series of such stages. From Equation 1.94, the energy difference between the two levels can be expressed as

$$\Delta E = \left(n_2^2 - n_1^2\right)\frac{\pi^2\hbar^2}{2m_cL^2} \tag{1.97}$$

where:
   L is the thickness of the well

For $n_2 = 3$ and $n_1 = 2$, Equation 1.97 becomes (Silfvast 2008)

$$\Delta E = (3^2 - 2^2)\frac{\pi^2\hbar^2}{2m_cL^2} \tag{1.98}$$

which is applicable to the $3 \to 2$ transition in GaInAs reported by Faist et al. (1994). As a footnote, it should be mentioned that the quantum well approach illustrated here

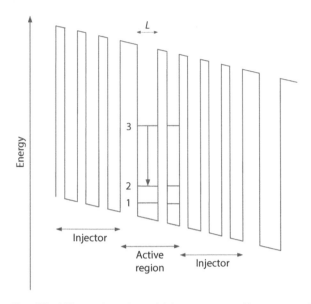

FIGURE 1.11 Simplified illustration of a multiple-quantum well structure relevant to quantum cascade lasers. An electron is injected from the *injector region* into the active region at $n_2 = 3$. Thus, a photon is emitted via the $3 \rightarrow 2$ transition. The electron continues to the next region where the process is repeated. By configuring a series of such stages, one electron can generate the emission of numerous photons. (Reproduced from Duarte, F.J., *Quantum Optics for Engineers*, CRC Press, New York, 2014. With permission from Taylor & Francis.)

also applies, with some modifications to the description of quantum wells in laser dye molecules (Schäfer 1990).

1.4.6.1   Example

Using $\Delta E = h\nu$, in Equation 1.97, it follows that the wavelength of emission is given by

$$\lambda = \left(n_2^2 - n_1^2 \right)^{-1} \frac{8 m_c c L^2}{h} \tag{1.99}$$

This expression can be used to provide an estimate of the emission wavelength for a QCL transition, provided the effective mass, relevant to the emitting semiconductor, and the thickness of the quantum well (L) are known (see Problem 1.6).

1.4.7   QUANTUM DOTS

Besides the multiple-quantum well configurations, other interesting semiconductor geometries include the quantum wire and the quantum dot. The quantum wire narrowly confines the electrons and holes in two directions (x,y). The quantum dot geometry severely confines the electrons in three dimensions (x,y,z). Under these circumstances, the quantized energy can be expressed as

$$E = E_c + \frac{k_x^2 \hbar^2}{2m_c} + \frac{k_y^2 \hbar^2}{2m_c} + \frac{k_x^2 \hbar^2}{2m_c} \qquad (1.100)$$

where:

k_x, k_y, k_z are defined according to Equation 1.93

The concept of quantum dot is not limited to semiconductor materials, but it also applies to nanoparticle gain media and nanoparticle core–shell gain media. For instance, for core–shell nanoparticles, the physics can be described with a Schrödinger equation of the form

$$\nabla^2 \Psi(r) - \frac{2m}{\hbar^2} \left[V(r) - E \right] \Psi(r) = 0 \qquad (1.101)$$

with the potential $V(r)$ defined by the core–shell geometry (Dong et al. 2013).

1.5   INTRODUCTION TO LASER RESONATORS AND LASER CAVITIES

A basic laser is composed of a gain medium, a mechanism to excite that medium, and an optical resonator and/or optical cavity. These optical resonators and optical cavities, known as laser resonators and laser cavities, are the optical systems that reflect radiation back to the gain medium and determine the amount of radiation to be emitted by the laser. In this section, a brief introduction to laser resonators is provided. In further chapters, Chapter 7 in particular, this subject is considered in more detail.

The most basic resonator, regardless of the method of excitation, is that composed of two mirrors aligned along a single optical axis as depicted in Figure 1.1. In this flat-mirror resonator, one of the mirrors is ~100% reflective, at the wavelength or wavelengths of interest, and the other mirror is partially reflective. The amount of reflectivity depends on the characteristics of the gain medium. The optimum reflectivity for the output coupler is often determined empirically. For a low-gain laser medium, this reflectivity can approach 99%, whereas for a high-gain laser medium, the reflectivity can be as low as 20%. In Figure 1.1, the gain medium is depicted with its output windows at an angle relative to the optical axis. If the angle of incidence of the laser emission on the windows is the Brewster angle, then the emission will be highly linearly polarized. If the windows are oriented as depicted in Figure 1.1, then the laser emission will be polarized parallel to the plane of incidence. An alternative to the flat-mirror approach is to use a pair of optically matched concave mirrors. Transverse and longitudinal excitation geometries are depicted in Figure 1.12. Further, in some resonators, the back mirror can be replaced by a diffraction grating as shown in Figure 1.13. This is often the case in tunable lasers.

These resonators might incorporate intracavity frequency-selective optical elements, such as Frabry–Perot etalons (Figure 1.13b), to narrow the emission linewidth. They can also include intracavity beam expanders to protect optics from optical damage and to be utilized in linewidth narrowing techniques as described in Chapter 4. Resonators that yield highly coherent or narrow-linewidth emission are often called oscillators and are considered in detail in Chapter 7.

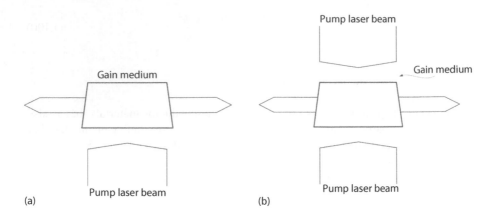

(a)   (b)

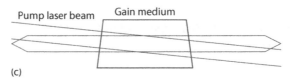

(c)

FIGURE 1.12 (a) Transverse laser excitation. (b) Transverse double-laser excitation. (c) Longitudinal laser excitation.

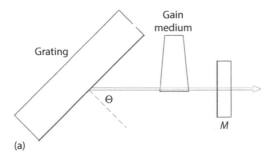

(a)

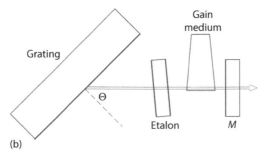

(b)

FIGURE 1.13 (a) Grating mirror resonator and (b) grating mirror resonator incorporating an intracavity etalon.

FIGURE 1.14 Cross section of a TEM$_{00}$ laser beam from a high-power narrow-linewidth dispersive laser oscillator. The spatial intensity profile of this beam is near-Gaussian. (Reprinted from *Opt. Commun.*, 117, Duarte, F.J., Solid-sate dispersive dye laser oscillator: Very compact cavity, 480–484, Copyright 1995b, with permission from Elsevier.)

The transverse mode structure in these resonators is approximately determined by the ratio (Siegman 1986)

$$N_F = \frac{w^2}{L\lambda} \tag{1.102}$$

known as the *Fresnel number*. Here, w is the beam waist at the gain region, L is the length of the cavity, and λ is the wavelength of emission. The lower this number, the better the beam quality of the emission, or the closer it will be to a single transverse mode designated by TEM$_{00}$. A TEM$_{00}$ is a clean beam (Duarte 1995b) with no spatial structure on it, as shown in Figure 1.14, and is generally round with a near-Gaussian intensity profile in the spatial domain. Thus, long lasers with relatively narrow beam waists tend to yield single-transverse-mode emission. As it will be examined in Chapters 4 and 7, an important part of laser cavity design consist in optimizing the dimensions of the beam waist to the cavity length to obtain TEM$_{00}$ emission and low beam divergences in compact configurations.

An additional class of linear laser resonators are the *unstable resonators*. These cavities depart from the flat-mirror design and incorporate curved mirrors as depicted in Figure 1.15. These mirror configurations are adopted from the field of reflective telescopes. A widely used design is a variation of the Newtonian telescope known as the Cassegrain telescope. In this configuration, the two mirrors have a high reflectivity. The advantages of unstable resonators include the use of large gain medium volumes and good transverse-mode discrimination. This topic will be considered further under the context of transfer ray matrices in Chapter 6. For a detailed treatment on the subject of unstable resonators, the reader should refer to Siegman (1986).

A further class of cavities are linear and ring laser resonators (Figure 1.16) developed for CW dye lasers (Hollberg 1990) and later applied to the generation of ultrashort pulses (Diels 1990; Diels and Rudolph 2006). A straightforward unidirectional ring resonator with an 8 shape is illustrated in Figure 1.16b. In these cavities,

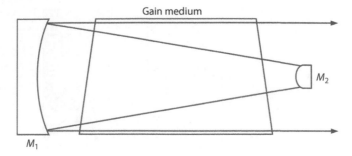

FIGURE 1.15   Basic unstable resonator laser cavity.

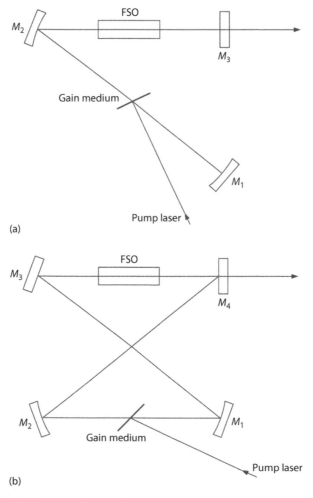

FIGURE 1.16   (a) Linear and (b) unidirectional eight-shaped ring dye laser cavities.

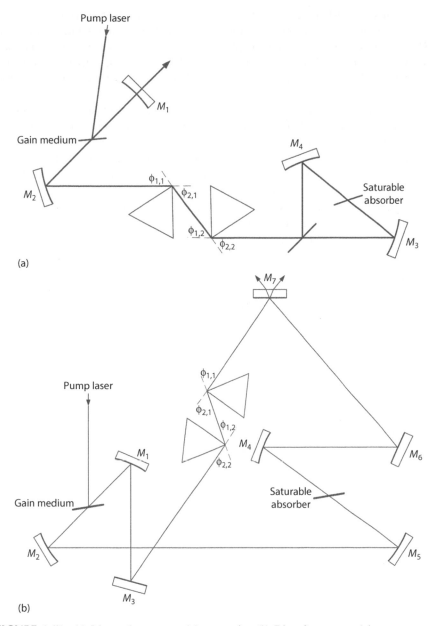

FIGURE 1.17   (a) Linear femtosecond laser cavity. (b) Ring femtosecond laser resonator. Both laser configurations include a saturable absorber and a multiple-prism pulse compressor.

the oscillation is in the form of a travelling wave that avoids the effect of *spatial hole burning* that causes the laser to oscillate in more than one longitudinal mode. Linear and ring resonators utilized in femtosecond laser configurations incorporate saturable absorbers as depicted in Figure 1.17. In the ring laser, a *collision* between two conterpropagating pulses occurs at the saturable absorber. This collision causes the two pulses interfere, thus creating a transient grating that shortens the emission pulse. This effect is known as colliding pulse mode locking (CPM) (Fork et al. 1981). The prisms in this cavity are deployed to provide negative dispersion and thus help in *pulse compression* as it will be described in Chapter 4. Although originally developed for dye lasers, these cavities are widely used with a variety of gain media.

Although most lasers do need efficient and well-designed optical resonators, some lasers have such high gain factors that they tend to emit laser like radiation sometimes called superradiant emission or superfluorescence with only one mirror, or even without external mirrors. This means that the intrinsic reflection factors from flat windows provide the necessary feedback for powerful emission albeit with poor coherence properties. More specifically, this emission tends to be broadband and highly divergent. One additional advantage of using inclined windows, when using high gain laser media, is to reduce parasitic reflections that tend to contribute to output noise. Some laser media that produce very high gains include copper vapor, molecular nitrogen, and laser dyes.

PROBLEMS

1.1 Show that in the steady state Equation 1.14 becomes Equation 1.18.

1.2 Show that in the steady state Equation 1.17 becomes Equation 1.21.

1.3 Show that by neglecting the triplet state from Equation 1.21, the gain can be expressed as Equation 1.23.

1.4 Starting from Equations 1.40 and 1.41, derive an expression for $|D_I|^2$ and show that it is equal to $|D_{II}|^2$.

1.5 Use Equation 1.53 to arrive at the expression for the transition cross section given in Equation 1.54.

1.6 For a $3 \rightarrow 2$ transition, use Equation 1.99 to estimate the emission wavelength of the GaInAs quantum cascade laser reported by Faist et al. (1994).

2 Dirac Optics

2.1 INTRODUCTION

Dirac introduced his enormously practical quantum *bra–ket* notation in 1939 as a "new notation for quantum mechanics" (Dirac 1939). Also, he discussed in his classic book *Principles of Quantum Mechanics*, first published in 1930 (Dirac 1978), the essence of interference as a one-photon phenomenon. He did so within a macroscopic framework using concepts such as "a beam of light consisting of a large number of photons," beam splitters," and "translational states of a photon" (Dirac 1978). That primordial and illuminating discussion surely qualifies Dirac as the father of quantum optics.

In 1965, Feynman discussed electron interference in two-slit thought experiments using probability amplitudes and Dirac's notation as tools (Feynman et al. 1965b). In 1991, Dirac's notation was applied to the propagation of coherent light in an *N*-slit interferometer (Duarte 1991). The emphasis of this chapter is to provide a straightforward introduction to the description of generalized interference using Dirac's notation. Here, the content, format, and style follow closely the initial introduction (Duarte 2003), although the discussion is extended using newer material (Duarte 2006) and relevant concepts adopted in a recent review (Duarte 2014).

2.2 DIRAC'S NOTATION IN OPTICS

The concept of the notation invented by Dirac can be explained by considering the propagation of a particle from plane s to plane x, as illustrated in Figure 2.1. According to the Dirac concept, there is a probability amplitude, denoted by $\langle x|s \rangle$, that quantifies such propagation. Historically, Dirac introduced the nomenclature of *ket vectors*, denoted by $|\ \rangle$, and *bra vectors*, denoted by $\langle\ |$, which are mirror images of each other. Thus, the probability amplitude is described by the *bra–ket* $\langle x|s \rangle$, which is a *complex number*.

It is important to note that in Dirac's notation the propagation from s to x is expressed in reverse by $\langle x|s \rangle$. In other words, the *starting* condition is at the *right* and the *final* condition is at the *left*. If the propagation of the photon is not directly from plane s to plane x, but it involves the passage through an intermediate plane j, as illustrated in Figure 2.2, then the probability amplitude describing such propagation is

$$\langle x|s \rangle = \langle x|j \rangle \langle j|s \rangle \tag{2.1}$$

If the photon from s must also propagate through planes j and k in its trajectory to x, as illustrated in Figure 2.3, then the probability amplitude is given by

$$\langle x|s \rangle = \langle x|k \rangle \langle k|j \rangle \langle j|s \rangle \tag{2.2}$$

FIGURE 2.1 Propagation from s to x is expressed as $\langle x|s\rangle$.

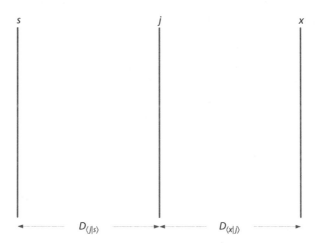

FIGURE 2.2 Propagation from s to x via an intermediate plane j is expressed as $\langle x|j\rangle \langle j|s\rangle$.

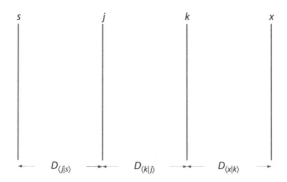

FIGURE 2.3 Propagation from s to x via two intermediate planes j and k is expressed as $\langle x|k\rangle \langle k|j\rangle \langle j|s\rangle$.

When at the intermediate plane of the case in Figure 2.2, a number of alternatives N are available to the passage of the photon, as depicted in Figure 2.4a; the overall probability amplitude must consider *every possible alternative*, which is expressed mathematically by a summation over j in the form of

$$\langle x|s \rangle = \sum_{j=1}^{N} \langle x|j \rangle \langle j|s \rangle \tag{2.3}$$

For the case of an additional intermediate plane with N alternatives, as illustrated in Figure 2.4b, the probability amplitude is written as

$$\langle x|s \rangle = \sum_{k=1}^{N} \sum_{j=1}^{N} \langle x|k \rangle \langle k|j \rangle \langle j|s \rangle \tag{2.4}$$

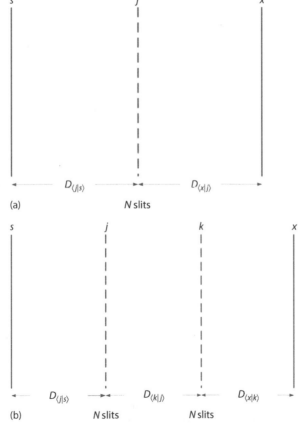

FIGURE 2.4 (a) Propagation from s to x via an array of N slits positioned at the intermediate plane j. (b) Propagation from s to x via an array of N slits positioned at the intermediate plane j and via an additional array of N slits positioned at k.

The addition of further intermediate planes, with N alternatives, can then be systematically incorporated in the notation. The Dirac notation, albeit originally applied to the propagation of a single photon or electron (Dirac 1978; Feynman et al. 1965a), also applies to describe the propagation of ensembles of coherent, or indistinguishable, photons, or indistinguishable quanta (Duarte 1991, 1993). This is in agreement with the interpretation that suggests that the principles of quantum mechanics are applicable to the description of macroscopic phenomena, which are not perturbed by observation (van Kampen 1988).

2.3   INTERFERENCE

As outlined by Feynman in his thought experiments, on two-slit electron interference, Dirac's notation offers a natural avenue to describe the propagation of electrons from a source to a detection plane via a pair of slits. This idea can be extended to the description of photon propagation from a source s to a screen detector x, via a transmission grating j comprised by N slits, as illustrated in Figure 2.5.

In the experimental scheme of Figure 2.5, a narrow-linewidth, or highly coherent, laser emits a Gaussian beam that is expanded in one dimension in the plane of propagation, or plane of incidence. Then, the central part of that expanded beam propagates through a wide illumination aperture, thus configuring the radiation source (s) that illuminates an array of N slits, or transmission grating (j). The interaction of the coherent radiation with the transmission grating (j) produces an interference signal at x. A crucial point here is that all the indistinguishable photons, or indistinguishable quanta, illuminate the array of N slits, or grating, simultaneously. If only one photon propagates, at any given time, then that individual photon illuminates the whole array of N slits simultaneously. The probability amplitude that describes the propagation from the source (s) to the detection plane (x), via the array of N slits (j), is given by (Duarte 1991, 1993)

$$\langle x|s \rangle = \sum_{j=1}^{N} \langle x|j \rangle \langle j|s \rangle$$

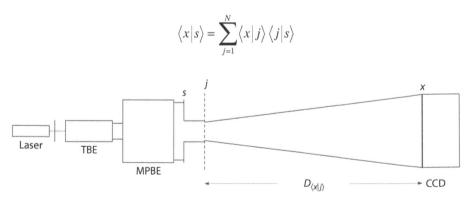

FIGURE 2.5 Optical architecture of the N-slit laser interferometer. Light from a TEM_{00} narrow-linewidth laser is transformed into an extremely elongated near-Gaussian source (s) to illuminate an array of N slits at j. Interaction of the coherent emission with the slit array produces interference at x. The j-to-x intra-interferometric distance is $D_{\langle x|j \rangle}$. This class of interferometric architecture was first introduced by Duarte (1991, 1993). (Reproduced from Duarte, F.J., et al., *J. Opt.*, 12, 015705, 2010. With permission from the Institute of Physics.)

According to Dirac (1978), the probability amplitudes can be represented by *wave functions of ordinary wave optics*. Thus, following Feynman et al. (1965a),

$$\langle j|s\rangle = \Psi(r_{j,s})e^{-i\theta_j} \tag{2.5}$$

$$\langle x|j\rangle = \Psi(r_{x,j})e^{-i\phi_j} \tag{2.6}$$

where:

  θ_j and ϕ_j are the *phase terms* associated with the incidence and diffraction waves, respectively

Using these expressions for the probability amplitudes, then, Equation 2.3 can be written as

$$\langle x|s\rangle = \sum_{j=1}^{N} \Psi(r_j)e^{-i\Omega_j} \tag{2.7}$$

where:

$$\Psi(r_j) = \Psi(r_{x,j})\Psi(r_{j,s}) \tag{2.8}$$

and

$$\Omega_j = (\theta_j + \phi_j) \tag{2.9}$$

The propagation probability can be obtained by expanding Equation 2.7 and multiplying the expansion by its complex conjugate. In other words, by performing the multiplication

$$\langle x|s\rangle \langle x|s\rangle^* = |\langle x|s\rangle|^2 \tag{2.10}$$

and using the identity

$$2\cos(\Omega_m - \Omega_j) = e^{-i(\Omega_m - \Omega_j)} + e^{i(\Omega_m - \Omega_j)} \tag{2.11}$$

we can write the generalized propagation probability in one dimension:

$$|\langle x|s\rangle|^2 = \sum_{j=1}^{N} \Psi(r_j) \sum_{m=1}^{N} \Psi(r_m)e^{i(\Omega_m - \Omega_j)} \tag{2.12}$$

which can be expressed as (Duarte and Paine 1989; Duarte 1991)

$$|\langle x|s\rangle|^2 = \sum_{j=1}^{N} \Psi(r_j)^2 + 2\sum_{j=1}^{N} \Psi(r_j) \left(\sum_{m=j+1}^{N} \Psi(r_m)\cos(\Omega_m - \Omega_j) \right) \tag{2.13}$$

Interference due to transmission in a two-dimensional transmission grating can be described considering the experimental setup depicted in Figure 2.6. Propagation of quanta occurs from s to x via a two-dimensional transmission grating j_{zy}, that is, j is

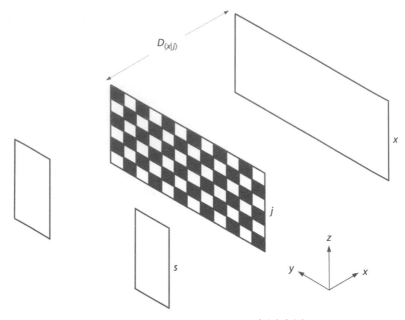

FIGURE 2.6 A two-dimensional representation of the $\langle x|j\rangle\langle j|s\rangle$ geometry.

replaced by a grid comprising j components in the y direction and j components in the z direction. Note that in the one-dimensional case, only the y component of j is present, which is written simply as j. The plane configured by the j_{zy} grid is orthogonal to the plane of propagation. Hence, for photon propagation from s to x, via j_{zy}, the probability amplitude is given by (Duarte 1995a)

$$\langle x|s\rangle = \sum_{j_z=1}^{N} \sum_{j_y=1}^{N} \langle x|j_{zy}\rangle \langle j_{zy}|s\rangle \qquad (2.14)$$

Now, if j is abstracted from j_{zy}, then the above equation can be expressed as

$$\langle x|s\rangle = \sum_{z=1}^{N} \sum_{y=1}^{N} \Psi(r_{zy})e^{-i\Omega_{zy}} \qquad (2.15)$$

and the corresponding probability is given by (Duarte 1995a)

$$\left|\langle x|s\rangle\right|^2 = \sum_{z=1}^{N} \sum_{y=1}^{N} \Psi(r_{zy}) \sum_{q=1}^{N} \sum_{p=1}^{N} \Psi(r_{qp})e^{i(\Omega_{qp}-\Omega_{zy})} \qquad (2.16)$$

For a three-dimensional transmission grating, it can be shown that (Duarte 1995a)

$$\left|\langle x|s\rangle\right|^2 = \sum_{z=1}^{N} \sum_{y=1}^{N} \sum_{x=1}^{N} \Psi(r_{zyx}) \sum_{q=1}^{N} \sum_{p=1}^{N} \sum_{r=1}^{N} \Psi(r_{qpr})e^{i(\Omega_{qpr}-\Omega_{zyx})} \qquad (2.17)$$

It is important to emphasize that the concepts described here apply to either the propagation of single photons and the propagation of ensembles of *coherent, indistinguishable*, or *monochromatic* photons or quanta. The application of quantum principles to the description of propagation of a large number of monochromatic, or indistinguishable, photons was already advanced by Dirac in his discussion of interference (Dirac 1978; Duarte 1998).

2.3.1 EXAMPLE

The generalized interference equation in one dimension (Equation 2.13) can be used to write explicit expressions for double-slit ($N = 2$), triple-slit ($N = 3$), quadruple-slit ($N = 4$), quintuple-slit ($N = 5$), sextuple-slit ($N = 6$), septuple-slit ($N = 7$) interference, and so on. For a seven-slit ($N = 7$) interference, we get

$$
\begin{aligned}
\left|\langle x|s\rangle\right|^2 &= \Psi(r_1)^2 + \Psi(r_2)^2 + \Psi(r_3)^2 + \Psi(r_4)^2 + \Psi(r_5)^2 + \Psi(r_6)^2 + \Psi(r_7)^2 \\
&+ 2\,\Big[\, \Psi(r_1)\Psi(r_2)\cos(\Omega_2 - \Omega_1) + \Psi(r_1)\Psi(r_3)\cos(\Omega_3 - \Omega_1) + \Psi(r_1)\Psi(r_4)\cos(\Omega_4 - \Omega_1) \\
&+ \Psi(r_1)\Psi(r_5)\cos(\Omega_5 - \Omega_1) + \Psi(r_1)\Psi(r_6)\cos(\Omega_6 - \Omega_1) + \Psi(r_1)\Psi(r_7)\cos(\Omega_7 - \Omega_1) \\
&+ \Psi(r_2)\Psi(r_3)\cos(\Omega_3 - \Omega_2) + \Psi(r_2)\Psi(r_4)\cos(\Omega_4 - \Omega_2) + \Psi(r_2)\Psi(r_5)\cos(\Omega_5 - \Omega_2) \\
&+ \Psi(r_2)\Psi(r_6)\cos(\Omega_6 - \Omega_2) + \Psi(r_2)\Psi(r_7)\cos(\Omega_7 - \Omega_2) + \Psi(r_3)\Psi(r_4)\cos(\Omega_4 - \Omega_3) \\
&+ \Psi(r_3)\Psi(r_5)\cos(\Omega_5 - \Omega_3) + \Psi(r_3)\Psi(r_6)\cos(\Omega_6 - \Omega_3) + \Psi(r_3)\Psi(r_7)\cos(\Omega_7 - \Omega_3) \\
&+ \Psi(r_4)\Psi(r_5)\cos(\Omega_5 - \Omega_4) + \Psi(r_4)\Psi(r_6)\cos(\Omega_6 - \Omega_4) + \Psi(r_4)\Psi(r_7)\cos(\Omega_7 - \Omega_4) \\
&+ \Psi(r_5)\Psi(r_6)\cos(\Omega_6 - \Omega_5) + \Psi(r_5)\Psi(r_7)\cos(\Omega_7 - \Omega_5) + \Psi(r_6)\Psi(r_7)\cos(\Omega_7 - \Omega_6) \,\Big]
\end{aligned}
\tag{2.18}
$$

2.3.2 GEOMETRY OF THE *N*-SLIT INTERFEROMETER

In addition to the generalized interferometric equations, it is important to consider the geometry of the transmission grating (j) in conjunction with the plane of interference (x) for the one-dimensional case as illustrated in Figure 2.7. According to the geometry, the phase difference term in Equation 2.13 can be expressed as (Duarte 1997)

$$
\cos\left[(\theta_m - \theta_j) \pm (\phi_m - \phi_j)\right] = \cos\left(|l_m - l_{m-1}|k_1 \pm |L_m - L_{m-1}|k_2\right)
\tag{2.19}
$$

where:

$$
k_1 = \frac{2\pi n_1}{\lambda_v}
\tag{2.20}
$$

and

$$
k_2 = \frac{2\pi n_2}{\lambda_v}
\tag{2.21}
$$

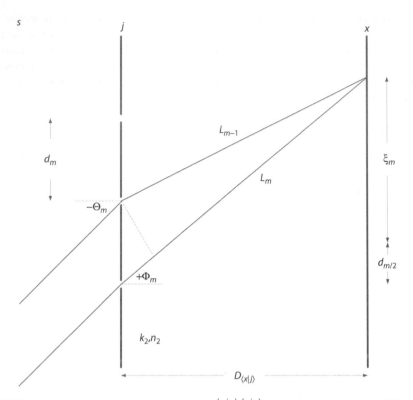

FIGURE 2.7 A detailed representation of the $\langle x|j\rangle\langle j|s\rangle$ geometry depicting the difference in path length and the angles of incidence and diffraction. (Reproduced from Duarte, F.J., *Am. J. Phys.*, 65, 637–640, 1997. With permission from the American Association of Physics Teachers.)

are the wavenumbers of the two optical regions defined in Figure 2.7. Here, $\lambda_1 = \lambda_v/n_1$ and $\lambda_2 = \lambda_v/n_2$, where λ_v is the vacuum wavelength and n_1 and n_2 are the corresponding indexes of refraction (Wallenstein and Hänsch 1974; Born and Wolf 1999). The phase differences can be expressed exactly via the following geometrical equations (Duarte 1993):

$$\left|L_m - L_{m-1}\right| = \frac{2\xi_m d_m}{\left|L_m + L_{m-1}\right|} \tag{2.22}$$

$$L_m^2 = D_{\langle x|j\rangle}^2 + \left(\xi_m + \frac{d_m}{2}\right)^2 \tag{2.23}$$

$$L_{m-1}^2 = D_{\langle x|j\rangle}^2 + \left(\xi_m - \frac{d_m}{2}\right)^2 \tag{2.24}$$

In this notation, ξ_m is the lateral displacement, on the x plane, from the projected median of d_m to the interference plane.

2.3.3   N-Slit Interferometer Experiment

The N-slit interferometer is illustrated in Figure 2.5. In practice, this interferometer can be configured with a variety of lasers including tunable lasers. However, one requirement is that the laser to be utilized must emit in the narrow-linewidth regime and in a single transverse mode (TEM_{00}) with a near-Gaussian profile. Ideally, the source should be a single-longitudinal-mode laser. The reason for this requirement is that narrow-linewidth lasers yield sharp, well-defined interference patterns close to those predicted theoretically for a single wavelength.

One particular configuration of the N-slit interferometer can be integrated by a TEM_{00} He–Ne laser ($\lambda = 632.82$ nm) with a beam 0.5 mm in diameter. It should be emphasized that this class of laser yields a smooth near-Gaussian beam profile and narrow-linewidth emission. The laser beam is then magnified, in two dimensions, by a ×20 Galilean telescope. Following the telescopic expansion, the beam is further expanded in one dimension by a ×5 multiple-prism beam expander. This optical arrangement yields an expanded smooth near-Gaussian beam approximately 50 mm wide. An option is to insert a convex lens prior to the multiple-prism expander. This produces an extremely elongated near-Gaussian beam of 20–30 μm at it maximum height by 50 mm in width (Duarte 1993). The beam propagation through this system can be accurately characterized using ray transfer matrices as discussed in Chapter 6 (Duarte 1995b). Also, as an option, at the exit of the multiple-prism beam expander, an aperture, a few millimeters wide, can be deployed. Thus, the source s can be either the exit prism of the multiple-prism beam expander or the wide aperture.

At this stage, it should be noted that for the illumination of two slits 50 μm in width, separated by 50 μm, the elongated Gaussian provides a nearly plane illumination profile, which is also approximately the case even if a larger number of slits, of these dimensions, are illuminated. For the particular case of a two-slit experiment, or Young's interference experiment, involving 50-μm slits separated by 50 μm and a grating-to-screen distance (a) of 10 cm, the interference signal is displayed in Figure 2.8a. The calculated interference using Equation 2.13 and assuming plane wave illumination is given in Figure 2.8b. The interference screen at x is a digital detector comprising a photodiode array, each 25 μm in width. For an array of $N = 100$ slits, each 30 μm in width and separated by 30 μm, the measured and calculated interferograms are shown in Figure 2.9. Here, the grating-to-digital detector distance is $D_{\langle x|j \rangle} = 75$ cm.

In practice, the transmission gratings are not perfect and offer an uncertainty in the dimension of the slits. The uncertainty in the slit dimensions of the grating, incorporating the 30-μm slits, used in this experiment was measured to be ≤2%. The theoretical interferogram for the grating comprising $N = 100$ slits, each 30.0 ± 0.6 μm wide and separated by 30.0 ± 0.6 μm, is given in Figure 2.10. Notice the slight symmetry deterioration.

When a wide slit is used to select the central portion of the elongated Gaussian beam, the interaction of the coherent laser beam with the slit results in diffraction prior to the illumination of the transmission grating. The interferometric equation (2.13) can be used to characterize this diffraction. This is done by dividing the wide

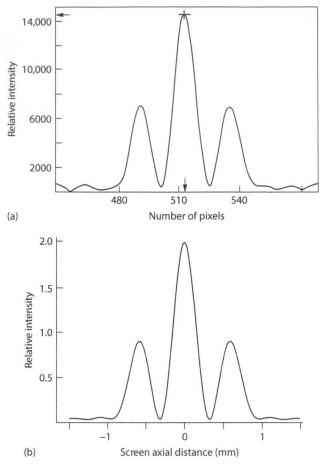

FIGURE 2.8 (a) Measured interferogram resulting from the interaction of coherent laser emission at $\lambda = 632.82$ nm and two ($N = 2$) slits 50 μm wide, separated by 50 μm. The j-to-x distance is $D_{\langle x|j \rangle} = 10$ cm. (b) Corresponding theoretical interferogram from Equation 2.13. Note that the *screen axial distance* refers to the distance at the interferometric plane that defines the spatial width of the interferogram and that is perpendicular to the propagation axis. (Reproduced from Duarte, F.J., *Quantum Optics for Engineers*, CRC Press, New York, 2014. With permission from Taylor & Francis.)

slit into hundreds of smaller slits. As an example, a 4-mm-wide aperture is divided into 800 slits, each 4 μm wide and separated by a 1-μm interslit distance. The calculated near-field diffraction pattern, for a distance on intra-interferometric distance of $D_{\langle x|j \rangle} = 10$ cm, is shown in Figure 2.11. Using this as the radiation source to illuminate the 100-slit grating, comprising 30 μm slits with an interslit distance of 30 μm (for $D_{\langle x|j \rangle} = 75$ cm), yields the theoretical interferogram displayed in Figure 2.12. Thus, the interferometric equation (2.13) can be used in a cascade approach from an illumination plane to an interference plane. This cascade approach can be applied to multiple intermediate planes. More on this approach is given in Chapter 10.

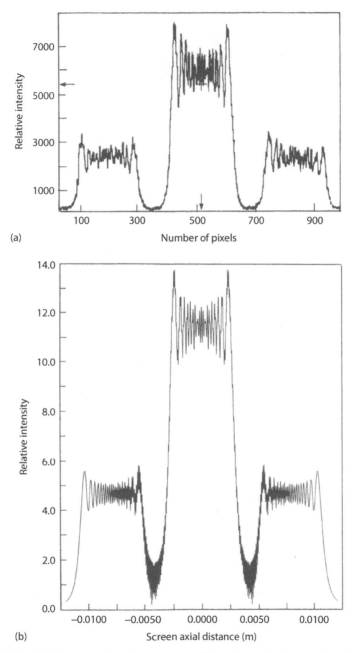

(a)

(b)

FIGURE 2.9 (a) Measured interferogram resulting from the interaction of coherent laser emission at $\lambda = 632.82$ nm and $N = 100$ slits 30 μm wide, separated by 30 μm. The j-to-x distance is $D_{\langle x|j \rangle} = 75$ cm. (b) Corresponding theoretical interferogram from Equation 2.13. (Reprinted from *Opt. Commun.*, 103, Duarte, F.J., On a generalized interference equation and interferometric measurements, 8–14, Copyright 1993, with permission from Elsevier.)

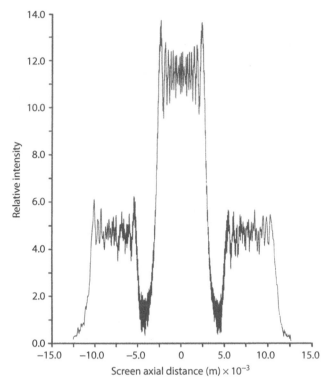

FIGURE 2.10   Theoretical interferometric/diffraction distribution using a $\leq 2\%$ uncertainty in the dimensions of the 30-μm slits. In this calculation, $N = 100$ and the j-to-x distance is $D_{\langle x|j\rangle} = 75\,\text{cm}$. A slight deterioration in the spatial symmetry of the distribution is evident. (Reprinted from *Opt. Commun.*, 103, Duarte, F.J., On a generalized interference equation and interferometric measurements, 8–14, Copyright 1993, with permission from Elsevier.)

2.4   GENERALIZED DIFFRACTION

Feynman, in his usual style, stated that "no one has ever been able to define the difference between interference and diffraction satisfactorily" (Feynman et al. 1965b; pp. 30–1). His point is well taken. In the discussion related to Figure 2.9, and its variants, reference was only made to interference. However, what we really have is interference in three diffraction orders: the zeroth, or central order, and the ± 1, or secondary orders. In other words, there is an interference pattern associated with each diffraction order. Physically, however, it is the same phenomenon. The interaction of coherent light with a set of slits, in the near field, gives rise to an interference pattern. As the distance $D_{\langle x|j\rangle}$ from j to x increases, the central interference pattern begins to give origin to secondary patterns that gradually separate from the central order at lower intensities. These are the ± 1 diffraction orders. This physical phenomenon, as one goes from the near to the far field, is illustrated in Figure 2.13. One of the beauties of the Dirac description of optics is the ability to continuously move from the near to the far field with a single mathematical description.

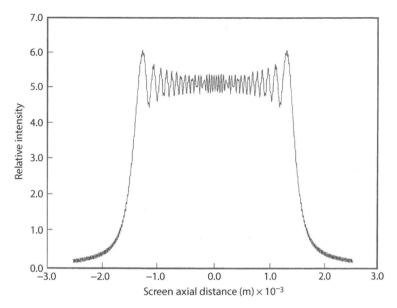

FIGURE 2.11   Theoretical near-field diffraction distribution produced by a 4-mm aperture illuminated at λ = 632.82 nm. The *j*-to-*x* distance is $D_{\langle x|j\rangle}$ = 10 cm. (Reprinted from *Opt. Commun.*, 103, Duarte, F.J., On a generalized interference equation and interferometric measurements, 8–14, Copyright 1993, with permission from Elsevier.)

The second interference–diffraction entanglement refers to the fact that our generalized interference equation can be naturally applied to describe a diffraction pattern produced by a single wide slit as shown in Figure 2.11. Under these circumstances, the wide slit is mathematically represented by a series of subslits.

The generalized description of diffraction given next includes the refinement and extension of the original presentation given by Duarte (1997, 2003) and includes the treatment of positive and negative diffraction as introduced by Duarte (2006), and follows the style of Duarte (2014).

The intimate relation between interference and diffraction has its origin in the interferometric equation itself (Duarte 2003):

$$\left|\langle x|s\rangle\right|^2 = \sum_{j=1}^{N}\Psi(r_j)^2 + 2\sum_{j=1}^{N}\Psi(r_j)\left[\sum_{m=j+1}^{N}\Psi(r_m)\cos(\Omega_m - \Omega_j)\right]$$

for it is the $\cos(\Omega_m - \Omega_j)$ term that gives rise to different diffraction orders while the interference follows the mechanics established the overall equation. Here, we revisit the geometry at the *N*-slit plane *j* and illustrate what is obviously seen in Figure 2.13: up on arrival to a slit, diffraction occurs symmetrically toward both sides as illustrated in Figures 2.14 through 2.17. Figure 2.14 depicts the usual description associated with incidence below the normal (−) and diffraction above the normal (+). Figure 2.15 illustrates the incidence above the normal (+) and the

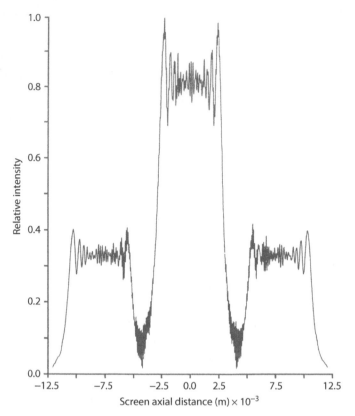

FIGURE 2.12 Theoretical interferometric distribution illustrating a cascade calculation. It incorporates diffraction-edge effects in the illumination. In this calculation, the width of the slits in the array is 30 μm separated by 30 μm, $N = 100$, and the j-to-x distance is $D_{\langle x|j \rangle} = 75$ cm. The aperture-to-grating distance is 10 cm. (Reprinted from *Opt. Commun.*, 103, Duarte, F.J., On a generalized interference equation and interferometric measurements, 8–14, Copyright 1993, with permission from Elsevier.)

diffraction above the normal (+) (Duarte 2006). For completeness, we also include the case of incidence below the normal (−) followed by diffraction below the normal (−) and incidence above the normal (+) followed by diffraction below the normal (−) (Figures 2.16 and 2.17). Thus, the equations describing the original geometry (Duarte 1997) are slightly modified to account for all the ± alternatives:

$$\cos\left[\pm(\theta_m - \theta_j) \pm (\phi_m - \phi_j)\right] = \cos\left(\pm \left|l_m - l_{m-1}\right| k_1 \pm \left|L_m - L_{m-1}\right| k_2\right) \qquad (2.25)$$

where:

$$k_1 = \frac{2\pi n_1}{\lambda_v} \qquad (2.26)$$

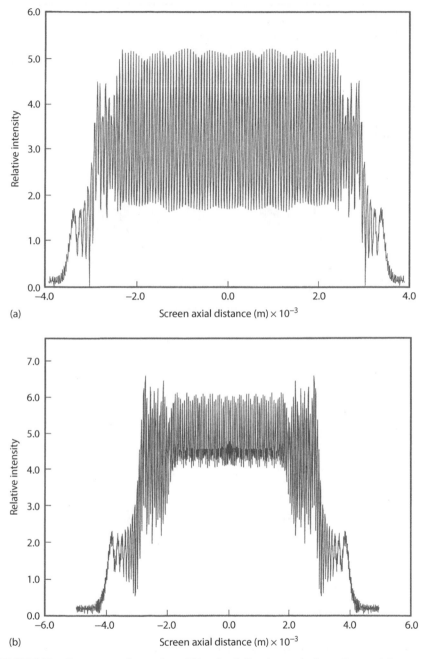

(a)

(b)

FIGURE 2.13 Emergence of secondary diffraction (±1) orders as the j-to-x distance is increased. (a) At a grating-to-screen distance of $D_{\langle x|j\rangle} = 5\,\mathrm{cm}$, the interferometric distribution is mainly part of a single order. At the boundaries, there is an incipient indication of emerging orders.

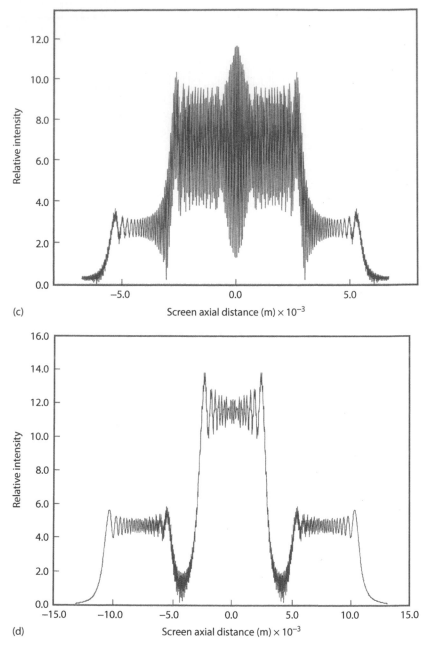

(c)   Screen axial distance (m) × 10⁻³

(d)   Screen axial distance (m) × 10⁻³

FIGURE 2.13 (Continued) (b) As the distance is increased to $D_{\langle x|j\rangle} = 10\,\text{cm}$, the presence of the emerging (±1) orders is more visible. (c) At a distance of $D_{\langle x|j\rangle} = 25\,\text{cm}$, the emerging ($\pm1$) orders give rise to an overall distribution with clear *shoulders*. (d) At a distance of $D_{\langle x|j\rangle} = 75\,\text{cm}$, the -1, 0, and $+1$ diffraction orders are clearly established. Notice the increase in the width of the distribution as the intra-interferometric distance $D_{\langle x|j\rangle}$ increases from 5 to 75 cm. The width of the slits is 30 μm separated by 30 μm, $N = 100$, and $\lambda = 632.82$ nm.

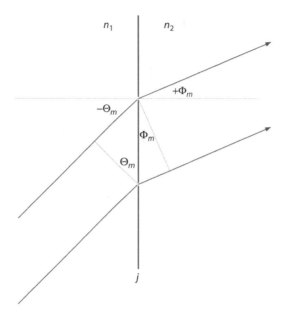

FIGURE 2.14 Incidence below the normal (−) and diffraction above the normal (+). (Reproduced from Duarte, F.J., *Quantum Optics for Engineers*, CRC Press, New York, 2014. With permission from Taylor & Francis.)

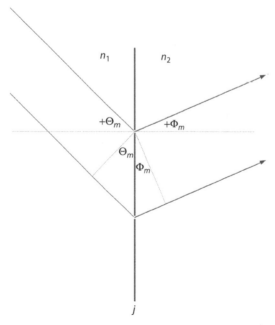

FIGURE 2.15 Incidence above the normal (+) and diffraction above the normal (+). (Reproduced from Duarte, F.J., *Quantum Optics for Engineers*, CRC Press, New York, 2014. With permission from Taylor & Francis.)

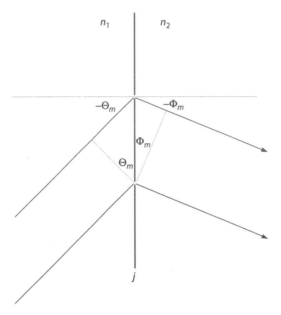

FIGURE 2.16   Incidence below the normal (−) followed by diffraction below the normal (−). (Reproduced from Duarte, F.J., *Quantum Optics for Engineers*, CRC Press, New York, 2014. With permission from Taylor & Francis.)

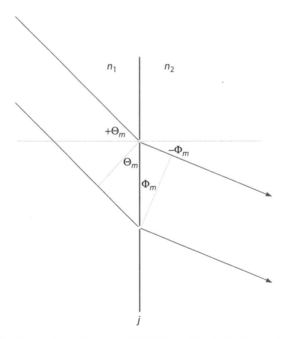

FIGURE 2.17   Incidence above the normal (+) followed by diffraction below the normal (−). (Reproduced from Duarte, F.J., *Quantum Optics for Engineers*, CRC Press, New York, 2014. With permission from Taylor & Francis.)

and

$$k_2 = \frac{2\pi n_2}{\lambda_v} \tag{2.27}$$

are the wavenumbers of the two optical regions defined in Figures 2.14 through 2.17. Here, as we saw earlier, $\lambda_1 = \lambda_v/n_1$ and $\lambda_2 = \lambda_v/n_2$, where λ_v is the vacuum wavelength and n_1 and n_2 are the corresponding indexes of refraction.

As mentioned earlier, the phase differences can be expressed exactly via the exact geometrical expressions (Duarte 1993):

$$\left| L_m - L_{m-1} \right| = \frac{2\xi_m d_m}{\left| L_m + L_{m-1} \right|} \tag{2.28}$$

$$L_m^2 = D_{\langle x|j \rangle}^2 + \left(\xi_m + \frac{d_m}{2} \right)^2 \tag{2.29}$$

$$L_{m-1}^2 = D_{\langle x|j \rangle}^2 + \left(\xi_m - \frac{d_m}{2} \right)^2 \tag{2.30}$$

From the geometry of Figure 2.7, we can write

$$\sin \Phi_m = \frac{\xi_m + (d_m/2)}{L_m} \tag{2.31}$$

and for the condition $D_{\langle x|j \rangle} \gg d_m$, we have $\left| L_m + L_{m-1} \right| \approx 2L_m$. Then using Equation 2.28, we have

$$\left| L_m - L_{m-1} \right| \approx d_m \sin \Phi_m \tag{2.32}$$

$$\left| l_m - l_{m-1} \right| \approx d_m \sin \Theta_m \tag{2.33}$$

where:
  Θ_m and Φ_m are the angles of incidence and diffraction, respectively. Given that maxima occur at

$$\left(\pm \left| l_m - l_{m-1} \right| n_1 \pm \left| L_m - L_{m-1} \right| n_2 \right) \frac{2\pi}{\lambda_v} = M\pi \tag{2.34}$$

then using Equations 2.32 and 2.33

$$d_m \left(\pm n_1 \sin \Theta_m \pm n_2 \sin \Phi_m \right) \frac{2\pi}{\lambda_v} = M\pi \tag{2.35}$$

where:
  $M = 0, 2, 4, 6, \dots$

For $n_1 = n_2$, we have $\lambda = \lambda_v$, and this equation reduces to the *generalized diffraction grating equation*:

$$d_m\left(\pm\sin\Theta_m\pm\sin\Phi_m\right)=m\lambda \qquad (2.36)$$

where:

$m = 0, 1, 2, 3,...$ are the various *diffraction orders*

A most important observation is due here: in our discussion on the interferometric equations, we have made explicit reference to the exact geometrical equations (2.28 through 2.30). However, in the derivation of Equations 2.35 and 2.36, we have used the approximation $D_{\langle x|j\rangle} \gg d_m$. Are we being consistent? The answer is yes. The exact equations (2.28 through 2.30) are used in the generalized interferometric equation (2.13), whereas the approximation $D_{\langle x|j\rangle} \gg d_m$ has been applied in the derivation of the generalized diffraction equation

$$d_m\left(\pm\sin\Theta_m\pm\sin\Phi_m\right)=m\lambda$$

that manifests itself in the *far field* as beautifully illustrated in Figure 2.13d. From Equation 2.36, it is clearly seen that beyond the zeroth order, m can take a series of $\pm$ values, that is, $m = \pm 1, \pm 2, \pm 3,...$.

2.4.1 POSITIVE DIFFRACTION

From the generalized diffraction equation (2.36) including both $\pm$ alternatives, the usual traditional diffraction equation can be stated as

$$d_m\left(\sin\Theta_m\pm\sin\Phi_m\right)=m\lambda \qquad (2.37)$$

which was previously derived starting from (Duarte 1997)

$$\cos\left[(\theta_m-\theta_j)\pm(\phi_m-\phi_j)\right]=\cos\left(|l_m-l_{m-1}|k_1\pm|L_m-L_{m-1}|k_2\right) \qquad (2.38)$$

From Equation 2.37, setting $\Theta_m = \Phi_m = \Theta$, the diffraction grating equation for Littrow configuration emerges from the well-known grating equation:

$$m\lambda = 2d_m\sin\Theta \qquad (2.39)$$

2.5 POSITIVE AND NEGATIVE REFRACTION

So far we have discussed interference and diffraction and we have seen how diffraction manifests itself as the interferometric distribution propagates toward the far field. An additional fundamental phenomenon in optics is *refraction*.

Refraction is the change in the geometrical path of a beam of light due to transmission from the original medium of propagation to a second medium with a different refractive index. For example, refraction is the bending of a ray of light caused due to propagation in a glass, or crystalline, prism.

If in the diffraction grating equation d_m is made very small relative to a given λ, diffraction ceases to occur and the only solution can be found for $m = 0$ (Duarte 1997).

That is, under these conditions, a grating made of grooves coated on a transparent substrate, such as optical glass, does not diffract but exhibits the refraction properties of the glass. For example, since the maximum value of $(\pm \sin \Theta_m \pm \sin \Phi_m)$ is 2, for a 5000-lines/mm transmission grating, let us say, no diffraction can be observed for the visible spectrum. Hence, for the condition $d_m \ll \lambda$, the diffraction grating equation can only be solved for

$$d_m(\pm n_1 \sin \Theta_m \pm n_2 \sin \Phi_m) \frac{2\pi}{\lambda_v} = 0 \qquad (2.40)$$

which leads to

$$(\pm n_1 \sin \Theta_m \pm n_2 \sin \Phi_m) = 0 \qquad (2.41)$$

For the case of incidence below the normal ($-$) and refraction above the normal ($+$) (Figure 2.14),

$$-n_1 \sin \Theta_m + n_2 \sin \Phi_m = 0 \qquad (2.42)$$

so that

$$n_1 \sin \Theta_m = n_2 \sin \Phi_m \qquad (2.43)$$

which is the well-known *equation of refraction*, also known as *Snell's law*. Under the present physical conditions, Θ_m is the angle of incidence and Φ_m becomes the *angle of refraction*. The same outcome is obtained for incidence above the normal ($+$) and refraction below the normal ($-$) (Figure 2.17).

For the case of incidence above the normal ($+$) and refraction above the normal ($+$) (Figure 2.15),

$$+n_1 \sin \Theta_m + n_2 \sin \Phi_m = 0 \qquad (2.44)$$

so that

$$n_1 \sin \Theta_m = -n_2 \sin \Phi_m \qquad (2.45)$$

which is the refraction law for *negative refraction*. The same outcome is obtained for incidence below the normal ($-$) and diffraction below the normal ($-$) (Figure 2.16).

2.6   REFLECTION

The discussion on interference, up to now, has involved an N-slit array or a transmission grating. It should be indicated that the arguments and physics apply equally well to a reflection interferometer (Duarte 2003), that is, to an interferometer incorporating a reflection, rather than a transmission, grating. Explicitly, if a mirror is placed at an infinitesimal distance immediately behind the N-slit array, as illustrated in Figure 2.18, then the transmission interferometer becomes a reflection interferometer. Under these circumstances, the equations

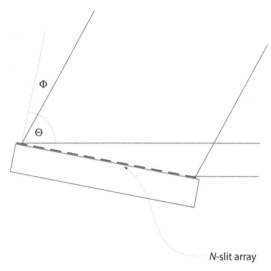

FIGURE 2.18   A reflection diffraction grating is formed by approaching a reflection surface at an infinitesimal distance to the array of N-slits.

$$d_m \left(\pm n_1 \sin \Theta_m \pm n_2 \sin \Phi_m \right) \frac{2\pi}{\lambda_v} = M\pi$$

and

$$d_m \left(\pm \sin \Theta_m \pm \sin \Phi_m \right) = m\lambda$$

apply in the reflection domain, with Θ_m being the incidence angle and Φ_m the diffraction angle in the *reflection* domain. For the case of $d_m \ll \lambda$ and $n_1 = n_2$, we then have

$$\left(\pm \sin \Theta_m \pm \sin \Phi_m \right) = 0 \qquad\qquad (2.46)$$

For incidence above the normal (+), and reflection below the normal (−),

$$+ \sin \Theta_m - \sin \Phi_m = 0 \qquad\qquad (2.47)$$

which means

$$\Theta_m = \Phi_m \qquad\qquad (2.48)$$

where:
   Θ_m is the angle of incidence
   Φ_m is the angle of reflection

This is known as the *law of reflection*.

2.7   THE CAVITY LINEWIDTH EQUATION

As previously outlined, starting from the generalized interferometric equation

$$\left|\langle x|s\rangle\right|^2 = \sum_{j=1}^{N} \Psi(r_j)^2 + 2\sum_{j=1}^{N} \Psi(r_j)\left[\sum_{m=j+1}^{N} \Psi(r_m)\cos(\Omega_m - \Omega_j)\right]$$

the generalized diffraction equation

$$d_m(\pm\sin\Theta_m \pm \sin\Phi_m) = m\lambda$$

is obtained, and for positive diffraction and $\Theta_m \approx \Phi_m(=\Theta)$, the diffraction grating equation

$$2d\sin\Theta = m\lambda$$

in Littrow configuration can be established.

Following the approach of Duarte (1992) and considering two slightly different wavelengths, an expression for the wavelength difference can be written as

$$\Delta\lambda = \frac{2d}{m}(\sin\Theta_1 - \sin\Theta_2) \qquad (2.49)$$

For $\Theta_1 \approx \Theta_2(=\Theta)$, this equation can be restated as

$$\Delta\lambda \approx \frac{2d}{m}\Delta\theta\left(1 - \frac{3\Theta^2}{3!} + \frac{5\Theta^4}{5!} - \frac{7\Theta^6}{7!} + \cdots\right) \qquad (2.50)$$

Differentiation of the grating equation leads to

$$\frac{\partial\theta}{\partial\lambda}\cos\Theta = \frac{m}{2d} \qquad (2.51)$$

and substitution into Equation 2.50 yields

$$\Delta\lambda \approx \Delta\theta\left(\frac{\partial\theta}{\partial\lambda}\right)^{-1}\left(1 - \frac{\Theta^2}{2!} + \frac{\Theta^4}{4!} - \frac{\Theta^6}{6!} + \cdots\right)(\cos\Theta)^{-1} \qquad (2.52)$$

which reduces to the well-known cavity linewidth equation (Duarte 1992):

$$\Delta\lambda \approx \Delta\theta\left(\frac{\partial\theta}{\partial\lambda}\right)^{-1} \qquad (2.53)$$

or

$$\Delta\lambda \approx \Delta\theta(\nabla_\lambda\theta)^{-1} \qquad (2.54)$$

where:

$$\nabla_\lambda \theta = (\partial\theta/\partial\lambda)$$

This equation has been used extensively to determine the emission linewidth in pulsed narrow-linewidth dispersive laser oscillators (Duarte 1990). It originates from the generalized N-slit interference equation and incorporates $\Delta\theta$ whose value can be determined either from the uncertainty principle or from the interferometric equation itself (see Chapter 3). This equation is well known in the field of classical spectrometers where it has been introduced using geometrical arguments (Robertson 1955). In addition to its technical and computational usefulness, Equation 2.53 or 2.54 or both illustrate the inherent interdependence between spectral and spatial coherence.

2.7.1   INTRODUCTION TO ANGULAR DISPERSION

Angular dispersion is an important quantity in optics that describes the ability for an optical element, such as a diffraction grating or prism, to geometrically separate a beam of light as a function of wavelength. As just seen, mathematically the angular dispersion is represented by the differential $(\partial\Theta/\partial\lambda)$. For spectrophotometers and wavelength meters based on dispersive elements, such as diffraction gratings and prism arrays, the dispersion should be as large as possible since it enables a higher wavelength spatial discrimination or higher spatial resolution. Further, in the case of dispersive laser oscillators, a high dispersion leads to the achievement of narrow-linewidth emission since the dispersive linewidth is given by

$$\Delta\lambda \approx \Delta\theta\left(\frac{\partial\theta}{\partial\lambda}\right)^{-1}$$

Next, two important forms of dispersion, directly relevant to narrow-linewidth laser oscillators, are introduced: diffraction grating dispersion and prismatic dispersion.

For a uniform diffraction grating, $d_m = d$, and the diffraction grating equation becomes

$$d(\sin\Theta \pm \sin\Phi) = m\lambda \tag{2.55}$$

The angular dispersion is calculated by differentiating Equation 2.55 so that

$$\frac{\partial\Theta}{\partial\lambda} = \frac{m}{d\cos\Theta} \tag{2.56}$$

or alternatively

$$\frac{\partial\Theta}{\partial\lambda} = \frac{(\sin\Theta \pm \sin\Phi)}{\lambda\cos\Theta} \tag{2.57}$$

For a prism deployed at minimum deviation, as illustrated in Figure 2.19, the following set of geometrical relations apply:

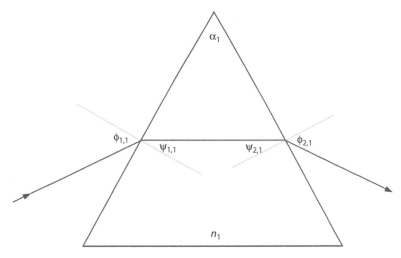

FIGURE 2.19   Single prism depicting refraction at minimum deviation.

$$\phi_1 + \phi_2 = \varepsilon + \alpha \tag{2.58}$$

$$\psi_1 + \psi_2 = \alpha \tag{2.59}$$

$$\sin\phi_1 = n\sin\psi_1 \tag{2.60}$$

$$\sin\phi_2 = n\sin\psi_2 \tag{2.61}$$

where:

ϕ_1 and ϕ_2 are the angles of incidence and emergence
ψ_1 and ψ_2 are the corresponding angles of refraction
α is the apex angle
ε is the angle of deviation
n is the index of refraction of the prism

Differentiating Equation 2.59 with respect to n, we obtain the identity

$$\frac{d\psi_1}{dn} = -\frac{d\psi_2}{dn} \tag{2.62}$$

and differentiating Equations 2.60 and 2.61, we obtain

$$\frac{d\phi_2}{dn} = \frac{\sin\psi_2}{\cos\phi_2} + \left(\frac{\cos\psi_2}{\cos\phi_2}\right)\tan\psi_1 \tag{2.63}$$

which is the result given by Born and Wolf (1999). The use of the identity

$$\frac{d\phi_2}{d\lambda} = \left(\frac{d\phi_2}{dn}\right)\left(\frac{dn}{d\lambda}\right) \tag{2.64}$$

provides the dispersion for a single prism (Duarte 1990):

$$\frac{d\phi_2}{d\lambda} = \left[\frac{\sin\psi_2}{\cos\phi_2} + \left(\frac{\cos\psi_2}{\cos\phi_2} \right) \tan\psi_1 \right] \frac{dn}{d\lambda} \tag{2.65}$$

which for orthogonal beam exit, that is, $\phi_2 \approx \psi_2 \approx 0$, becomes

$$\frac{d\phi_2}{d\lambda} \approx \tan\psi_1 \frac{dn}{d\lambda} \tag{2.66}$$

The generalized multiple-prism dispersion, for both positive and negative diffraction, is considered in detail in Chapter 4.

2.8   DIRAC AND THE LASER

In his seminal discussion on interference, in 1930, Dirac introduced the concept of interference as a one-photon phenomenon (Dirac 1978). Explicitly, he stated, "Each photon then interferes only with itself. Interference between two different photons never occurs" (p. 9). This concept is central to explain the physics of the N-slit interferometer since it is a single photon that illuminates the whole array of N slits, or grating, simultaneously. In the case of a monochromatic laser beam or an ensemble of *indistinguishable* photons, or indistinguishable quanta, all the indistinguishable photons illuminate the array of N slits, or grating, simultaneously. In the past, this concept has been the source of some controversy due to a misunderstanding of the Dirac interpretation, which implies that *indistinguishable photons*, regardless of source of origin, are the *same photons*.

As explained in Chapter 11, interference *a là Dirac*, utilizing indistinguishable photons, via

$$\left| \langle x | s \rangle \right|^2 = \sum_{j=1}^{N} \Psi(r_j)^2 + 2 \sum_{j=1}^{N} \Psi(r_j) \left[\sum_{m=j+1}^{N} \Psi(r_m) \cos(\Omega_m - \Omega_j) \right]$$

yields spatially sharp, well-defined interferograms with a high degree of visibility near unity. Due to the integrated and cumulative nature of the detection process (Duarte 2004, 2008), interferograms from semicoherent sources produce interferograms with broad features, lower spatial definition, and decreased visibility (see Figure 11.11). At the other extreme, plainly distinguishable photons, or photons of distinct different wavelengths (such as a blue photon and a red photon), do not interfere with each other.

The Dirac discussion on the interference of photons goes even further. It begins with reference to *a beam of roughly monochromatic light*, and then prior to his dictum on interference, he wrote about *a beam of light having a large number of photons*. It is this beam of light that in his discussion is divided into two components and is subsequently *made to interference*. In present terms, this is no different than the description of interference due to the interaction of a high-power narrow-linewidth laser beam with a two-beam interferometer (Duarte 1998). In other words, in 1930 Dirac provided perhaps the earliest physical description of a laser beam, laser interference, and hence, laser optics.

PROBLEMS

2.1 Show that substitution of Equations 2.5 and 2.6 into Equation 2.3 leads to Equation 2.7.

2.2 Show that Equation 2.12 can be expressed as Equation 2.13.

2.3 From the geometry of Figure 2.7, derive Equations 2.22 through 2.24.

2.4 Write an equation for $|\langle x|s \rangle|^2$ in the case relevant to Figure 2.12, that is, an N-slit grating illuminated by a wide single slit. Assume that the single wide slit can be represented by an array of N subslits.

2.5 Show that, for orthogonal beam exit, Equation 2.65 reduces to Equation 2.66.

3 The Uncertainty Principle in Optics

3.1 APPROXIMATE DERIVATION OF THE UNCERTAINTY PRINCIPLE

Heisenberg's uncertainty principle (Heisenberg 1927) is of fundamental importance to optics and to laser optics in particular. Here, optical arguments are applied to outline an approximate derivation of the uncertainty principle.

3.1.1 THE WAVE CHARACTER OF PARTICLES

The quantum energy of a particle is given by the well-known quantum energy equation

$$E = h\nu \tag{3.1}$$

where:
  h is Planck's constant

Equating this to the relativistic energy $E = mc^2$ and using the identity $\lambda = c/\nu$, an expression for the momentum of a particle can be given as

$$p = \frac{h}{\lambda} \tag{3.2}$$

which, using the identity

$$k = \frac{2\pi}{\lambda} \tag{3.3}$$

can be restated as

$$p = \hbar k \tag{3.4}$$

This momentum equation was applied to particles, such as electrons, by de Broglie (Haken 1981). Thus, wave properties such as frequency and wavelength were attributed to the motion of particles.

3.1.2 THE DIFFRACTION IDENTITY AND THE UNCERTAINTY PRINCIPLE

In his discussion of position and momentum, Feynman considers the relative uncertainty in the wavelength that can be measured with a given grating (Feynman et al., 1965). Using such approach, he arrives at an expression for the resolving power of a diffraction grating: $\Delta\lambda/\lambda$. His discussion is then extended to include the length of a wave train in order to derive an identity for $\Delta\lambda/\lambda^2$. Here, the origin of this expression is examined from an interferometric perspective.

The generalized one-dimensional interferometric equation derived using Dirac's notation is given by (Duarte 1991)

$$\left|\langle x|s\rangle\right|^2 = \sum_{j=1}^{N}\Psi(r_j)^2 + 2\sum_{j=1}^{N}\Psi(r_j)\left[\sum_{m=j+1}^{N}\Psi(r_m)\cos(\Omega_m - \Omega_j)\right] \tag{3.5}$$

The interference term in this equation can be expressed as (Duarte 1997)

$$\cos\left[(\theta_m - \theta_j)\pm(\phi_m - \phi_j)\right] = \cos\left(\left|l_m - l_{m-1}\right|k_1 \pm \left|L_m - L_{m-1}\right|k_2\right) \tag{3.6}$$

where:

$$k_1 = \frac{2\pi n_1}{\lambda_v}$$

and

$$k_2 = \frac{2\pi n_2}{\lambda_v}$$

are the wave numbers for the two optical regions, defined in Figure 3.1 (see Chapter 2 for further details). Also, $\left|l_m - l_{m-1}\right|$ and $\left|L_m - L_{m-1}\right|$ are the corresponding path differences. Since it can be shown that *maxima* can occur at

$$\left(\left|l_m - l_{m-1}\right|n_1 \pm \left|L_m - L_{m-1}\right|n_2\right)\frac{2\pi}{\lambda_v} = M\pi \tag{3.7}$$

where:
  $M = 0, \pm 2, \pm 4, \pm 6, \ldots$

it can be shown that

$$d_m\left(n_1\sin\Theta_m \pm n_2\sin\Phi_m\right)\frac{2\pi}{\lambda_v} = M\pi \tag{3.8}$$

which, for $n_1 = n_2 = 1$ and $\lambda = \lambda_v$, reduces to

$$d_m(\sin\Theta_m \pm \sin\Phi_m)k = M\pi \tag{3.9}$$

that can be expressed as the well-known grating equation

$$d_m(\sin\Theta_m \pm \sin\Phi_m) = m\lambda \tag{3.10}$$

where:
  $m = 0, \pm 1, \pm 2, \pm 3, \ldots$

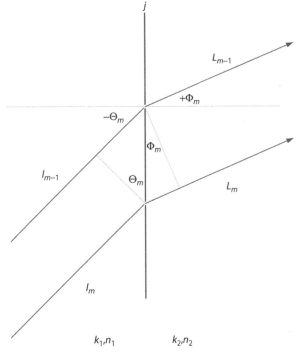

FIGURE 3.1 Boundary at a transmission grating depicting corresponding path differences. Further details on the geometry of the transmission grating are given in Chapter 2.

For a grating deployed in the reflection domain, and at Littrow configuration, $\Theta_m = \Phi_m = \Theta$ so that the grating equation reduces to

$$2d \sin\Theta = m\lambda \qquad (3.11)$$

Using this equation, one can consider an expanded light beam incident on a reflection grating as illustrated in Figure 3.2. For an infinitesimal change in wavelength

$$\lambda_1 = \frac{2d}{m}\left(\frac{\Delta x_1}{l}\right) \qquad (3.12)$$

$$\lambda_2 = \frac{2d}{m}\left(\frac{\Delta x_2}{l}\right) \qquad (3.13)$$

where:
 l is the grating length
 Δx is the path difference

From the geometry of Figure 3.2 and the corresponding grating equation for the Littrow configuration (Equation 3.11),

$$\frac{2d}{m} = \frac{l\lambda}{\Delta x} \qquad (3.14)$$

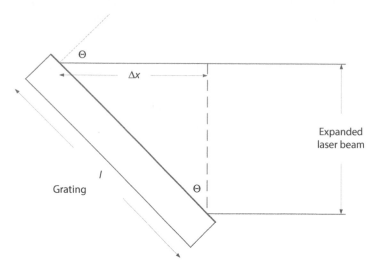

FIGURE 3.2 Path differences in a diffraction grating of the reflective class in Littrow configuration. From the geometry, $sin\ \Theta = (\Delta x/l)$.

Then, it follows that

$$\Delta\lambda = \frac{l\lambda}{\Delta x}\left(\frac{\Delta x_1 - \Delta x_2}{l}\right)$$ (3.15)

In order to distinguish between a maximum and a minimum, the *difference* in path differences should be equal to a single wavelength, that is,

$$(\Delta x_1 - \Delta x_2) \approx \lambda$$ (3.16)

Hence, Equation 3.15 reduces to the well-known diffraction identity

$$\Delta\lambda \approx \frac{\lambda^2}{\Delta x}$$ (3.17)

which in turn leads to

$$\Delta v \approx \frac{c}{\Delta x}$$ (3.18)

Assuming a grating composed of uniform and equally spaced slits of width d, separated themselves by a distance d, the total length of the grating can be stated as $l \approx 2Nd$, where N is the total number of slits. Hence, the path difference can be written as $\Delta x \approx mN\lambda$ so that Equation 3.17 can also be expressed as

$$\frac{\Delta\lambda}{\lambda^2} \approx \frac{1}{mN\lambda}$$ (3.19)

and multiplication by λ yields

$$\frac{\Delta\lambda}{\lambda} \approx \frac{1}{mN}$$ (3.20)

which is known as the *resolving power* of a diffraction grating. Now, considering $p = \hbar k$ for two slightly different wavelengths leads to

$$p_1 - p_2 = \frac{h(\lambda_1 - \lambda_2)}{\lambda_1 \lambda_2} \tag{3.21}$$

and since the difference between λ_1 and λ_2 is infinitesimal, this equation can be restated as

$$\Delta p \approx h\frac{\Delta \lambda}{\lambda^2} \tag{3.22}$$

Substitution of Equation 3.22 into 3.17 leads to

$$\Delta p \Delta x \approx h \tag{3.23}$$

which is known as *Heisenberg's uncertainty principle* (Dirac 1978). This approach offers a geometrical perspective on the foundations of the uncertainty principle. It should also be noted that many authors also express the uncertainty principle in its minimum form where the right-hand side is given by $\hbar/2$ rather than h:

$$\Delta p \Delta x \approx \frac{\hbar}{2} \tag{3.24}$$

This minimal version of the uncertainty principle, in addition to a generalized version, is discussed in Duarte (2014).

3.1.3 ALTERNATIVE VERSIONS OF THE UNCERTAINTY PRINCIPLE

In addition to $\Delta p\,\Delta x \approx h$, the uncertainty principle can be expressed in various useful versions. Assuming an independent derivation, and using $p = \hbar k$, it can be expressed in its wavelength-spatial form:

$$\Delta \lambda\,\Delta x \approx \lambda^2 \tag{3.25}$$

which can also be stated in its frequency-spatial version:

$$\Delta \nu\,\Delta x \approx c \tag{3.26}$$

which are Equations 3.17 and 3.18 in an alternative format. Using $E = mc^2$, the uncertainty principle can also be written in its energy–time form:

$$\Delta E\,\Delta t \approx h \tag{3.27}$$

which, using the quantum energy $E = h\nu$, can be transformed to its frequency–time version:

$$\Delta \nu\,\Delta t \approx 1 \tag{3.28}$$

This result is very important in the area of pulsed lasers and implies that a laser emitting a pulse of a given duration Δt has a minimum spectral linewidth Δv. It also implies that by measuring the width of the spectral emission the duration of that pulse can be determined. The time

$$\Delta t \approx \frac{1}{\Delta v} \tag{3.29}$$

is also known as the *coherence time*. From this time, the *coherence length* can be defined as

$$\Delta x \approx \frac{c}{\Delta v} \tag{3.30}$$

which is an alternative form of Equation 3.26. This equation has gigantic physical implications as it illustrates the fact that a single photon, with an infinitesimal narrow linewidth Δv, can have an enormous coherence length.

3.2   APPLICATIONS OF THE UNCERTAINTY PRINCIPLE IN OPTICS

In this section, some useful applications of the uncertainty principle in beam propagation and intracavity optics are considered.

3.2.1   BEAM DIVERGENCE

If the uncertainty principle is assumed to be derived from independent and rigorous methods, then it can be used to derive some useful identities in optics. For example, starting from

$$\Delta p \, \Delta x \approx h$$

the application of $p = \hbar k$ yields

$$\Delta k \, \Delta x \approx 2\pi \tag{3.31}$$

which leads directly to

$$\Delta \lambda \approx \frac{\lambda^2}{\Delta x}$$

For a diffraction-limited beam traveling in the z-direction

$$k_x = k \sin \theta \tag{3.32}$$

and/or a very small angle θ, we can write

$$k_x \approx k\theta \tag{3.33}$$

Using $\Delta k_x \Delta x \approx 2\pi$, it is readily seen that the beam has an angular divergence given by

$$\Delta\theta \approx \frac{\lambda}{\Delta x} \qquad (3.34)$$

This equation indicates that the angular spread of a propagating beam of wavelength λ is inversely proportional to its original width, that is, narrower beams exhibit a larger angular spread or divergence (see Figure 3.3). Also it states that the light of shorter wavelength experiences less beam divergence, which is a well-known experimental fact in laser physics (see Figure 3.4).

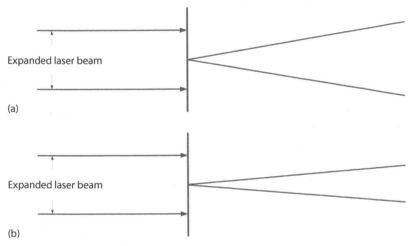

FIGURE 3.3   Beam divergence for two different apertures at wavelength λ. (a) An expanded laser beam is incident on a microhole of diameter $2w$. (b) The same laser beam is incident on a microhole of diameter $4w$.

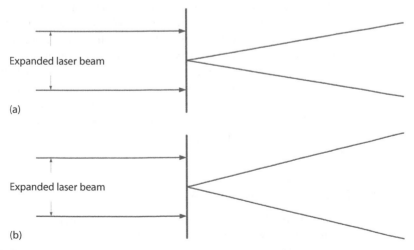

FIGURE 3.4   Beam divergence for (a) $\lambda = 450$ nm and (b) $\lambda = 650$ nm. In both cases, an expanded laser beam is incident on identical microholes of diameter $2w$.

Equation 3.34 has the same form of the classical equation for beam divergence (Duarte 1990):

$$\Delta\theta = \frac{\lambda}{\pi w}\left[1+\left(\frac{L_{\mathcal{R}}}{B}\right)^2+\left(\frac{AL_{\mathcal{R}}}{B}\right)^2\right]^{1/2} \tag{3.35}$$

where:

   w is the beam waist
   $L_{\mathcal{R}} = (\pi w^2/\lambda)$ is the Rayleigh length
   A and B are spatial propagation parameters defined in Chapter 6

For well-chosen experimental conditions, the second and third terms in the parentheses of Equation 3.35 approach zero so that

$$\Delta\theta \approx \frac{\lambda}{\pi w} \tag{3.36}$$

and the beam divergence is said to approach its *diffraction limit*. The equivalence of Equations 3.34 and 3.36 is self evident.

The generalized interference equation

$$\left|\langle x|s\rangle\right|^2 = \sum_{j=1}^{N}\Psi(r_j)^2 + 2\sum_{j=1}^{N}\Psi(r_j)\left[\sum_{m=j+1}^{N}\Psi(r_m)\cos(\Omega_m-\Omega_j)\right]$$

was previously used to derive $\Delta\lambda \approx \lambda^2/\Delta x$, which eventually leads to $\Delta p\Delta x \approx h$ and subsequently leads to an expression to the *diffraction limited* beam divergence.

3.2.1.1   Example

Here, we estimate the divergence of a coherent beam as it propagates in space using the one-dimensional generalized interferometric equation. For instance, at $\lambda = 632.82$ nm, a beam with an original dimension of $2w_0 = 200\ \mu m$ can be estimated via Equation 3.5 to spread to $2w \approx 1.89$ mm (full width at half maximum [FWHM]) following a propagation of 0.5 m using an approximate graphical method (see Figure 3.5). The same beam can be estimated to increase further to $2w \approx 18.9$ mm (FWHM) following a propagation of 5 m. Using the dimensions of the beam waist w, and simple geometry, this yields an approximate beam divergence of $\Delta\theta \approx 1.9$ mrad. Alternatively, using Equation 3.36, the diffraction-limited beam divergence can be determined directly (for $w_0 = 100$ nm and $\lambda = 632.82$ nm) as $\Delta\theta \approx 2.01$ mrad.

3.2.2   BEAM DIVERGENCE AND ASTRONOMY

One application of the beam divergence equation relates to the angular resolution limit of telescopes used in astronomical observations. Reflection telescopes such as the Newtonian and Cassegrain telescopes are depicted in Figure 3.6 and discussed further in Chapter 6. The angular resolution that can be accomplished with these

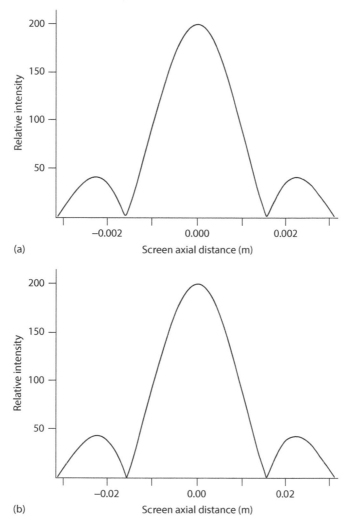

FIGURE 3.5 Beam divergence determined from the generalized interference equation. (a) Beam profile following propagation through a distance of 0.5 m. (b) Beam profile following propagation through a distance of 5 m. Here, $w_0 = 100$ μm and $\lambda = 632.82$ nm. Using the difference in beam waist w over the propagation distance, the beam divergence is determined to be $\Delta\theta \approx 1.9$ mrad.

telescopes under ideal conditions is approximately described by the diffraction limit given in Equation 3.36. That is, the smallest angular discrimination, or resolution limit, of a telescope with a diameter $D = 2w$ is given by

$$\Delta\theta \approx \frac{2\lambda}{\pi D} \tag{3.37}$$

This equation indicates that the two avenues to increase the angular resolution of a telescope are either to observe at shorter wavelengths or to increase the diameter

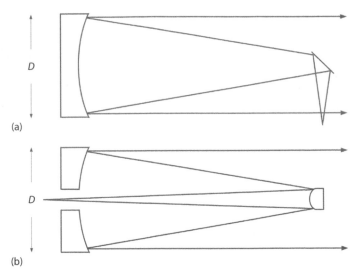

(a)

(b)

FIGURE 3.6 Reflection telescopes used in astronomical observations: (a) Newtonian telescope; (b) Cassegranian telescope.

of the telescope. The latter option has been adopted by optical designers that have built very large telescopes. At $\lambda = 500$ nm, the angular resolution for a telescope, with $D = 10$ m, is $\Delta\theta \approx 3.18 \times 10^{-8}$ rad. For a diameter of $D = 100$ m, the angular resolution at the same wavelength becomes $\Delta\theta \approx 3.18 \times 10^{-9}$ rad. Furthermore, for an hypothetical diameter of $D = (1000/\pi)$ m, the angular resolution at $\lambda = 500$ nm becomes $\Delta\theta \approx 1.00 \times 10^{-9}$ rad. In addition to better angular resolutions, large aperture telescopes provide increased signal since the area of collection increases substantially. In future, these ultralarge telescopes might become important in the search of exoplanets and in the study of fundamental physics phenomena during the very early stages of the universe.

Albeit great progress has been made in the construction of very large segmented mirrors, it might be a long time before telescopes of the extraordinary diameters mentioned here are built. In the meantime, interferometric telescopic arrays (see Chapter 11) can be used to significantly increase the light collection area and resolving power of individual telescopes. Such an array is the Very Large Telescope located at the Atacama Desert in northern Chile, which is composed of four telescopes, each telescope with a diameter of 8.2 m. As mentioned in Chapter 11, the first interferometric telescope was the Hanbury Brown–Twiss interferometer that, as emphasized by Feynman, is indeed a quantum optics instrument. We should also keep in mind that, as explicitly explained in this chapter, the angular resolution of telescopes can be explained either via quantum interferometric principles or via the uncertainty principle itself. Thus, it should not be surprising to see other quantum concepts adopted in the improvement of telescopic resolution.

3.2.2.1   Laser Guide Star

An important feature of large and very large terrestrial telescopes is the use of *laser guide star* techniques, in conjunction with adaptive optics, to correct for atmospheric turbulence. Atmospheric turbulence can seriously affect the optical homogeneity of the transmission medium as described and discussed in Chapter 10 from an interferometric perspective.

In guide star applications, the laser is tuned to the well-known atomic sodium $3^2S_{1/2} - 3^2P_{3/2}$ transition at $\lambda = 588.9963$ nm (also known as the sodium D_2 line; Mitchel and Zemansky 1971) and its beam is projected 90–100 km above the planetary surface. Emission from the sodium guide star induced by the illumination from the laser is collected by a Cassegrain telescope and used to characterize the atmospheric-induced distortions resulting from turbulence. The spatial information thus collected is used to drive the adaptive optics and compensate for atmospheric turbulence. In this application, three most important parameters of the laser are optical power, laser linewidth (Δv), and beam divergence ($\Delta\theta$). More specifically, the laser should yield a diffraction-limited beam and single-longitudinal-mode emission (see Chapter 7).

Laser development for guide star applications originated with work on high-power tunable dye lasers (Everett 1989; Primmerman et al. 1991) and includes both pulsed and continuous-wave (CW) tunable lasers (Bass et al. 1992; Pique and Farinotti 2003). Alternatives to direct emission tunable organic dye lasers are frequency-doubled diode-pumped fiber Raman lasers (Feng et al. 2009; Henry et al. 2013).

The use of Amici prisms, or compounds prisms, for atmospheric dispersion corrections in astronomical applications is discussed in Chapter 4.

3.2.2.2   Example

Consider the propagation of a diffraction-limited laser beam over a distance of 95 km at $\lambda = 589$ nm for guide star applications in astronomy. A multiple-prism grating tunable laser oscillator (Duarte and Piper 1984; Duarte 1999) can be designed to provide single-longitudinal-mode oscillation at $\lambda \approx 589$ nm in a laser with a beam divergence, which is nearly diffraction limited. Thus, its divergence is characterized by $\Delta\theta \approx (\lambda/\pi w)$. For a beam waist $w \approx 125$ μm, the corresponding beam divergence becomes $\Delta\theta \approx 1.5$ mrad. If the beam of this oscillator were to be propagated over a distance of 95 km, it would illuminate a circle of nearly 285 m in diameter. This would yield a very weak power density. If, however, the beam is expanded without introducing further divergence by a factor of $M \approx 200$, then the new beam divergence becomes $\Delta\theta \approx (\lambda/M\pi w)$. Now, as a consequence of the beam expansion, the beam diameter at the required propagation distance becomes about 1.42 m, which is close to the dimensions needed to achieve the necessary power densities with existing high-power tunable lasers. It should be mentioned that the beam magnification factors of about 200 are not difficult to attain. Further, for this type of application, the narrow-linewidth oscillator emission is augmented at amplifier stages, where the beam also increases its dimensions, thus reducing the requirements on M.

3.3   THE INTERFEROMETRIC EQUATION
AND THE UNCERTAINTY PRINCIPLE

Previously, it has been shown that from the Dirac principle

$$\langle x|s \rangle = \sum_{j=1}^{N} \langle x|j \rangle \langle j|s \rangle$$

the generalized interferometric equation

$$\left| \langle x|s \rangle \right|^2 = \sum_{j=1}^{N} \Psi(r_j)^2 + 2 \sum_{j=1}^{N} \Psi(r_j) \left[\sum_{m=j+1}^{N} \Psi(r_m) \cos(\Omega_m - \Omega_j) \right]$$

can be obtained (Duarte 1991, 1993). This equation, originally derived while considering single-photon propagation (Duarte 1991), has also been shown to accurately predict interferometric distributions generated by the interaction of ensembles of indistinguishable photons with N-slit arrays (Duarte 1993). Furthermore, as shown by Duarte (1992), and explained in Chapter 2, the phase term of this generalized interferometric equation can be used to derive the cavity linewidth equation:

$$\Delta\lambda \approx \Delta\theta(\nabla_\lambda\theta)^{-1}$$

where:
   $\Delta\theta$ is the beam divergence
   $\nabla_\lambda\theta = (\partial\theta/\partial\lambda)$, which is the intracavity angular dispersion

This equation has been used extensively to determine the emission linewidth in pulsed narrow-linewidth dispersive laser oscillators (Duarte 2003). As shown earlier in this chapter, the beam divergence term

$$\Delta\theta \approx \frac{\lambda}{\pi w}$$

can be distilled from the uncertainty principle

$$\Delta p \, \Delta x \approx h$$

which itself can also be obtained in an approximate approach, starting from the *phase term* of the interferometric equation, thus emphasizing a fundamental connection between these two principles. Besides, as shown by example, an estimate of $\Delta\theta$ can be obtained directly from the interferometric distributions provided via the interferometric equation. The phase term of the interferometric equation is in itself created by the interaction of probability amplitudes via the mechanics of the basic Dirac quantum principle

$$\langle x|s \rangle = \sum_{j=1}^{N} \langle x|j \rangle \langle j|s \rangle$$

where the probability amplitudes are represented by *wave functions of ordinary wave optics* (Dirac 1978):

$$\langle j|s\rangle = \Psi(r_{j,s})e^{-i\theta_j}$$

$$\langle x|j\rangle = \Psi(r_{x,j})e^{-i\phi_j}$$

This exposition indicates that interference, as described by the Dirac notation, has a central and crucial position at the foundations of quantum mechanics and its significance might even be deeper than the significance of the uncertainty principle itself. The importance of interference to quantum mechanical principles was first highlighted by Dirac (1978) and then by Feynman et al. (1965).

3.3.1 QUANTUM CRYPTOGRAPHY

Quantum secure communications and quantum cryptography are subjects of intense research activity. In this regard, it should be mentioned that Heisenberg's uncertainty principle is often associated with the essence of quantum cryptography (Bennett et al. 1992).

An alternative approach to describe the mechanics of quantum cryptography is to use the probability amplitude of entangled polarizations of quanta traveling in opposite directions (Ekert 1991). The probability amplitude for entangled polarizations of quanta traveling in opposite directions was first introduced in the form of (Ward 1949)

$$|s\rangle = \left(|x\rangle_1|y\rangle_2 - |y\rangle_1|x\rangle_2\right) \tag{3.38}$$

which following straightforward normalization becomes

$$|\Psi_-\rangle = \frac{1}{\sqrt{2}}\left(|x\rangle_1|y\rangle_2 - |y\rangle_1|x\rangle_2\right) \tag{3.39}$$

which is the ubiquitous probability amplitude depicting quantum entanglement given in numerous research papers and textbooks (see, e.g., Mandel and Wolf 1995). This probability amplitude can be derived via Ward's argument (Ward 1949), a Hamiltonian approach (Duarte 2014), or via a neat and transparent approach from an *N*-slit interferometric perspective (Duarte 2013, 2014) again utilizing the Dirac principle:

$$\langle x|s\rangle = \sum_{j=1}^{N}\langle x|j\rangle\langle j|s\rangle$$

Briefly, the probability amplitude for a single photon to propagate from a source *s* to the interferometric plane configured by a detector *d* via apertures 1 and 2 is given by $\langle d|s\rangle$

$$\langle d|s\rangle = \langle d|2\rangle\langle 2|s\rangle + \langle d|1\rangle\langle 1|s\rangle \tag{3.40}$$

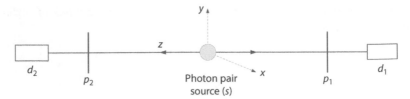

FIGURE 3.7   Generic experimental configuration of a source emitting two indistinguishable quanta traveling in opposite directions with entangled polarizations. This experimental arrangement is a simplified rendition of the original configuration introduced by Pryce and Ward.

For an experiment with a central photon source emitting indistinguishable quanta toward *identical detectors* $(d_1 = d_2 = d)$ positioned in opposite directions via polarization analyzers p_1 and p_2 (see Figure 3.7), the corresponding probability amplitude can be described as

$$\langle d|s \rangle = \langle d|p_2 \rangle \langle p_2|s \rangle + \langle d|p_1 \rangle \langle p_1|s \rangle \tag{3.41}$$

In abstract form, this probability amplitude can be expressed as

$$|s \rangle = |p_2 \rangle \langle p_2|s \rangle + |p_1 \rangle \langle p_1|s \rangle \tag{3.42}$$

Defining

$$|C_1 \rangle = |p_1 \rangle \langle p_1|s \rangle \tag{3.43}$$

and

$$|C_2 \rangle = |p_2 \rangle \langle p_2|s \rangle \tag{3.44}$$

Equation 3.42 can be expressed as

$$|s \rangle = |C_2 \rangle + |C_1 \rangle \tag{3.45}$$

Using the Dirac identity applicable to the state of several similar particles

$$|X \rangle = |a_1 \rangle |b_2 \rangle |c_3 \rangle \dots |g_n \rangle \tag{3.46}$$

for quanta 1 and quanta 2, we can write

$$|C_2 \rangle = |x \rangle_1 |y \rangle_2 \tag{3.47}$$

and

$$|C_1 \rangle = |y \rangle_1 |x \rangle_2 \tag{3.48}$$

Substituting back in Equation 3.45, and following normalization, leads to

$$|\Psi_+\rangle = \frac{1}{\sqrt{2}}\left(|x\rangle_1|y\rangle_2 + |y\rangle_1|x\rangle_2\right) \tag{3.49}$$

and its linear combination

$$|\Psi_-\rangle = \frac{1}{\sqrt{2}}\left(|x\rangle_1|y\rangle_2 - |y\rangle_1|x\rangle_2\right)$$

An extension of the Ward probability amplitude is the probability amplitude for n entangled particles propagating in two different directions (Duarte 2013):

$$|\Psi\rangle = \frac{1}{\sqrt{2}}\left(|a\rangle_1|b\rangle_2|c\rangle_3 \cdots |g\rangle_n \pm |a'\rangle_1|b'\rangle_2|c'\rangle_3 \cdots |g'\rangle_n\right) \tag{3.50}$$

For N path alternatives, the overall probability amplitude is given by

$$|s\rangle = \sum_{j=1}^{N}|C_{N+1-j}\rangle \tag{3.51}$$

where the *first* term of the path probability amplitude series can be written as

$$C_{N+1-j} = \prod_{m=1}^{n} |\gamma\rangle_m \tag{3.52}$$

These are a series of probability amplitudes starting at C_{N+1-j}, with $j = 1$, and ending at C_1, which is reached when $j = N$. Here, n is the total number of quanta, which is an even number since quanta participate in pairs. For each pair $(1, 2), (3, 4)...(n-1, n)$, γ represents a pair of orthogonal polarization alternatives such as x and y, and ϕ and ϕ'.

For an even number of paths, the normalized probability amplitudes, designated by C_R, where R is a Roman numeral, have the form:

$$C_R = \frac{1}{\sqrt{N}} \sum_{m=1}^{N} (\pm)C_m \tag{3.53}$$

However, for $N = 3$,

$$C_I = \frac{1}{\sqrt{3}}\left(C_3 + C_2 + iC_1\right) \tag{3.54}$$

$$C_{II} = \frac{1}{\sqrt{3}}\left(C_3 - C_2 + iC_1\right) \tag{3.55}$$

$$C_{III} = \frac{1}{\sqrt{3}}\left(iC_3 + jC_2 + kC_1\right) \tag{3.56}$$

where:
i, j, and k are defined as Hamilton's quaternions

A detailed discussion of the Ward probability amplitude for entangled quanta and its corresponding experimental arrangement is given by Duarte (2014).

3.3.1.1 Example

For four particles ($n = 4$), that is, *two pairs*, and four paths ($N = 4$), as illustrated in Figure 3.8,

$$|s\rangle = |C_4\rangle + |C_3\rangle + |C_2\rangle + |C_1\rangle \tag{3.57}$$

or

$$|s\rangle = |C_1\rangle + |C_2\rangle + |C_3\rangle + |C_4\rangle \tag{3.58}$$

so that

$$|s\rangle = |x\rangle_1 |y\rangle_2 |\phi\rangle_3 |\phi'\rangle_4 + |y\rangle_1 |x\rangle_2 |\phi'\rangle_3 |\phi\rangle_4$$
$$+ |\phi\rangle_1 |\phi'\rangle_2 |x\rangle_3 |y\rangle_4 + |\phi'\rangle_1 |\phi\rangle_2 |y\rangle_3 |x\rangle_4 \tag{3.59}$$

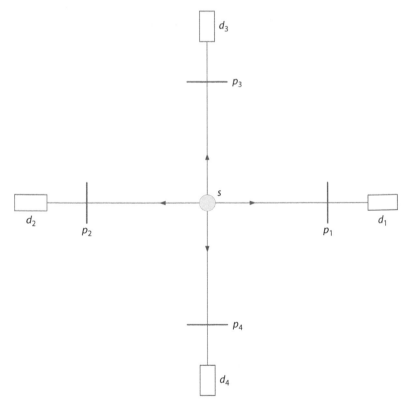

FIGURE 3.8 Experimental configuration of a source emitting *two pairs* ($n = 4$) of indistinguishable quanta traveling in four different directions ($N = 4$) with entangled polarizations. In the theoretical development, it is assumed that all detectors are identical ($d_1 = d_2 = d_3 = d_4 = d$).

$$C_1 = |x\rangle_1 |y\rangle_2 |\phi\rangle_3 |\phi'\rangle_4 \qquad (3.60)$$

$$C_2 = |y\rangle_1 |x\rangle_2 |\phi'\rangle_3 |\phi\rangle_4 \qquad (3.61)$$

$$C_3 = |\phi\rangle_1 |\phi'\rangle_2 |x\rangle_3 |y\rangle_4 \qquad (3.62)$$

$$C_4 = |\phi'\rangle_1 |\phi\rangle_2 |y\rangle_3 |x\rangle_4 \qquad (3.63)$$

The normalization condition

$$1 = |C_I|^2 + |C_{II}|^2 + |C_{III}|^2 + |C_{IV}|^2 \qquad (3.64)$$

requires that

$$C_I = \frac{1}{\sqrt{4}} \left(C_1 + C_2 + C_3 + C_4 \right) \qquad (3.65)$$

$$C_{II} = \frac{1}{\sqrt{4}} \left(C_1 + C_2 - C_3 - C_4 \right) \qquad (3.66)$$

$$C_{III} = \frac{1}{\sqrt{4}} \left(C_1 - C_2 + C_3 - C_4 \right) \qquad (3.67)$$

$$C_{IV} = \frac{1}{\sqrt{4}} \left(C_1 - C_2 - C_3 + C_4 \right) \qquad (3.68)$$

which lead to the following explicit normalized probability amplitudes

$$
|\Psi\rangle = \frac{1}{\sqrt{4}} \left(|x\rangle_1 |y\rangle_2 |\phi\rangle_3 |\phi'\rangle_4 + |y\rangle_1 |x\rangle_2 |\phi'\rangle_3 |\phi\rangle_4 \right.
$$
$$
\left. + |\phi\rangle_1 |\phi'\rangle_2 |x\rangle_3 |y\rangle_4 + |\phi'\rangle_1 |\phi\rangle_2 |y\rangle_3 |x\rangle_4 \right) \qquad (3.69)
$$

$$
|\Psi\rangle = \frac{1}{\sqrt{4}} \left(|x\rangle_1 |y\rangle_2 |\phi\rangle_3 |\phi'\rangle_4 + |y\rangle_1 |x\rangle_2 |\phi'\rangle_3 |\phi\rangle_4 \right.
$$
$$
\left. - |\phi\rangle_1 |\phi'\rangle_2 |x\rangle_3 |y\rangle_4 - |\phi'\rangle_1 |\phi\rangle_2 |y\rangle_3 |x\rangle_4 \right) \qquad (3.70)
$$

$$
|\Psi\rangle = \frac{1}{\sqrt{4}} \left(|x\rangle_1 |y\rangle_2 |\phi\rangle_3 |\phi'\rangle_4 - |y\rangle_1 |x\rangle_2 |\phi'\rangle_3 |\phi\rangle_4 \right.
$$
$$
\left. + |\phi\rangle_1 |\phi'\rangle_2 |x\rangle_3 |y\rangle_4 - |\phi'\rangle_1 |\phi\rangle_2 |y\rangle_3 |x\rangle_4 \right) \qquad (3.71)
$$

$$|\Psi\rangle = \frac{1}{\sqrt{4}}\left(|x\rangle_1 |y\rangle_2 |\phi\rangle_3 |\phi'\rangle_4 - |y\rangle_1 |x\rangle_2 |\phi'\rangle_3 |\phi\rangle_4 \right.$$

$$\left. - |\phi\rangle_1 |\phi'\rangle_2 |x\rangle_3 |y\rangle_4 + |\phi'\rangle_1 |\phi\rangle_2 |y\rangle_3 |x\rangle_4 \right)$$

(3.72)

This is a complete symmetrical configuration where the number of path alternatives ($N = 4$) is matched by the number of particles ($n = 4$), which corresponds to *two pairs* of particles propagating via *two pairs* of alternative paths. Once this mechanics is established, one can readily extend the multiparticle probability amplitude to other similar higher symmetrical configurations.

PROBLEMS

3.1 Calculate the diffraction-limited beam divergence at FWHM for (1) a laser beam with a $w = 100$ μm radius at $\lambda = 590$ nm and (2) a laser beam with a $w = 500$ μm radius at $\lambda = 590$ nm.

3.2 Repeat the calculations of Problem 3.1 for $\lambda = 308$ nm. Comment.

3.3 Calculate the single-pass dispersive cavity linewidth for a high-power tunable laser yielding a diffraction-limited beam divergence, $w = 100$ μm in radius, at $\lambda = 590$ nm. Assume that an appropriate beam expander illuminates a 3300 lines/mm grating deployed in the first order. The grating has a 50 mm length perpendicular to the grooves.

3.4 (a) For a high-power pulsed laser delivering a $\Delta v \approx 350$ MHz laser linewidth, estimate its shortest possible pulse width at FWHM. (b) For a laser emitting $\Delta t \approx 10$ fs pulses, estimate its broadest possible spectral width in nanometers centered around $\lambda = 600$ nm.

3.5 Show that Equation 3.34 follows from Equation 3.33.

3.6 Estimate the angular resolution of a single 8.2 m diameter telescope at $\lambda = 500$ m.

3.7 Write explicit normalized probability amplitudes for $N = 3$ propagation channels and $n = 6$ quanta, that is, three pairs of photons with entangled polarizations.

4 The Physics of Multiple-Prism Optics

4.1 INTRODUCTION

Multiple-prism arrays were first introduced by Newton (1704) in his book *Opticks*. In this visionary volume, Newton reported on arrays of nearly isosceles prisms in additive and compensating configurations to control the propagation path and the dispersion of light. Further, he also illustrated slight beam expansion in a single isosceles prism. In his treatise, Newton does provide a written, qualitative, description of light dispersion in a sequence of prisms, thus providing the foundations for prismatic spectrometers and related instrumentation. In this chapter, a mathematical description of light dispersion in generalized multiple-prism arrays is given. Although the theory was originally developed to quantify the phenomenon of intracavity linewidth narrowing in multiple-prism grating tunable laser oscillators (Duarte and Piper 1982), it is applicable beyond the domain of tunable lasers to optics in general. The theory considers single-pass as well as multiple-pass dispersion in generalized multiple-prism arrays and is useful to *both* linewidth narrowing and pulse compression. As a pedagogical tool, reduction of the generalized formulae to single-prism calculations is included in addition to a specific design of a particular practical high-magnification multiple-prism beam expander.

In order to facilitate the flow of information, and to maintain the focus on the physics, some of the mathematical steps involved in the derivations are not included. For further details, references to the original work are cited. The mathematical notation is consistent with the notation used in the original literature.

Besides a technical interest in quantifying multiple-prism arrays for laser, astronomical, and analytical instrumentation, the generalized multiple-prism dispersion theory is of interest from an intrinsic perspective. In the case of prismatic optics, the word *dispersion* really means the ranking or ordering of colors according to their wavelength. In other words, the phenomenon that is studied is the transition of white light, or broadband radiation, to highly ordered radiation according to its frequency. The multiple-prism dispersion theory enables the quantification of a transition from an initial state of entropy to a lower state of entropy—a transition that is rare in nature.

This chapter is a revised version of the original chapter published in the first edition of *Tunable Laser Optics* (Duarte 2003) and includes additional material on Amici prisms or compound prisms, generalized multiple-prism dispersion due to positive and negative refraction, and higher order phase derivatives for multiple-prism laser pulse compression.

4.2 GENERALIZED MULTIPLE-PRISM DISPERSION

Multiple-prism arrays are widely used in optics in a variety of applications such as (1) intracavity beam expanders in narrow-linewidth tunable laser oscillators, (2) extracavity beam expanders, (3) pulse compressors in ultrafast lasers, and (4) dispersive elements in spectrometers. Albeit multiple-prism arrays were first introduced by Newton (1704), a mathematical description of their dispersion had to wait a long time until their application as intracavity beam expanders in narrow-linewidth tunable lasers (Duarte and Piper 1982). Generalized multiple-prism arrays are depicted in Figure 4.1.

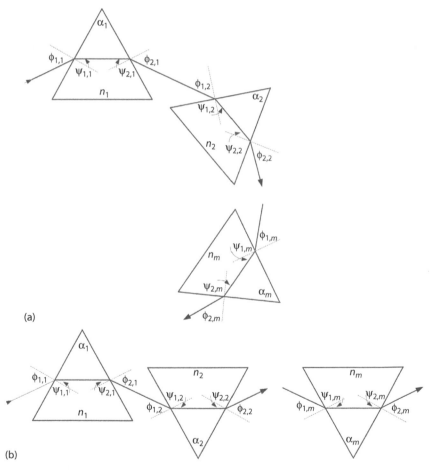

FIGURE 4.1 Generalized multiple-prism arrays: (a) Additive configuration; (b) compensating configuration. Depiction of these generalized prismatic configurations was introduced by Duarte and Piper. (Reproduced from Duarte, F.J., *Quantum Optics for Engineers*, CRC Press, New York, 2014. With permission from Taylor & Francis; Data from Duarte, F.J., and Piper, J.A., *Am. J. Phys.*, 51, 1132–1134, 1983.)

Considering positive and negative refraction for the mth prism of the arrangements, the angular relations are given by (Duarte and Piper 1982; Duarte 2006)

$$\phi_{1,m} + \phi_{2,m} = \varepsilon_m \pm \alpha_m \tag{4.1}$$

$$\psi_{1,m} + \psi_{2,m} = \alpha_m \tag{4.2}$$

$$\sin\phi_{1,m} = \pm n_m \sin\psi_{1,m} \tag{4.3}$$

$$\sin\phi_{2,m} = \pm n_m \sin\psi_{2,m} \tag{4.4}$$

As illustrated in Figure 4.1, $\phi_{1,m}$ and $\phi_{2,m}$ are the angles of incidence and emergence, respectively, and $\psi_{1,m}$ and $\psi_{2,m}$ are the corresponding angles of refraction at the mth prism. In these equations, the positive sign (+) relates to positive refraction, whereas the negative sign (−) indicates negative refraction.

Differentiating Equations 4.3 and 4.4 and using

$$\frac{d\psi_{1,m}}{dn} = -\frac{d\psi_{2,m}}{dn} \tag{4.5}$$

the single-pass dispersion following the mth prism is given by (Duarte and Piper 1982, 1983; Duarte 2006)

$$\nabla_\lambda\phi_{2,m} = \pm\mathcal{H}_{2,m}\nabla_\lambda n_m \pm (k_{1,m}k_{2,m})^{-1}\left[\mathcal{H}_{1,m}\nabla_\lambda n_m(\pm)\nabla_\lambda\phi_{2,(m-1)}\right] \tag{4.6}$$

where:
$$\nabla_\lambda = \partial/\partial\lambda$$

and the following geometrical identities apply:

$$k_{1,m} = \frac{\cos\psi_{1,m}}{\cos\phi_{1,m}} \tag{4.7}$$

$$k_{2,m} = \frac{\cos\phi_{2,m}}{\cos\psi_{2,m}} \tag{4.8}$$

$$\mathcal{H}_{1,m} = \frac{\tan\phi_{1,m}}{n_m} \tag{4.9}$$

$$\mathcal{H}_{2,m} = \frac{\tan\phi_{2,m}}{n_m} \tag{4.10}$$

The $k_{1,m}$ and $k_{2,m}$ factors represent the physical beam expansion experienced at the mth prism by the incidence and emergence beams, respectively. In Equation 4.6, the signs in parentheses, that is, ($\pm$), refer to deployment in either a positive (+) or a compensating (−) configuration, while the simple $\pm$ indicates either positive or negative refraction (Duarte 2006).

The generalized *single-pass dispersion equation* indicates that the cumulative dispersion at the mth prism, namely, $\nabla_\lambda \phi_{2,m}$, is a function of the geometry of the mth prism, the position of the light beam relative to this prism, the material of this prism, and the cumulative dispersion up to the previous prism, $\nabla_\lambda \phi_{2,(m-1)}$ (Duarte and Piper 1982, 1983). For completeness, it is useful to mention that the development of dispersive equations for both positive and negative refraction arises naturally from an interferometric perspective (Duarte 1993, 2006).

The generalized single-pass dispersion equation for positive refraction is

$$\nabla_\lambda \phi_{2,m} = \mathcal{H}_{2,m} \nabla_\lambda n_m + (k_{1,m} k_{2,m})^{-1} \left[\mathcal{H}_{1,m} \nabla_\lambda n_m \pm \nabla_\lambda \phi_{2,(m-1)} \right] \tag{4.11}$$

For the special case of orthogonal beam exit, that is, $\phi_{2,m} = 0$ and $\psi_{2,m} = 0$, we have $\mathcal{H}_{2,m} = 0$, $k_{2,m} = 1$, and Equation 4.11 reduces to

$$\nabla_\lambda \phi_{2,m} = (k_{1,m})^{-1} \left[\mathcal{H}_{1,m} \nabla_\lambda n_m \pm \nabla_\lambda \phi_{2,(m-1)} \right] \tag{4.12}$$

For an array of r identical isosceles or equilateral prisms deployed symmetrically in an additive configuration for positive refraction, so that $\phi_{1,m} = \phi_{2,m}$, the cumulative dispersion reduces to (Duarte 1990a)

$$\nabla_\lambda \phi_{2,r} = r \nabla_\lambda \phi_{2,1} \tag{4.13}$$

The generalized single-pass dispersion equation for positive refraction (Equation 4.11) can also be restated in a more practical and explicit notation (Duarte 1985, 1989, 1990a):

$$\nabla_\lambda \phi_{2,r} = \sum_{m=1}^{r} (\pm 1) \mathcal{H}_{1,m} \left(\prod_{j=m}^{r} k_{1,j} \prod_{j=m}^{r} k_{2,j} \right)^{-1} \nabla_\lambda n_m$$

$$+ (M_1 M_2)^{-1} \sum_{m=1}^{r} (\pm 1) \mathcal{H}_{2,m} \left(\prod_{j=1}^{m} k_{1,j} \prod_{j=1}^{m} k_{2,j} \right) \nabla_\lambda n_m \tag{4.14}$$

where:

$$M_1 = \prod_{j=1}^{r} k_{1,j} \tag{4.15}$$

and

$$M_2 = \prod_{j=1}^{r} k_{2,j} \tag{4.16}$$

are the respective beam expansion factors. For the important practical case of r right-angle prisms designed for orthogonal beam exit (i.e., $\phi_{2,m} = \psi_{2,m} = 0$), Equation 4.14 reduces to

$$\nabla_\lambda \phi_{2,r} = \sum_{m=1}^{r} (\pm 1)\mathcal{H}_{1,m} \left(\prod_{j=m}^{r} k_{1,j} \right)^{-1} \nabla_\lambda n_m \tag{4.17}$$

If in addition the prisms have identical apex angles ($\alpha_1 = \alpha_2 = \alpha_3 = \cdots = \alpha_m$) and are configured to have the same angle of incidence ($\phi_{1,1} = \phi_{1,2} = \phi_{1,3} = \cdots = \phi_{1,m}$), then Equation 4.17 can be written as (Duarte 1985)

$$\nabla_\lambda \phi_{2,r} = \tan \psi_{1,1} \sum_{m=1}^{r} (\pm 1) \left(\frac{1}{k_{1,m}} \right)^{m-1} \nabla_\lambda n_m \tag{4.18}$$

Further, if the angle of incidence for all prisms is Brewster's angle, then the single-pass dispersion reduces to the elegant expression:

$$\nabla_\lambda \phi_{2,r} = \sum_{m=1}^{r} (\pm 1) \left(\frac{1}{n_m} \right)^{m} \nabla_\lambda n_m \tag{4.19}$$

4.2.1 DOUBLE-PASS GENERALIZED MULTIPLE-PRISM DISPERSION

The evaluation of intracavity dispersion in tunable laser oscillators incorporating multiple-prism beam expanders requires the assessment of the double-pass, or return-pass, dispersion. The double-pass dispersion of multiple-prism beam expanders was derived by thinking of the return pass as a mirror image of the first light passage as illustrated in Figure 4.2. The return-pass dispersion corresponds to the dispersion experienced by the return light beam at the first prism. Thus, it is given by $\partial \phi'_{1,m} / \partial \lambda = \nabla_\lambda \phi'_{1,m}$, where the prime character indicates return pass (Duarte and Piper 1982, 1984):

$$\nabla_\lambda \phi'_{1,m} = \mathcal{H}'_{1,m} \nabla_\lambda n_m + \left(k'_{1,m} k'_{2,m} \right) \left[\mathcal{H}'_{2,m} \nabla_\lambda n_m \pm \nabla_\lambda \phi'_{1,(m+1)} \right] \tag{4.20}$$

where:

$$k'_{1,m} = \frac{\cos \psi'_{1,m}}{\cos \phi'_{1,m}} \tag{4.21}$$

$$k'_{2,m} = \frac{\cos \phi'_{2,m}}{\cos \psi'_{2,m}} \tag{4.22}$$

$$\mathcal{H}'_{1,m} = \frac{\tan \phi'_{1,m}}{n_m} \tag{4.23}$$

$$\mathcal{H}'_{2,m} = \frac{\tan \phi'_{2,m}}{n_m} \tag{4.24}$$

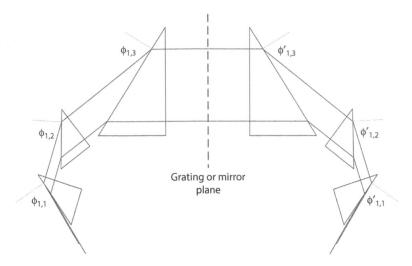

FIGURE 4.2 Multiple-prism beam expander geometry in additive configuration and its mirror image. A dispersive analysis through the multiple-prism array and its mirror image is equivalent to a double-pass, or return-pass, analysis. This type of description for multiple-prism grating assemblies was first introduced by Duarte and Piper. (Reproduced from Duarte, F.J., *Quantum Optics for Engineers*, CRC Press, New York, 2014. With permission from Taylor & Francis; Data from Duarte, F.J., and Piper, J.A., *Opt. Commun.*, 43, 303–307, 1982.)

Here, $\nabla_\lambda \phi'_{1,(m+1)}$ provides the cumulative single-pass multiple-prism dispersion and the dispersion of the diffraction grating, that is,

$$\nabla_\lambda \phi'_{1,(m+1)} = (\nabla_\lambda \Theta_G \pm \nabla_\lambda \phi_{2,r}) \tag{4.25}$$

where:

$\nabla_\lambda \Theta_G$ is the grating dispersion

If the grating is replaced by a mirror, then we simply have the prismatic contribution and

$$\nabla_\lambda \phi'_{1,(m+1)} = \nabla_\lambda \phi_{2,r} \tag{4.26}$$

Defining $\nabla_\lambda \phi'_{1,m} = \nabla \Phi_P$, where Φ stands for return pass and P for multiple prism, or a plurality of prisms, the explicit version of the generalized double-pass dispersion for a multiple-prism mirror system is given by (Duarte 1985, 1989)

$$\nabla_\lambda \Phi_P = 2 M_1 M_2 \sum_{m=1}^{r} (\pm 1) \mathcal{H}_{1,m} \left(\prod_{j=m}^{r} k_{1,j} \prod_{j=m}^{r} k_{2,j} \right)^{-1} \nabla_\lambda n_m$$

$$+ 2 \sum_{m=1}^{r} (\pm 1) \mathcal{H}_{2,m} \left(\prod_{j=1}^{m} k_{1,j} \prod_{j=1}^{m} k_{2,j} \right) \nabla_\lambda n_m \tag{4.27}$$

For the case of r right-angle prisms designed for orthogonal beam exit (i.e., $\phi_{2,m} = \psi_{2,m} = 0$), Equation 4.27 reduces to

$$\nabla_\lambda \Phi_P = 2M_1 \sum_{m=1}^{r} (\pm 1) \mathcal{H}_{1,m} \left(\prod_{j=m}^{r} k_{1,j} \right)^{-1} \nabla_\lambda n_m \qquad (4.28)$$

which can also be expressed as (Duarte 1985)

$$\nabla_\lambda \Phi_P = 2 \sum_{m=1}^{r} (\pm 1) \left(\prod_{j=1}^{m} k_{1,j} \right) \tan \psi_{1,m} \nabla_\lambda n_m \qquad (4.29)$$

If the angle of incidence for all prisms in the array is made equal to the Brewster angle, this equation simplifies further to (Duarte 1990a)

$$\nabla_\lambda \Phi_P = 2 \sum_{m=1}^{r} (\pm 1)(n_m)^{m-1} \nabla_\lambda n_m \qquad (4.30)$$

4.2.2 Multiple Return-Pass Generalized Multiple-Prism Dispersion

Here we consider a multiple-prism grating or multiple-prism mirror assembly as illustrated in Figures 4.2 and 4.3. The light beam enters the first prism of the array, it is then expanded, and it is either diffracted back or reflected back into the

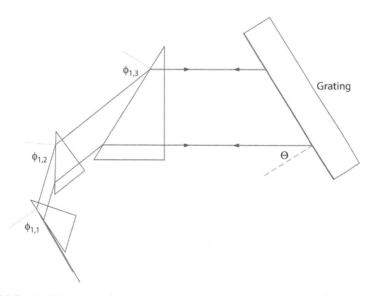

FIGURE 4.3 Additional perspective on the multiple-prism grating assembly used to perform the dispersive multiple-return pass analysis as explained by Duarte and Piper. The multiple-prism configuration can be either additive or compensating and can be composed of any number of prisms. (Data from Duarte, F.J., and Piper, J.A., *Opt. Commun.*, 43, 303–307, 1982.)

multiple-prism array. In a dispersive laser oscillator, this process goes forth and back many times, thus giving rise to the concept of intracavity double pass or intracavity return pass. For the first return pass toward the first prism in the array, the dispersion is given by

$$\nabla_\lambda \phi_{2,m} = \mathcal{H}_{2,m} \nabla_\lambda n_m + (k_{1,m} k_{2,m})^{-1} \left[\mathcal{H}_{1,m} \nabla_\lambda n_m \pm \nabla_\lambda \phi_{2,(m-1)} \right]$$

If N denotes the number of passes toward the grating or reflecting element and $2N$ denotes the number of return passes toward the first prism in the sequence, we have (Duarte and Piper 1984)

$$(\nabla_\lambda \phi_{2,m})_N = \mathcal{H}_{2,m} \nabla_\lambda n_m + (k_{1,m} k_{2,m})^{-1} \left\{ \mathcal{H}_{1,m} \nabla_\lambda n_m \pm \left[\nabla_\lambda \phi_{2,(m-1)} \right]_N \right\} \quad (4.31)$$

and

$$(\nabla_\lambda \phi'_{1,m})_{2N} = \mathcal{H}'_{1,m} \nabla_\lambda n_m + \left(k'_{1,m} k'_{2,m} \right) \left\{ \mathcal{H}'_{2,m} \nabla_\lambda n_m \pm \left[\nabla_\lambda \phi'_{1,(m+1)} \right]_{2N} \right\} \quad (4.32)$$

For the first prism of the array, $[\nabla_\lambda \phi_{2,(m-1)}]_N$ (with $N = 3, 5, 7, \ldots$) in Equation 4.31 is replaced by $(\nabla_\lambda \phi'_{1,1})_{2N}$ (with $N = 1, 2, 3, \ldots$). Likewise for the last prism of the assembly, $[\nabla_\lambda \phi'_{1,(m+1)}]_{2N}$ (with $N = 1, 2, 3, \ldots$) in Equation 4.32 is replaced by $[\nabla_\lambda \Theta_G + (\nabla_\lambda \phi_{2,r})_N]$ (with $N = 1, 3, 5, \ldots$). In the case of the grating being replaced by a mirror, this expression becomes simply $(\nabla_\lambda \phi_{2,r})_N$ (with $N = 1, 3, 5, \ldots$). Thus, the multiple return-pass dispersion for a multiple-prism grating assembly is given by (Duarte and Piper 1984)

$$(\nabla_\lambda \theta)_R = \left(RM \nabla_\lambda \Theta_G + R \nabla_\lambda \Phi_P \right) \quad (4.33)$$

where:
  R is the number of return passes
  M is the overall beam magnification of the multiple-prism beam expander

This equation illustrates the very important fact that in the return-pass dispersion of a multiple-prism grating assembly, the dispersion of a grating is multiplied by the factor RM. Once again, if the grating is replaced by a mirror, that is, $\nabla_\lambda \Theta_G = 0$, the dispersion reduces to

$$(\nabla_\lambda \theta)_R = R \nabla_\lambda \Phi_P \quad (4.34)$$

which implies that the multiple-prism intracavity dispersion increases linearly as a function of R. In practice, the quantity R has a limit imposed by experimental conditions and is usually $R \approx 3$ for pulsed high-power narrow-linewidth tunable organic lasers emitting in the nanosecond regime (Duarte and Piper 1984; Duarte 2001a) (see Section 4.3.1).

4.2.3 SINGLE-PRISM EQUATIONS

Using Equation 4.11 for $m = 1$ yields

$$\nabla_\lambda \phi_{2,m} = \mathcal{H}_{2,1} \nabla_\lambda n_1 + (k_{1,1} k_{2,1})^{-1} (\mathcal{H}_{1,m} \nabla_\lambda n_1) \tag{4.35}$$

which can also be expressed as (Duarte 1990a)

$$\nabla_\lambda \phi_{2,1} = \left(\frac{\sin \psi_{2,1}}{\cos \phi_{2,1}} \right) \nabla_\lambda n_1 + \left(\frac{\cos \psi_{2,1}}{\cos \phi_{2,1}} \right) \tan \psi_{1,1} \nabla_\lambda n_1 \tag{4.36}$$

Division by $\nabla_\lambda n$ yields the result given for $\nabla_n \phi_{2,1}$ by Born and Wolf (1999). For orthogonal beam exit ($\phi_{2,1} \approx \psi_{2,1} \approx 0$), Equation 4.36 simplifies to

$$\nabla_\lambda \phi_{2,1} \approx \tan \psi_{1,1} \nabla_\lambda n \tag{4.37}$$

which is the result given by Wyatt (1978). For a single prism designed for orthogonal beam exit and deployed at Brewster's angle of incidence, Equation 4.19 becomes

$$\nabla_\lambda \phi_{2,1} = \left(\frac{1}{n} \right) \nabla_\lambda n \tag{4.38}$$

a result that also follows from Equation 4.37.

For a double-pass analysis, for $m = 1$, Equation 4.32 becomes

$$\nabla_\lambda \phi'_{1,1} = \mathcal{H}'_{1,1} \nabla_\lambda n_1 + \left(k'_{1,1} k'_{2,1} \right) \left(\mathcal{H}'_{2,1} \nabla_\lambda n_1 \pm \nabla_\lambda \phi_{2,1} \right) \tag{4.39}$$

which for orthogonal beam exit reduces to

$$\nabla_\lambda \Phi_P = 2(k_{1,1} \tan \psi_{1,1}) \nabla_\lambda n \tag{4.40}$$

For incidence at the Brewster angle, this equation simplifies further to

$$\nabla_\lambda \Phi_P = 2 \nabla_\lambda n \tag{4.41}$$

Reduction from Equation 4.40 to 4.41 implies that at the Brewster angle of incidence, the beam magnification

$$k_{1,1} = n \tag{4.42}$$

and

$$\tan \psi_{1,1} = \frac{1}{n} \tag{4.43}$$

4.3   MULTIPLE-PRISM DISPERSION LINEWIDTH NARROWING

The cavity linewidth equation is derived, in Chapter 3, from interferometric principles as (Duarte 1992)

$$\Delta\lambda = \Delta\theta(\nabla_\lambda\theta)^{-1} \tag{4.44}$$

where:
  $\Delta\theta$ is the beam divergence
  $\nabla_\lambda\theta$ is the overall intracavity dispersion

The message from this equation is that for narrow-linewidth emission we need to minimize the beam divergence and to increase the intracavity dispersion. The multiple-return-pass cavity linewidth for dispersive oscillator configurations, as illustrated in Figures 4.4 and 4.5, is given by (Duarte 2001b)

$$\Delta\lambda = \Delta\theta_R \left(MR\nabla_\lambda\Theta_G + R\nabla_\lambda\Phi_P \right)^{-1} \tag{4.45}$$

and the multiple-return-pass beam divergence (see Chapter 6) is given by

$$\Delta\theta_R = \frac{\lambda}{\pi w}\left[1+\left(\frac{L_{\mathcal{R}}}{B_R}\right)^2+\left(\frac{A_R L_{\mathcal{R}}}{B_R}\right)^2\right]^{1/2} \tag{4.46}$$

where:
  w is the beam waist
  $L_{\mathcal{R}} = (\pi w^2/\lambda)$ is the Rayleigh length
  A_R and B_R are multiple-pass propagation matrix coefficients

For an optimized multiple-prism grating laser oscillator, $\Delta\theta_R$ approaches its diffraction limit following a few return passes.

4.3.1   MECHANICS OF LINEWIDTH NARROWING IN OPTICALLY PUMPED PULSED LASER OSCILLATORS

The factor R in the multiple-return-pass linewidth equation is related to the total intracavity transit time from the beginning of the excitation pulse to the onset of laser oscillation. Thus, R is determined experimentally by measuring the time delay $\delta\tau$ between the leading edge of the pump pulse and the leading edge of the laser emission pulse. This perspective is consistent with the mechanics of recording single-pulse interferograms, which provide the laser linewidth at the early stages of oscillation that is broader than at subsequent times of emission. That is, although the linewidth narrowing continues throughout the duration of the laser emission, the time-integrated recording process, using photographic means for instance, yields information as broad as that recorded at the onset of the laser oscillation. Once $\delta\tau$ has been measured, the number return intracavity passes is given by

$$R = \frac{\delta\tau}{\Delta t} \tag{4.47}$$

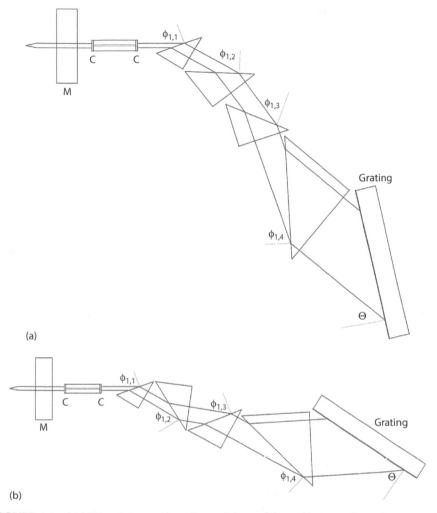

(a)

(b)

FIGURE 4.4   (a) MPL grating semiconductor laser oscillators incorporating a (+, +, +, −) compensating multiple-prism configuration (a) and a (+, −, +, −) compensating multiple-prism configuration (b). (Reproduced from Duarte, F.J., *Tunable Laser Applications*, 2nd edn., Chapter 5, CRC Press, New York, 2009. With permission from Taylor & Francis.)

where:
  Δt is the time taken by the emission to cover twice the cavity length or $\Delta t = 2L/c$
    so that

$$R = \delta\tau\left(\frac{c}{2L}\right) \tag{4.48}$$

In sum, from the beginning of the pulsed laser excitation, the R factor contributes to reduce both the beam divergence and the laser linewidth. As indicated by Equation 4.46, the beam divergence can only decrease toward its diffraction limit, whereas

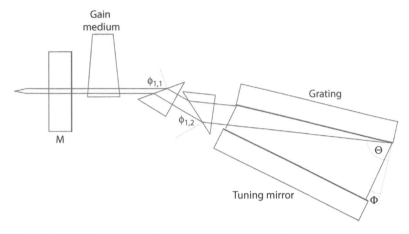

FIGURE 4.5 HMPGI grating solid-state organic laser oscillator incorporating a compensating (+, −) double-prism configuration. (Reprinted from *Opt. Laser Technol.*, 29, Duarte, F.J., Multiple-prism near-grazing-incidence grating solid-state dye-laser oscillator, 513–519, Copyright 1997, with permission from Elsevier.)

the laser linewidth decreases linearly as a function of R. In turn, R is a finite number that can be as low as $R \approx 3$ for high-power short-pulse excitation. It is also necessary to remember that Equation 4.45 is the dispersive cavity linewidth equation, that is, it quantifies the ability of a dispersive oscillator cavity to narrow the frequency transmission window of the cavity. Once the dispersion linewidth limits emission to a single longitudinal mode, the actual linewidth of that mode is determined by the dynamics of the laser. For further information on this subject, the reader should refer to Duarte and Piper (1984) and Duarte (2001b).

4.3.2 Design of Zero-Dispersion Multiple-Prism Beam Expanders

In practice, the dispersion of the grating multiplied by the beam expansion, that is, $M(\nabla_\lambda \Theta_G)$, amply dominates the overall intracavity dispersion. Thus, it is desirable to remove the dispersion component originating from the multiple-prism beam expander so that

$$\Delta\lambda \approx \Delta\theta_R \left(MR\nabla_\lambda\Theta_G \right)^{-1} \tag{4.49}$$

At this stage, we go back to Equation 4.29 and set the requirement $\nabla_\lambda \Phi_P = 0$ so that (Duarte 1985)

$$\nabla_\lambda \Phi_P = 2 \sum_{m=1}^{r} (\pm 1) \left(\prod_{j=1}^{m} k_{1,j} \right) \tan\psi_{1,m} \nabla_\lambda n_m = 0$$

For a double-prism expander yielding zero dispersion,

$$(k_{1,1}) \tan\psi_{1,1} = (k_{1,1}) k_{1,2} \tan\psi_{1,2} \tag{4.50}$$

For a three-prism expander yielding zero dispersion,

$$(k_{1,1} + k_{1,1}k_{1,2})\tan\psi_{1,1} = (k_{1,1}k_{1,2})k_{1,3}\tan\psi_{1,3} \tag{4.51}$$

For a four-prism expander yielding zero dispersion,

$$(k_{1,1} + k_{1,1}k_{1,2} + k_{1,1}k_{1,2}k_{1,3})\tan\psi_{1,1} = (k_{1,1}k_{1,2}k_{1,3})k_{1,4}\tan\psi_{1,4} \tag{4.52}$$

For a five-prism expander yielding zero dispersion,

$$(k_{1,1} + k_{1,1}k_{1,2} + k_{1,1}k_{1,2}k_{1,3} + k_{1,1}k_{1,2}k_{1,3}k_{1,4})\tan\psi_{1,1} = (k_{1,1}k_{1,2}k_{1,3}k_{1,4})k_{1,5}\tan\psi_{1,5} \tag{4.53}$$

The beauty of these zero-dispersion multiple-prism configurations is that the multiple-prism expander contributes to greatly augment the intracavity dispersion, while the tuning characteristics of the laser are exclusively determined by the diffraction grating.

4.3.2.1 Example

To illustrate the design of a quasi-achromatic multiple-prism beam expander, a case of practical interest, with $M \approx 100$, and a four-prism expander deployed in a compensating configuration, similar to that outlined in Figure 4.4a, is considered. The compensating configuration selected is $(+, +, +, -)$; that is, the additive dispersion of the first three prisms is subtracted by the fourth prism, thus yielding zero overall dispersion. Multiple-prism beam expanders deployed intracavity in tunable lasers employ right-angle prisms designed for orthogonal beam exit, that is, $\phi_{2,m} \approx \psi_{2,m} \approx 0$. Thus, we apply Equation 4.52:

$$(k_{1,1} + k_{1,1}k_{1,2} + k_{1,1}k_{1,2}k_{1,3})\tan\psi_{1,1} = (k_{1,1}k_{1,2}k_{1,3})k_{1,4}\tan\psi_{1,4}$$

If the first three prisms are identical, then we have

$$\tan\psi_{1,1} = \tan\psi_{1,2} = \tan\psi_{1,3} \tag{4.54}$$

A particular wavelength of interest in tunable lasers is $\lambda = 590$ nm. At this wavelength, for a material such as optical crown glass, $n = 1.5167$. For $M \approx 100$, we select

$$k_{1,1}k_{1,2}k_{1,3} = 64$$

so that

$$k_{1,1} = k_{1,2} = k_{1,3} = 4$$

This means that for the first three prisms

$$\phi_{1,1} = \phi_{1,2} = \phi_{1,3} = 79.01°$$

and

$$\psi_{1,1} = \psi_{1,2} = \psi_{1,3} = 40.33°$$

which also becomes the apex angle of the first three prisms. Using Equation 4.52, it is found that for the fourth prism, $\phi_{1,4} = 59.39°$, $\psi_{1,4} = 34.57°$, and $k_{1,4} = 1.62$. This yields an overall beam magnification factor of $M = 64 \times 1.62 \approx 103.68$. Normally, the beam waist w in a well-designed high-power multiple-prism grating laser oscillator is about 100 μm. This means that the expanded beam at the exit surface of the fourth prism is $2wM \approx 20.7$ mm. Thus, this particular multiple-prism beam expander can be composed of three small prisms with a hypotenuse of 17 mm, albeit the first two can be even smaller. The larger fourth prism requires a hypotenuse of 27 mm. The thickness of all prisms can be 10 mm. The surfaces of these prisms are usually polished to yield a surface flatness of λ/4 or better.

4.3.2.2 Example

Optimized compact high-power solid-state multiple-prism laser oscillators have been demonstrated to yield single-longitudinal-mode oscillation at $\Delta v \approx 350$ MHz, at pulses $\Delta t \approx 3$ ns, near the limit allowed by Heisenberg's uncertainty principle (Duarte 1999). The oscillator, illustrated in Figure 4.6, requires the use of a small fused silica double-prism beam expander with $M \approx 42$ and $\phi_{2,m} \approx \psi_{2,m} \approx 0$ at λ = 590 nm. Thus we use Equation 4.29 or 4.50 to obtain (Duarte 2000)

$$\tan \psi_{1,1} = k_{1,2} \tan \psi_{1,2} \qquad (4.55)$$

For $n = 1.4583$ and $\phi_{1,1} = 88.60°$, we get $\psi_{1,1} \approx 43.28°$ and $k_{1,1} \approx 29.80$. With these initial parameters, Equation 4.55 yields for the second prism $\phi_{1,2} \approx 53.93°$, $\psi_{1,2} \approx 33.66°$, and $k_{1,2} \approx 1.41$. Therefore, the overall intracavity beam expansion becomes

$$M = k_{1,1}k_{1,2} \approx 42.13$$

For a beam waist of $w = 100$ μm, this implies $2wM \approx 8.43$ mm. These dimensions require the first prism to have a hypotenuse of ~8 mm and the second prism a

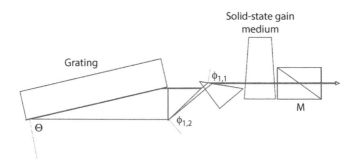

FIGURE 4.6 Optimized multiple-prism grating tunable organic laser oscillator incorporating a compensating (+, −) double-prism configuration. This oscillator demonstrated single-transverse-mode beam characteristics and single-longitudinal-mode emission at $\Delta v \approx 350$ MHz and $\Delta t \approx 3$ ns. (Reproduced from Duarte, F.J., *Appl. Opt.*, 38, 6347–6349, 1999. With permission from the Optical Society.)

hypotenuse of ~10 mm. In this particular oscillator, this intracavity beam expansion is used to illuminate a 3300 lines/mm grating deployed at an angle of incidence ~77° in Littrow configuration (Duarte 1999).

Shay and Duarte (2009) describe the design of a zero-dispersion five-prism beam expander for fused silica at $\lambda = 1550$ nm ($n = 1.44402$), yielding an overall beam expansion of $M \approx 987$.

4.4 DISPERSION OF AMICI, OR COMPOUND, PRISMS

Amici prisms, direct-vision prisms, and compound prisms are terms that refer to multiple-prism assemblies designed not to deviate the light beam. In this regard, such prisms are essentially multiple-prism arrays where the prisms are deployed, in compensating configurations, right next to each other without space in between. One of the main applications of Amici prisms is in the field of astronomy for the correction of atmospheric dispersion in adaptive optics systems for very large telescopes (Wynne 1997). The description of the dispersion calculation of Amici prisms given here follows a review given by Duarte (2013). The expression applied here is the single-pass generalized multiple-prism dispersion equation for positive refraction, that is, Equation 4.11:

$$\nabla_\lambda \phi_{2,m} = \mathcal{H}_{2,m} \nabla_\lambda n_m + (k_{1,m} k_{2,m})^{-1} \left[\mathcal{H}_{1,m} \nabla_\lambda n_m \pm \nabla_\lambda \phi_{2,(m-1)} \right]$$

For a single prism ($m = 1$), as illustrated in Figure 4.7, this equation reduces to Equation 4.35:

$$\nabla_\lambda \phi_{2,1} = \mathcal{H}_{2,1} \nabla_\lambda n_1 + (k_{1,1} k_{2,1})^{-1} (\mathcal{H}_{1,m} \nabla_\lambda n_1)$$

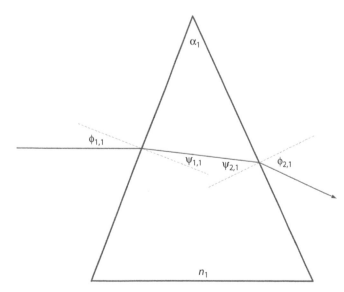

FIGURE 4.7 Single dispersive prism, made of fused silica, with a negative dispersion of $\nabla_\lambda \phi_{2,1} \approx -0.0386$ for the D_2 line of sodium at $\lambda = 588.9963$ nm. The calculations indicate a severe beam deviation.

For a double-prism configuration ($m = 2$) in a compensating mode, as illustrated in Figure 4.8, the dispersion equation becomes

$$\nabla_\lambda \phi_{2,2} = \mathcal{H}_{2,2} \nabla_\lambda n_m + (k_{1,2}k_{2,2})^{-1}(\mathcal{H}_{1,2}\nabla_\lambda n_m \pm \nabla_\lambda \phi_{2,1}) \qquad (4.56)$$

and for a three-prism ($m = 3$) compensating configuration, as illustrated in Figure 4.9, the dispersion equation becomes

$$\nabla_\lambda \phi_{2,3} = \mathcal{H}_{2,3} \nabla_\lambda n_m + (k_{1,3}k_{2,3})^{-1}(\mathcal{H}_{1,3}\nabla_\lambda n_m \pm \nabla_\lambda \phi_{2,2}) \qquad (4.57)$$

thus explicitly illustrating the iterative nature of the generalized multiple-prism equations.

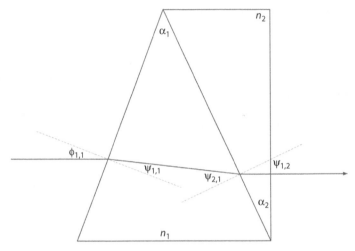

FIGURE 4.8 Double-prism Amici configuration comprising a fused silica prism and a higher diffractive index Schott SF10 prism. Here, $\nabla_\lambda \phi_{2,2} \approx -0.0330$ at $\lambda = 588.9963$ nm and the exit beam is in the same direction, albeit still displaced, as the incident beam. (Drawing not to scale.)

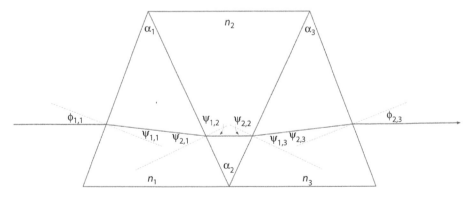

FIGURE 4.9 Three-prism Amici, direct-vision, configuration. The first and third prisms are made of fused silica ($n_1 = n_3 = 1.458413$), whereas the second prism, positioned at the center of the configuration, is made of Schott SF10 ($n_2 = 1.728093$). As explained in the text, this compound prism is formed by unfolding the double-prism configuration shown in Figure 4.8. Here, the overall dispersion is $\nabla_\lambda \phi_{2,3} \approx +0.0720$ at $\lambda = 588.9963$ nm, and the exit beam is collinear with the incident beam as discussed in the text. (Drawing not to scale.)

4.4.1 EXAMPLE

Here, the design and evaluation of dispersion for a two-prism and a three-prism Amici configuration is demonstrated following the example given by Duarte (2013). Selecting a fused silica prism ($n_1 = 1.458413$ at $\lambda = 588.9963$ nm, also known as the sodium D_2 line), as the first dispersive stage, with the following angular dimensions $\alpha_1 = 45°$, $\phi_{1,1} = 20.66°$, $\psi_{1,1} = 14.00°$, $\psi_{2,1} = 31.00°$, and $\phi_{2,1} = 48.69°$ (see Figure 4.7), Equation 4.35 reduces to

$$\nabla_\lambda \phi_{2,1} \approx 1.1039 \ \nabla_\lambda n_1$$

so that, using $\nabla_\lambda n_1 \approx -0.034986 \ \mu m^{-1}$, the dispersion is

$$\nabla_\lambda \phi_{2,1} \approx 1.1039 \ \nabla_\lambda n_1 \approx -0.0386 \tag{4.58}$$

Notice that the quantity $\nabla_\lambda \phi_{2,1}$ has units of radians μm^{-1}. This prism deviates the incident light beam. To create a nondeviating beam path, a second prism is introduced, in compensating configuration, adjoined to the first prism, thus creating a compound or Amici prism, as illustrated in Figure 4.8. For the second prism, a higher refractive index material, in this case, Schott SF10 ($n_2 = 1.728093$ at $\lambda = 588.9963$ nm) with an apex angle $\alpha_2 = 25.76°$, is selected. For this compensating prism, the angular quantities are $\phi_{2,1} = \phi_{1,2} = 48.69°$, $\psi_{1,2} = 25.76°$, and $\psi_{2,2} = \phi_{2,2} = 0$. Using Equations 4.56 and 4.58, with $\nabla_\lambda n_2 \approx -0.127157 \ \mu m^{-1}$, the dispersion for the two-prism Amici configuration becomes

$$\nabla_\lambda \phi_{2,2} \approx 0.4826 \ \nabla_\lambda n_2 - 0.8092 \ \nabla_\lambda n_1 \approx -0.0331 \tag{4.59}$$

in units of radians μm^{-1}. This compound prism is illustrated in Figure 4.8 where it is observed that albeit the propagation of the beam is straight, the exit beam is slightly displaced relative to its position at the incidence surface.

This displacement is corrected using a colinear three-prism transmission configuration as illustrated in Figure 4.9. The two-prism Amici configuration, of Figure 4.8, is unfolded, thus creating a three-prism direct vision configuration (Duarte 2013). This multiple-prism assembly is also known as a three-prism Amici configuration (Jenkins and White 1957) and is illustrated in Figure 4.9. For this particular example, $\alpha_1 = 45°$, $\phi_{1,1} = 20.66°$, $\psi_{1,1} = 14.00°$, $\psi_{2,1} = 31.00°$, $\alpha_2 = 51.53°$, $\psi_{1,2} = \psi_{2,2} = 25.76°$, $\alpha_3 = 45°$, $\psi_{1,3} = 31.00°$, $\psi_{2,3} = 14.00°$, and $\phi_{2,3} = 20.66°$. Using Equation 4.56, the dispersion for the new double-prism configuration becomes

$$\nabla_\lambda \phi_{2,2} \approx 1.316872 \ \nabla_\lambda n_2 - 1.103930 \ \nabla_\lambda n_1 \tag{4.60}$$

Then, from Equation 4.57, and noting that $m = 1$ and $m = 3$ are identical so that $\nabla_\lambda n_3 = \nabla_\lambda n_1$, the estimated cumulative single-pass dispersion becomes

$$\nabla_\lambda \phi_{2,3} \approx 1.763263 \ \nabla_\lambda n_1 - 1.051694 \ \nabla_\lambda n_2 \approx +0.0720 \tag{4.61}$$

A single prism provides a negative dispersion, as seen in Equation 4.58, while deviating the original path of the incident beam. The two-prism Amici configuration slightly modifies this negative dispersion, while making the light beam propagate in the same direction as the incident beam, although the two beams remain displaced relative to each other as shown in Figure 4.8. Unfolding the two-prism arrangement to create a three-prism assembly as illustrated in Figure 4.9 yields a three-prism Amici configuration that makes the incident and exit beam collinear. Furthermore, the negative dispersion that characterized the single prism and the double prism is converted to a positive dispersion with greater magnitude. The quantitative conversion from negative dispersion ($\nabla_\lambda\phi_{2,2} \approx -0.0331$) to positive dispersion ($\nabla_\lambda\phi_{2,3} \approx +0.0720$) is consistent with the qualitative observations of Jenkins and White (1957).

Here, it has been shown quantitatively, via the generalized multiple-prism dispersion theory, that although these Amici, or compound, prisms do provide an exit beam collinear with the incident beam, they are also quite dispersive, hence their application in the correction of atmospheric dispersion in astronomical optics (Duarte 2013).

4.5 MULTIPLE-PRISM DISPERSION AND PULSE COMPRESSION

Prismatic pulse compression was demonstrated for the first time by Dietel et al. (1983) yielding a pulse duration of 53 fs. A prism compensating pair was introduced by Diels et al. (1985) obtaining a pulse duration of 85 fs. Two compensating prism pairs, as pulse compressors, were demonstrated by Fork et al. (1984) who reported pulses of 65 fs. Generalized dispersion equations applicable to pulse compression were introduced by Duarte and Piper (1982) and Duarte (1987a). An extension of this theory to arbitrary higher order phase derivatives was introduced by Duarte (2009). A double-prism and a quadruple-prism pulse compression configurations are depicted in Figures 4.10 and 4.11, respectively. Various other configurations are also possible as discussed by Duarte (1995a).

The material in this section is based on the original description given by Duarte (2003), and the new higher derivatives disclosed by Duarte (2009) and a recent review on the subject (Duarte 2013).

From the uncertainty relation

$$\Delta v \, \Delta t \approx 1 \qquad\qquad (4.62)$$

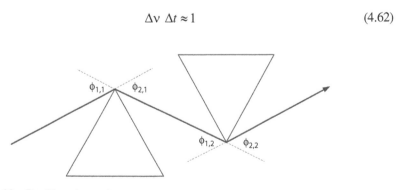

FIGURE 4.10 Double-prism pulse compressor.

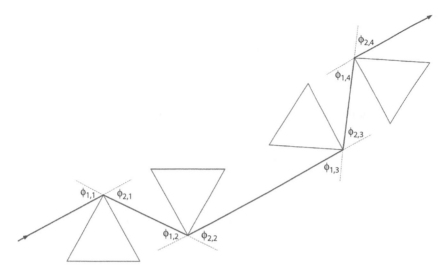

FIGURE 4.11 Four-prism pulse compressor obtained by symmetrically unfolding the double-prism configuration.

it is immediately apparent that the generation of ultrashort time pulses (Δt) requires the simultaneous generation of a very wide spectral distribution (Δv). From the cavity linewidth equation

$$\Delta\lambda = \Delta\theta(\nabla_\lambda\theta)^{-1}$$

that can be expressed in the frequency domain as

$$\Delta v = \left(\frac{c}{\lambda^2}\right)\Delta\theta(\nabla_\lambda\theta)^{-1} \tag{4.63}$$

it is also clear that the generation of very wide spectral emission requires *the least amount of intracavity dispersion.*

Pulse compression in ultrashort pulse, or femtosecond, lasers requires the control of the first, second, and third derivatives of the intracavity dispersion (see, e.g., Diels and Rudolph 2006). Using the identity (Duarte 1987a)

$$\nabla_n\phi_{2,m} = \nabla_\lambda\phi_{2,m}(\nabla_\lambda n_m)^{-1} \tag{4.64}$$

the generalized single-pass dispersion equation

$$\nabla_\lambda\phi_{2,m} = \mathcal{H}_{2,m}\nabla_\lambda n_m + (k_{1,m}k_{2,m})^{-1}\left[\mathcal{H}_{1,m}\nabla_\lambda n_m \pm \nabla_\lambda\phi_{2,(m-1)}\right]$$

can be restated as (Duarte 2009)

$$\nabla_n\phi_{2,m} = \mathcal{H}_{2,m} + (\mathcal{M})^{-1}\left[\mathcal{H}_{1,m} \pm \nabla_n\phi_{2,(m-1)}\right] \tag{4.65}$$

where:

$$(\mathcal{M})^{-1} = k_{1,m}^{-1} k_{2,m}^{-1} \tag{4.66}$$

Thus, the second derivative of the refraction angle, corresponding to $r = 2$, or the first derivative of the dispersion $\nabla_n \phi_{2,m}$, is given by (Duarte 1987a)

$$
\begin{aligned}
\nabla_n^2 \phi_{2,m} &= \nabla_n \mathcal{H}_{2,m} \\
&+ \left(\nabla_n \mathcal{M}^{-1} \right) \left[\mathcal{H}_{1,m} \pm \nabla_n \phi_{2,(m-1)} \right] \\
&+ \left(\mathcal{M}^{-1} \right) \left[\nabla_n \mathcal{H}_{1,m} \pm \nabla_n^2 \phi_{2,(m-1)} \right]
\end{aligned}
\tag{4.67}
$$

The second derivative of the dispersion $\nabla_n \phi_{2,m}$, corresponding to $r = 3$, is given by (Duarte 2009)

$$
\begin{aligned}
\nabla_n^3 \phi_{2,m} &= \nabla_n^2 \mathcal{H}_{2,m} \\
&+ \left(\nabla_n^2 \mathcal{M}^{-1} \right) \left[\mathcal{H}_{1,m} \pm \nabla_n \phi_{2,(m-1)} \right] \\
&+ 2 \left(\nabla_n \mathcal{M}^{-1} \right) \left[\nabla_n \mathcal{H}_{1,m} \pm \nabla_n^2 \phi_{2,(m-1)} \right] \\
&+ \left(\mathcal{M}^{-1} \right) \left[\nabla_n^2 \mathcal{H}_{1,m} \pm \nabla_n^3 \phi_{2,(m-1)} \right]
\end{aligned}
\tag{4.68}
$$

The third derivative of the dispersion $\nabla_n \phi_{2,m}$, corresponding to $r = 4$, is given by (Duarte 2009)

$$
\begin{aligned}
\nabla_n^4 \phi_{2,m} &= \nabla_n^3 \mathcal{H}_{2,m} \\
&+ \left(\nabla_n^3 \mathcal{M}^{-1} \right) \left[\mathcal{H}_{1,m} \pm \nabla_n \phi_{2,(m-1)} \right] \\
&+ 3 \left(\nabla_n^2 \mathcal{M}^{-1} \right) \left[\nabla_n \mathcal{H}_{1,m} \pm \nabla_n^2 \phi_{2,(m-1)} \right] \\
&+ 3 \left(\nabla_n \mathcal{M}^{-1} \right) \left[\nabla_n^2 \mathcal{H}_{1,m} \pm \nabla_n^3 \phi_{2,(m-1)} \right] \\
&+ \left(\mathcal{M}^{-1} \right) \left[\nabla_n^3 \mathcal{H}_{1,m} \pm \nabla_n^4 \phi_{2,(m-1)} \right]
\end{aligned}
\tag{4.69}
$$

The fourth derivative of the dispersion $\nabla_n \phi_{2,m}$, corresponding to $r = 5$, is given by (Duarte 2009)

$$\nabla_n^5 \phi_{2,m} = \nabla_n^4 \mathcal{H}_{2,m} + \left(\nabla_n^4 \mathcal{M}^{-1}\right)\left[\mathcal{H}_{1,m} \pm \nabla_n \phi_{2,(m-1)}\right]$$

$$+ 4\left(\nabla_n^3 \mathcal{M}^{-1}\right)\left[\nabla_n \mathcal{H}_{1,m} \pm \nabla_n^2 \phi_{2,(m-1)}\right]$$

$$+ 6\left(\nabla_n^2 \mathcal{M}^{-1}\right)\left[\nabla_n^2 \mathcal{H}_{1,m} \pm \nabla_n^3 \phi_{2,(m-1)}\right] \tag{4.70}$$

$$+ 4\left(\nabla_n \mathcal{M}^{-1}\right)\left[\nabla_n^3 \mathcal{H}_{1,m} \pm \nabla_n^4 \phi_{2,(m-1)}\right]$$

$$+ \left(\mathcal{M}^{-1}\right)\left[\nabla_n^4 \mathcal{H}_{1,m} \pm \nabla_n^5 \phi_{2,(m-1)}\right]$$

The fifth derivative of the dispersion $\nabla_\lambda \phi_{2,m}$, corresponding to $r = 6$, is given by

$$\nabla_n^6 \phi_{2,m} = \nabla_n^5 \mathcal{H}_{2,m} + \left(\nabla_n^5 \mathcal{M}^{-1}\right)\left[\mathcal{H}_{1,m} \pm \nabla_n \phi_{2,(m-1)}\right]$$

$$+ 5\left(\nabla_n^4 \mathcal{M}^{-1}\right)\left[\nabla_n \mathcal{H}_{1,m} \pm \nabla_n^2 \phi_{2,(m-1)}\right]$$

$$+ 10\left(\nabla_n^3 \mathcal{M}^{-1}\right)\left[\nabla_n^2 \mathcal{H}_{1,m} \pm \nabla_n^3 \phi_{2,(m-1)}\right] \tag{4.71}$$

$$+ 10\left(\nabla_n^2 \mathcal{M}^{-1}\right)\left[\nabla_n^3 \mathcal{H}_{1,m} \pm \nabla_n^4 \phi_{2,(m-1)}\right]$$

$$+ 5\left(\nabla_n \mathcal{M}^{-1}\right)\left[\nabla_n^4 \mathcal{H}_{1,m} \pm \nabla_n^5 \phi_{2,(m-1)}\right]$$

$$+ \left(\mathcal{M}^{-1}\right)\left[\nabla_n^5 \mathcal{H}_{1,m} \pm \nabla_n^6 \phi_{2,(m-1)}\right]$$

The sixth derivative of the dispersion $\nabla_n \phi_{2,m}$, corresponding to $r = 7$, is given by (Duarte 2013)

$$\nabla_n^7 \phi_{2,m} = \nabla_n^6 \mathcal{H}_{2,m} + \left(\nabla_n^6 \mathcal{M}^{-1}\right)\left[\mathcal{H}_{1,m} \pm \nabla_n \phi_{2,(m-1)}\right]$$

$$+ 6\left(\nabla_n^5 \mathcal{M}^{-1}\right)\left[\nabla_n \mathcal{H}_{1,m} \pm \nabla_n^2 \phi_{2,(m-1)}\right]$$

$$+ 15\left(\nabla_n^4 \mathcal{M}^{-1}\right)\left[\nabla_n^2 \mathcal{H}_{1,m} \pm \nabla_n^3 \phi_{2,(m-1)}\right]$$

$$+ 20\left(\nabla_n^3 \mathcal{M}^{-1}\right)\left[\nabla_n^3 \mathcal{H}_{1,m} \pm \nabla_n^4 \phi_{2,(m-1)}\right] \tag{4.72}$$

$$+ 15\left(\nabla_n^2 \mathcal{M}^{-1}\right)\left[\nabla_n^4 \mathcal{H}_{1,m} \pm \nabla_n^5 \phi_{2,(m-1)}\right]$$

$$+ 6\left(\nabla_n \mathcal{M}^{-1}\right)\left[\nabla_n^5 \mathcal{H}_{1,m} \pm \nabla_n^6 \phi_{2,(m-1)}\right]$$

$$+ \left(\mathcal{M}^{-1}\right)\left[\nabla_n^6 \mathcal{H}_{1,m} \pm \nabla_n^7 \phi_{2,(m-1)}\right]$$

Hence, it can be seen that from the second term on, the numerical factors can be predetermined from Pascal's triangle relative to N, where $(N + 1)$ is the order of the derivative (Duarte 2009). Observing the progression of the higher derivatives, a generalized expression is found to be (Duarte 2013)

$$\nabla_n^r \phi_{2,m} = \nabla_n^{r-1} \mathcal{H}_{2,m} + (\mathcal{M})^{-1} \left(\nabla_n + \zeta \right)^{r-1} \tag{4.73}$$

where:

$$\zeta^s = \nabla_n^s \mathcal{H}_{1,m} \pm \nabla_n^{s+1} \phi_{2,(m-1)} \tag{4.74}$$

$$\zeta^0 = 1 = \mathcal{H}_{1,m} \pm \nabla_n \phi_{2,(m-1)} \tag{4.75}$$

Here, when writing the expansion in r, in Equation 4.73, the term $\zeta^0 = 1$ must be included as defined in Equation 4.75. Also, in Equation 4.74, the maximum value of the exponent is $s = (r - 1)$.

These equations represent the complete description of the generalized multiple-prism dispersion theory applicable to pulse compression prismatic arrays in femtosecond, or ultrashort, pulse lasers and nonlinear optics. Exact numerical calculations to determine $\nabla_n \phi_{2,m}$ and $\nabla_n^2 \phi_{2,m}$, for $m = 1, 2, 3, 4$, were performed by Duarte (1990b). In these calculations, the angle of incidence was deviated by minute amounts from the Brewster angle of incidence. Duarte (2009) provides the exact values, as a function of the refractive index n, for $\nabla_n \phi_{2,m}$, $\nabla_n^2 \phi_{2,m}$, and $\nabla_n^3 \phi_{2,m}$. Simplifying assumptions include incidence at the Brewster angle of incidence, prisms of identical isosceles geometry, and made of the same material with refractive index $n_m = n$.

Furthermore, these equations have been applied to yield specific numerical results (Duarte 1987a, 1990b): for example, Osvay et al. (2004, 2005) have used the lower order derivatives, given here, in practical femtosecond lasers to determine dispersions and laser pulse durations for double-prism compressors with excellent agreement between theory and experiments. Table 4.1 provides the refractive index and dispersions for some prism materials used in prismatic pulse compressors.

4.5.1 Example

A four-prism compressor (Fork et al. 1984) is formed by unfolding the double-prism configuration about a symmetry axis perpendicular to the exit beam depicted in Figure 4.10, as illustrated in Figure 4.11. For exactly balanced compensating prism arrays composed of two pairs of compensating prisms, it can be shown that (Duarte 1987a, 2009)

$$\nabla_n \phi_{2,1} = \nabla_n \phi_{2,3} = 2 \tag{4.76}$$

$$\nabla_n^2 \phi_{2,1} = \nabla_n^2 \phi_{2,3} = (4n - 2n^{-3}) \tag{4.77}$$

$$\nabla_n^3 \phi_{2,1} = \nabla_n^3 \phi_{2,3} = (24n^2 + 8n^0 - 12n^{-2} + 6n^{-4} + 6n^{-6}) \tag{4.78}$$

TABLE 4.1
Refractive Index and Its Derivatives for Prism Materials Used in Pulse Compression

Material	λ (μm)	n (λ)	$\nabla_\lambda n$ (μm^{-1})	$\nabla_\lambda^2 n$ (μm^{-2})	Reference
Quartz	0.62	1.457	−0.03059	0.1267	Fork et al. (1984)
Fused silica[a]	0.62	1.45740	−0.03059	0.1267	
	0.80	1.45332	−0.01728	0.0399	
BK7	0.62	1.51554	−0.03613	0.15509	Duarte (1995b)
SF10[a]	0.62	1.72447	−0.10751	0.55988	
F2	0.62	1.61747	−0.07357	0.34332	Diels (1990)
LaSF9	0.62	1.84629	−0.11189	0.57778	Duarte (1995b)
	0.80	1.83257	−0.05201	0.18023	Duarte (1995b)
ZnSe[b]	0.62	2.586	−0.698	5.068	Duarte (1995b)
	0.80	2.511	−0.246	1.163	Duarte (1995b)

[a] Calculated using data in Chapter 13.
[b] Calculated using data from Marple (1964).

$$\nabla_n \phi_{2,2} = \nabla_n \phi_{2,4} = 0 \tag{4.79}$$

$$\nabla_n^2 \phi_{2,2} = \nabla_n^2 \phi_{2,4} = 0 \tag{4.80}$$

$$\nabla_n^3 \phi_{2,2} = \nabla_n^3 \phi_{2,4} = 0 \tag{4.81}$$

A six-prism pulse compressor has been used in semiconductor laser pulse compression by Pang et al. (1992). Certainly, the equations provided here can be applied to prismatic pulse compressors comprising any number of prisms and deployed in any suitable geometry.

4.6 APPLICATIONS OF MULTIPLE-PRISM ARRAYS

So far, the use of multiple-prism arrays as intracavity beam expanders in narrow-linewidth tunable laser oscillators and as pulse compressors in ultrafast lasers has been examined. However, as discussed by Duarte (2000), multiple-prism arrays as one-dimensional telescopes, or beam expanders, have found a plethora of alternative applications:

1. Extracavity double-prism beam expander to correct the ellipticity of laser beams generated by semiconductor lasers (Maker and Ferguson 1989). These expanders make use of the prism pairs first introduced by Brewster (1813).
2. Generalized laser beam shaping devices as discussed by Duarte (1987b, 1995c).

3. Beam expanding devices for optical computing (Lohmann and Stork 1989).
4. One-dimensional beam expanders yielding *extremely elongated* Gaussian beams, with height-to-width ratios in the 1:1000 to 1:3000 range for microscopy applications (Duarte 1987a, 1993, 1995b). This includes characterization of transmission and reflection imaging surfaces (see Chapter 10). Note that this type of illumination is also known in the literature as *light sheet illumination* and *selective plane illumination*.
5. Multiple-prism beam expansion for N-slit interferometry (Duarte 1991, 1993).
6. One-dimensional multiple-prism telescopes in conjunction with digital detectors (Duarte 1993, 1995b). This includes applications in conjunction with transmission grating in areas of wavelength measurements (Duarte 1995b) and secure optical communications (Duarte 2002; Duarte et al. 2010, 2011).
7. One-dimensional beam expansion for application in laser printers for imaging applications, also known as *laser sensitometers* (Duarte 2001a; Duarte et al. 2005).
8. Beam expansion for imaging systems used in astronomy, particularly in the subfield of astrometry (Sirat et al. 2005).
9. Multiple-prism assemblies for Amici prisms for dispersion correction in direct vision instrumentation for astronomy applications (Wynne 1997).

PROBLEMS

4.1 Show that Equation 4.11 can also be expressed in its explicit form of Equation 4.14.

4.2 Show that for orthogonal beam exit Equation 4.14 reduces to Equation 4.17.

4.3 Show that for prism with ($\alpha_1 = \alpha_2 = \alpha_3 = ... = \alpha_m$) and deployed for Brewster's angle of incidence, Equation 4.17 can be restated as Equation 4.19.

4.4 Design a three-prism beam expander with $\nabla_\lambda \Phi_P = 0$ for $M \approx 70$. Assume $\lambda = 590$ nm, $n = 1.5167$ and $w = 100$ μm.

4.5 Show that for $m = 1$, orthogonal beam exit, and Brewster's angle of incidence, Equation 4.17 reduces to Equation 4.38.

4.6 Expand Equation 4.73 for $r = 7$ to arrive at Equation 4.72.

5 Polarization

5.1 INTRODUCTION

This chapter is a revised version of Chapter 5 on classical *linear polarization* published in Duarte (2003) and includes some new subject matter on polarization as discussed in Duarte (2014).

5.2 MAXWELL EQUATIONS

Maxwell equations are of fundamental importance since they describe the whole of classical electromagnetic phenomena. From a classical perspective, light can be described as waves of electromagnetic radiation. As such, Maxwell equations are very useful to illustrate a number of the characteristics of light including polarization. It is customary to just state these equations without derivation. Since our goal is simply to apply them, the usual approach will be followed. However, for those interested, it is mentioned that a derivation by Dyson (1990) attributed to Feynman is available in the literature. *Maxwell equations* in the rationalized metric system are given by (Feynman et al. 1965)

$$\nabla \cdot \boldsymbol{B} = 0 \tag{5.1}$$

$$\nabla \cdot \boldsymbol{E} = \frac{\rho}{\varepsilon_0} \tag{5.2}$$

$$c^2 \nabla \times \boldsymbol{B} = \frac{\partial \boldsymbol{E}}{\partial t} + \frac{\boldsymbol{j}}{\varepsilon_0} \tag{5.3}$$

$$\nabla \times \boldsymbol{E} = -\frac{\partial \boldsymbol{B}}{\partial t} \tag{5.4}$$

These equations illustrate, with succinct beauty, the unique coexistence of the electric field and the magnetic field in nature. The first two equations give the value of the given flux through a closed surface, whereas the second two equations give the value of a line integral around a loop. In this notation,

$$\nabla = \left(\frac{\partial}{\partial x}, \frac{\partial}{\partial y}, \frac{\partial}{\partial z} \right) \tag{5.5}$$

where:
$\boldsymbol{E}$ is the electric vector
$\boldsymbol{B}$ is the magnetic induction
ρ is the electric charge density

101

j is the electric current density
ε_0 is the *permittivity of free space*
c is the speed of light (see Chapter 13)

In addition to Maxwell equations, the following identities are useful:

$$j = \sigma E \tag{5.6}$$

$$D = \varepsilon E \tag{5.7}$$

$$B = \mu H \tag{5.8}$$

where:
D is the electric displacement
H is the magnetic vector
σ is the specific conductivity
ε is the dielectric constant (or permittivity)
μ is the magnetic permeability

In the Gaussian systems of units, Maxwell equations are given in the form of (see, e.g., Born and Wolf 1999)

$$\nabla \cdot B = 0 \tag{5.9}$$

$$\nabla \cdot E = 4\pi\rho \tag{5.10}$$

$$\nabla \times H = \frac{1}{c}\left(\frac{\partial D}{\partial t} + 4\pi\ j \right) \tag{5.11}$$

$$\nabla \times E = -\frac{1}{c}\ \frac{\partial B}{\partial t} \tag{5.12}$$

It should be noted that many authors in the field of optics prefer to use Maxwell equations in the Gaussian system of units. As explained by Born and Wolf (1999), in this system E, D, j, and ρ are measured in electrostatic units, whereas H and B are measured in electromagnetic units.

For the case of no charges or currents, that is, $j = 0$ and $\rho = 0$, and a homogeneous medium, Maxwell equations and the given identities can be applied in conjunction with the vector identity

$$\nabla \times \nabla \times E = \nabla\ \nabla \cdot E - \nabla^2 E \tag{5.13}$$

to obtain wave equations of the form (Born and Wolf 1999):

$$\nabla^2 E - \frac{\varepsilon\mu}{c^2}\frac{\partial^2 E}{\partial t^2} = 0 \tag{5.14}$$

This leads to an expression for the velocity of propagation:

$$v = c(\varepsilon\mu)^{-1/2} \tag{5.15}$$

Comparison of this expression with the law of positive refraction, derived in Chapter 4, leads to what is known as Maxwell's formula (Born and Wolf 1999):

$$n = (\varepsilon\mu)^{1/2} \tag{5.16}$$

where:
  n is the refractive index

It is useful to note that in vacuum

$$c^2 = (\varepsilon_0\mu_0)^{-1} \tag{5.17}$$

in the rationalized metric system, where μ_0 is the *permeability of free space* (Lorrain and Corson 1970). The values of fundamental constants are listed in Chapter 13.

5.3 POLARIZATION AND REFLECTION

Following the convention of Born and Wolf (1999), we consider a reflection boundary, depicted in Figure 5.1, and a plane of incidence established by the incidence ray and the normal to the reflection surface. Here, the reflected component $\mathcal{R}_\parallel$ is parallel to the plane of incidence and the reflected component $\mathcal{R}_\perp$ is perpendicular to the plane of incidence.

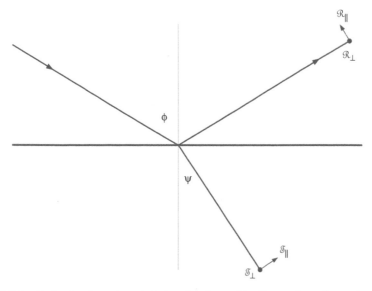

FIGURE 5.1 Reflection boundary defining the plane of incidence. Sometimes, the plane of incidence is also referred to as the plane of propagation.

For the case of $\mu_1 = \mu_2 = 1$, Born and Wolf (1999) consider the electric, and magnetic, vectors as complex plane waves. In this approach, the incident electric vector is represented by the equations of the form:

$$E_x^{(i)} = -A_{\parallel} \cos\phi (e^{-i\tau_i})$$ (5.18)

$$E_y^{(i)} = -A_{\perp} (e^{-i\tau_i})$$ (5.19)

$$E_z^{(i)} = -A_{\parallel} \sin\phi (e^{-i\tau_i})$$ (5.20)

where:
 $A_{\parallel}$ and $A_{\perp}$ are complex amplitudes
 τ_i is the usual plane wave phase factor

Using corresponding equations for E and H for transmission and reflection in conjunction with Maxwell's relation, with $\mu = 1$, and the law for positive refraction, Born and Wolf (1999) derive the *Fresnel formulae*:

$$\mathfrak{I}_{\parallel} = \left[\frac{2\sin\psi\cos\phi}{\sin(\phi+\psi)\cos(\phi-\psi)} \right] A_{\parallel}$$ (5.21)

$$\mathfrak{I}_{\perp} = \left[\frac{2\sin\psi\cos\phi}{\sin(\phi+\psi)} \right] A_{\perp}$$ (5.22)

$$\mathfrak{R}_{\parallel} = \left[\frac{\tan(\phi-\psi)}{\tan(\phi+\psi)} \right] A_{\parallel}$$ (5.23)

$$\mathfrak{R}_{\perp} = \left[\frac{\sin(\phi-\psi)}{\sin(\phi+\psi)} \right] A_{\perp}$$ (5.24)

Using these equations, the transmissivity and reflectivity for both polarizations can be expressed as

$$\mathfrak{I}_{\parallel} = \left[\frac{(\sin 2\phi \sin 2\psi)}{\sin^2(\phi+\psi)\cos^2(\phi-\psi)} \right]$$ (5.25)

$$\mathfrak{I}_{\perp} = \left[\frac{(\sin 2\phi \sin 2\psi)}{\sin^2(\phi+\psi)} \right]$$ (5.26)

$$\mathfrak{R}_{\parallel} = \left[\frac{\tan^2(\phi-\psi)}{\tan^2(\phi+\psi)} \right]$$ (5.27)

$$\mathfrak{R}_{\perp} = \left[\frac{\sin^2(\phi-\psi)}{\sin^2(\phi+\psi)} \right]$$ (5.28)

and

$$\mathcal{R}_{\parallel} + \mathcal{T}_{\parallel} = 1 \qquad (5.29)$$

$$\mathcal{R}_{\perp} + \mathcal{T}_{\perp} = 1 \qquad (5.30)$$

Using these expressions for transmissivity and reflectivity, the degree of polarization, $\mathcal{P}$, is defined as (Born and Wolf 1999)

$$\mathcal{P} = \frac{(\mathcal{R}_{\parallel} - \mathcal{R}_{\perp})}{(\mathcal{R}_{\parallel} + \mathcal{R}_{\perp})} \qquad (5.31)$$

The usefulness of these equations is self-evident once $\mathcal{R}_{\parallel}$ is calculated, as a function of angle of incidence (Figure 5.2), for fused silica at $\lambda \approx 590$ nm ($n = 1.4583$). Here, we see that $\mathcal{R}_{\parallel} = 0$ at 55.5604°. At this angle, ($\phi + \psi$) becomes 90° so that $\tan(\phi + \psi)$ approaches infinity, thus causing $\mathcal{R}_{\parallel} = 0$. This particular ϕ is known as the *Brewster angle* (ϕ_B) and has a very important role in laser optics. Since at $\phi = \phi_B$ the angle of refraction becomes $\psi = (90 - \phi)$ degrees and the law of positive refraction takes the form of

$$\tan \phi_B = n \qquad (5.32)$$

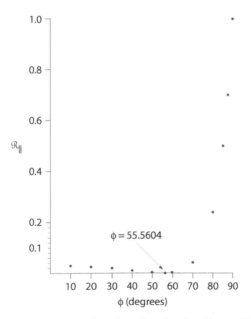

FIGURE 5.2   Reflection intensity as a function of angle of incidence. The angle at which the reflection vanishes is known as the Brewster angle.

For orthogonal, or normal, incidence, the difference between the two polarizations vanishes. Using the law of positive refraction and the appropriate trigonometric identities, in Equations 5.25 through 5.28, it can be shown that (Born and Wolf 1999)

$$\mathcal{R} = \left[\frac{(n-1)}{(n+1)}\right]^2 \tag{5.33}$$

and

$$\mathcal{I} = \left[\frac{4n}{(n+1)}\right]^2 \tag{5.34}$$

5.3.1 PLANE OF INCIDENCE

The discussion in Section 5.3 uses parameters such as $\mathcal{R}_{\parallel}$ and $\mathcal{R}_{\perp}$. In this convention, $\parallel$ means parallel to the plane of incidence and $\perp$ means perpendicular, orthogonal, or normal to the plane of incidence. The plane of incidence is defined, following Born and Wolf (1999), in Figure 5.1. This plane can also be referred to as the plane of propagation.

On more explicit terms, let us consider a laser beam propagating on a plane parallel to the surface of an optical table. If that beam is made to illuminate the hypotenuse of a right-angle prism, whose triangular base is parallel to the surface of the table, then the plane of incidence is established by the incident laser beam and perpendicular to the hypotenuse of the prism. In other words, in this case, the plane of incidence is parallel to the surface of the optical table. Moreover, if that prism is allowed to expand the transmitted beam, as discussed later in this chapter, then the beam expansion is parallel to the plane of incidence.

The linear polarization of a laser can often be orthogonal to an external plane of incidence. When this is the case, and the maximum transmission of the laser through external optics is desired, either the laser is rotated by $\pi/2$ about its axis of propagation, or a collinear $\pi/2$ polarization rotator is used as discussed later in this chapter.

5.4   JONES CALCULUS

Jones calculus is a matrix approach to describe, in a unified form, both linear and circular polarization. It was introduced by Jones (1947) and a good review of the subject was given by Robson (1974). Here, the salient features of the Jones calculus are described without derivation. This presentation follows a review given by Duarte (2014).

The electric field can be expressed in complex terms, in x and y coordinates, in vector form:

$$\begin{pmatrix} E_{0x} \\ E_{0y} \end{pmatrix} = \begin{pmatrix} E_0 e^{i\phi_x} \\ E_0 e^{i\phi_y} \end{pmatrix} \tag{5.35}$$

In this notation, linear polarization in the x-direction is represented by

$$\begin{pmatrix} E_{0x} \\ E_{0y} \end{pmatrix} = \begin{pmatrix} 1 \\ 0 \end{pmatrix} \tag{5.36}$$

whereas linear polarization in the y-direction is described by

$$\begin{pmatrix} E_{0x} \\ E_{0y} \end{pmatrix} = \begin{pmatrix} 0 \\ 1 \end{pmatrix} \tag{5.37}$$

Subsequently,

$$\begin{pmatrix} E_{0x} \\ E_{0y} \end{pmatrix} = \frac{1}{\sqrt{2}} \begin{pmatrix} 1 \\ \pm 1 \end{pmatrix} \tag{5.38}$$

describes diagonal (or oblique) polarization at a $\pi/4$ angle, relative to the x-axis (+) or the y-axis (−).

Circular polarization is described by the vector

$$\begin{pmatrix} E_{0x} \\ E_{0y} \end{pmatrix} = \frac{1}{\sqrt{2}} \begin{pmatrix} 1 \\ \pm i \end{pmatrix} \tag{5.39}$$

where:
  $+i$ applies to right circularly polarized light
  $-i$ applies to left circularly polarized light

Figure 5.3 illustrates the various polarization alternatives.

Jones calculus introduces 2×2 matrices to describe the optical elements transforming the polarization of the incidence radiation in the following format:

$$\begin{pmatrix} a_{11} & a_{12} \\ a_{21} & a_{22} \end{pmatrix} \begin{pmatrix} A_x \\ A_y \end{pmatrix} = \begin{pmatrix} a_{11}A_x + a_{12}A_y \\ a_{21}A_x + a_{22}A_y \end{pmatrix} \tag{5.40}$$

where the 2×2 matrix, on the left-hand side, represents the optical element, whereas the polarization vector multiplying this matrix corresponds to the incident radiation. The vector, on the right-hand side, describes the polarization of the resulting radiation.

Useful Jones matrices include the matrix for transmission of linearly polarized light in the x-direction:

$$\begin{pmatrix} a_{11} & a_{12} \\ a_{21} & a_{22} \end{pmatrix} = \begin{pmatrix} 1 & 0 \\ 0 & 0 \end{pmatrix} \tag{5.41}$$

and the y-direction

$$\begin{pmatrix} a_{11} & a_{12} \\ a_{21} & a_{22} \end{pmatrix} = \begin{pmatrix} 0 & 0 \\ 0 & 1 \end{pmatrix} \tag{5.42}$$

For light linearly polarized at a $\pi/4$ angle, the matrix becomes

$$\begin{pmatrix} a_{11} & a_{12} \\ a_{21} & a_{22} \end{pmatrix} = \frac{1}{2} \begin{pmatrix} 1 & 1 \\ 1 & 1 \end{pmatrix} \tag{5.43}$$

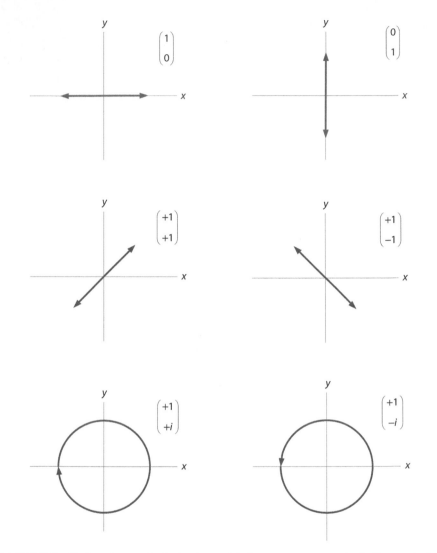

FIGURE 5.3   Various forms of polarization and their vector representation in Jones calculus.

The generalized polarization Jones matrix for linearly polarized light at an angle θ to the x-axis is given by

$$\begin{pmatrix} a_{11} & a_{12} \\ a_{21} & a_{22} \end{pmatrix} = \begin{pmatrix} \cos^2\theta & \sin\theta\cos\theta \\ \sin\theta\cos\theta & \sin^2\theta \end{pmatrix} \qquad (5.44)$$

The right circular polarizer is described as

$$\begin{pmatrix} a_{11} & a_{12} \\ a_{21} & a_{22} \end{pmatrix} = \frac{1}{2}\begin{pmatrix} 1 & -i \\ i & 1 \end{pmatrix} \qquad (5.45)$$

whereas the left circular polarizer is described as

$$\begin{pmatrix} a_{11} & a_{12} \\ a_{21} & a_{22} \end{pmatrix} = \frac{1}{2}\begin{pmatrix} 1 & i \\ -i & 1 \end{pmatrix} \tag{5.46}$$

The generalized rotation matrix for birefringent rotators is given by (Robson 1974)

$$\begin{pmatrix} a_{11} & a_{12} \\ a_{21} & a_{22} \end{pmatrix} = \begin{pmatrix} \cos\theta & \sin\theta \\ -e^{i\delta}\sin\theta & e^{i\delta}\cos\theta \end{pmatrix} \tag{5.47}$$

where:
δ is the phase angle
α is the rotation angle about the z-axis

For a quarter-wave plate, $\delta = \pi/2$, the rotation matrix becomes

$$\begin{pmatrix} a_{11} & a_{12} \\ a_{21} & a_{22} \end{pmatrix} = \begin{pmatrix} \cos\theta & \sin\theta \\ -i\sin\theta & i\cos\theta \end{pmatrix} \tag{5.48}$$

5.4.1 EXAMPLE

A laser beam linearly polarized in the x-direction is sent through an optical element that allows the transmission of y-polarization only, thus using Equations 5.36 and 5.42

$$\begin{pmatrix} 0 & 0 \\ 0 & 1 \end{pmatrix}\begin{pmatrix} 1 \\ 0 \end{pmatrix} = \begin{pmatrix} 0 \\ 0 \end{pmatrix} \tag{5.49}$$

so that no light is transmitted following the y-polarizer, as can be verified by a simple experiment.

5.5 POLARIZING PRISMS

There are two avenues to induce polarization using prisms. The first involves simple reflection as characterized by Fresnel's equations and straightforward refraction. This approach is valid for windows, prisms, or multiple-prism arrays, made from homogeneous optical materials such as optical glass or fused silica. The second approach involves double refraction in crystalline transmission media exhibiting birefringence.

5.5.1 TRANSMISSION EFFICIENCY IN MULTIPLE-PRISM ARRAYS

Here, attention is focused on the polarization characteristics of multiple-prism arrays. First, we illustrate the use of transmission loss equations, which include the corresponding Fresnel coefficient, to evaluate the overall loss via an arbitrary number of incidence and exit surfaces. Second, an example involving a double-prism beam expander is given where the transmission losses, for both components of polarization, are calculated.

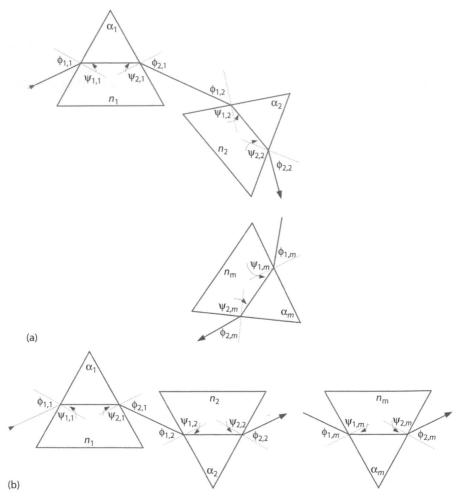

FIGURE 5.4 Generalized multiple-prism array in additive configuration (a) and compensating configuration (b). Depiction of multiple-prism arrays in this form was introduced by Duarte and Piper. (Data from Duarte F.J., and Piper, J.A., *Am. J. Phys.*, 51, 1132–1134, 1983.)

For a generalized multiple-prism array, as shown in Figure 5.4, the cumulative reflection losses at the incidence surface of the mth prism are given by (Duarte et al. 1990)

$$L_{1,m} = L_{2,(m-1)} + [1 - L_{2,(m-1)}]\mathcal{R}_{1,m} \qquad (5.50)$$

whereas the losses at the mth exit surface are given by

$$L_{2,m} = L_{1,m} + (1 - L_{1,m})\mathcal{R}_{2,m} \qquad (5.51)$$

where:

$\mathcal{R}_{1,m}$ and $\mathcal{R}_{2,m}$ are given by either $\mathcal{R}_{\parallel}$ or $\mathcal{R}_{\perp}$

In practice, the optics is deployed so that the polarization of the propagation beam is parallel to the plane of incidence, meaning that the reflection coefficient is given by $\mathfrak{R}_\parallel$. It should be noted that these equations apply not just to prisms but also to optical wedges and any homogeneous optical element, with an input and exit surface, used in the transmission domain.

5.5.2 INDUCED POLARIZATION IN A DOUBLE-PRISM BEAM EXPANDER

Polarization induction in multiple-prism beam expanders should be apparent once the reflectivity equations are combined with the transmission equations (5.50 and 5.51). In this section, this effect is made clear by considering the transmission efficiency, for both components of polarization, of a simple double-prism beam expander as illustrated in Figure 5.5. This beam expander is a modified version of one described by Duarte (2003) and consists of two identical prisms made of fused silica, with $n = 1.4583$ at $\lambda \approx 590$ nm and an apex angle of $42.7098°$. Both prisms are deployed to yield identical magnifications and for orthogonal beam exit. This implies that $\phi_{1,1} = \phi_{1,2} = 81.55°$, $\psi_{1,1} = \psi_{1,2} = 42.7098°$, $\phi_{2,1} = \phi_{2,2} = 0°$, and $\psi_{2,1} = \psi_{2,2} = 0°$.

Thus, for radiation polarized parallel to the plane of incidence,

$$L_{1,1} = \mathfrak{R}_{1,1} = 0.3008$$

$$L_{2,1} = L_{1,1}$$

$$L_{1,2} = L_{2,1} + (1 - L_{2,1})\mathfrak{R}_{1,2} = 0.5111$$

$$L_{2,2} = L_{1,2}$$

whereas for radiation polarized perpendicular to the plane of incidence

$$L_{1,1} = \mathfrak{R}_{1,1} = 0.5758$$

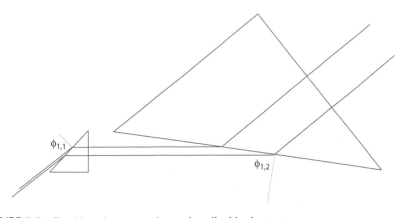

FIGURE 5.5 Double-prism expander as described in the text.

$$L_{2,1} = L_{1,1}$$

$$L_{1,2} = L_{2,1} + (1 - L_{2,1})\mathcal{R}_{1,2} = 0.8200$$

$$L_{2,2} = L_{1,2}$$

and also

$$k_{1,1} = k_{1,2} = 5.0005$$

$$M = k_{1,1}k_{1,2} = 25.0045$$

Thus, for this particular beam expander, the cumulative reflection losses are 51.11% for light polarized parallel to the plane of incidence, whereas they increase to 82.00% for radiation polarized perpendicular to the plane of incidence. This example helps to illustrate the fact that multiple-prism beam expanders exhibit a clear polarization preference. It is easy to see that the addition of further stages of beam magnification leads to increased discrimination. When incorporated in frequency-selective dispersive laser cavities, these beam expanders contribute significantly toward the emission of laser emission polarized parallel to the plane of propagation where the polarization preference is reinforced by multiple-return passes.

See Chapter 4 for a generalized description of multiple-prism dispersion. Chapter 7 describes the use of multiple-prism arrays in laser oscillators.

5.6   DOUBLE-REFRACTION POLARIZERS

These are crystalline prism pairs that exploit the birefringence effect in crystals. In birefringent materials, the dielectric constant, ε, is different in each of the x, y, and z directions so that the propagation velocity is different in each direction:

$$v_a = c(\varepsilon_x)^{-1/2} \tag{5.52}$$

$$v_b = c(\varepsilon_y)^{-1/2} \tag{5.53}$$

$$v_c = c(\varepsilon_z)^{-1/2} \tag{5.54}$$

Since polarization of a transmission medium is determined by the D vector, it is possible to describe the polarization characteristics in each direction. Further, it can be shown that there are two different velocities for the refracted radiation in any given direction (Born and Wolf 1999). As a consequence of the law of refraction, these two velocities lead to two different propagation paths in the crystal and give origin to the *ordinary* and *extraordinary* rays. In other words, the two velocities lead to *double refraction*.

Of particular interest in this class of polarizers are those known as the Nicol prism, the Rochon prism, the Glan–Foucault prism, the Glan–Thompson prism, and

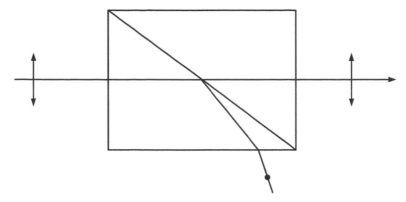

FIGURE 5.6 Generic Glan–Thompson polarizer. The beam polarized parallel to the plane of incidence is transmitted while the complementary component is deviated (drawing not to scale).

the Wollaston prism. According to Bennett and Bennett (1978), a Glan–Foucault prism pair is an air-spaced Glan–Thompson prism pair. In Glan-type polarizers, the extraordinary ray is transmitted from the first to the second prism in the propagation direction of the incident beam. However, the diagonal surfaces of the two prisms are predetermined to induce total internal reflection for the ordinary ray (see Figure 5.6). Glan-type polarizers are very useful since they can be oriented to discriminate in favor of either polarization component with negligible beam deviation. Normally, these polarizers are made of either quartz or calcite. Commercially available calcite Glan–Thompson polarizers with a useful aperture of 10 mm provide extinction ratios of $\sim 5 \times 10^{-5}$. It should be noted that Glan-type polarizers are used in straightforward propagation applications as well as intracavity elements. For instance, the tunable single-longitudinal-mode laser oscillator depicted in Figure 5.7 incorporates a Glan–Thompson polarizer as an output coupler. In this particular polarizer, the inner window is antireflection coated, whereas the outer window is coated for partial reflectivity to act as an output coupler mirror. The laser emission from multiple-prism grating oscillators is highly polarized parallel to the plane of incidence by the interaction of the intracavity flux with the multiple-prism expander and the grating. The function of the polarizer output coupler here is to provide further discrimination against unpolarized single-pass amplified spontaneous emission. These dispersive tunable laser oscillators yield extremely low levels of broadband amplified spontaneous emission measured to be in the $10^{-7} - 10^{-6}$ range (Duarte 1995, 1999).

The Wollaston prism, illustrated in Figure 5.8, is usually fabricated of either crystalline quartz or calcite. These prisms are assembled from two matched and complementary right-angle prisms whose crystalline optical axes are oriented orthogonal to each other. These prisms are widely used as beam splitters of beams with orthogonal polarizations. The beam separation provided by calcite is significantly greater than that achievable with crystalline quartz. Also, for both materials, the beam separation is wavelength dependent.

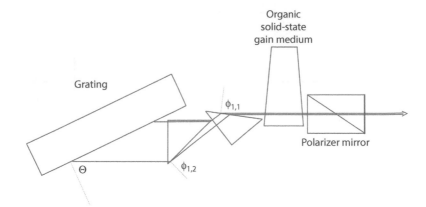

FIGURE 5.7 Solid-state MPL grating dye laser oscillator, yielding single-longitudinal-mode emission, incorporating a Glan–Thompson polarizer output coupler. The reflective coating is applied to the outer surface of the polarizer. (Reprinted from *Opt. Commun.*, 117, Duarte, F.J., Solid-state dispersive dye laser oscillator: Very compact cavity, 480–484, Copyright 1995, with permission from Elsevier.)

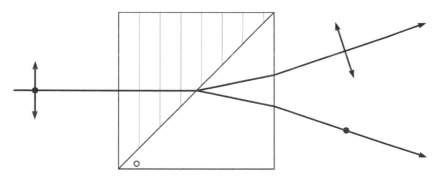

FIGURE 5.8 Generic Wollaston prism. The lines and circle represent the direction of the crystalline optical axis of the prism components (drawing not to scale).

The use of these prisms in quantum cryptography optical configurations is outlined in the work of Duarte (2014). In those optical configurations, a Wollaston prism is used after an electro-optical polarization rotator (such as a Pockels cell) to spatially separate photons corresponding to orthogonal polarizations. For a description of electro-optical polarization rotators, see the work of Saleh and Teich (1991).

5.7 INTENSITY CONTROL OF LASER BEAMS USING POLARIZATION

A very simple, and yet powerful, technique to control and/or attenuate the intensity of linearly polarized laser beams involves the transmission of the laser beam through a prism pair such as a Glan–Thompson polarizer followed by rotation of the polarizer (Duarte 2001). The essence of this technique is illustrated in Figure 5.9. In this

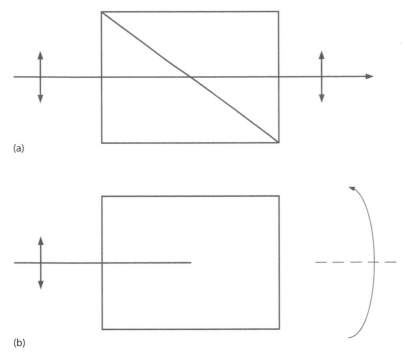

(a)

(b)

FIGURE 5.9 Attenuation of polarized laser beams using a Glan–Thompson polarizer: (a) polarizer set for ~100% transmission; (b) clockwise rotation of the polarizer, about the axis of propagation by $\pi/2$, yielding ~0% transmission. The amount of transmitted light can be varied continuously by rotating the polarizer in the $0 \leq \theta \leq \pi/2$ range. (Data from Duarte, F.J., Laser sensitometer using a multiple-prism beam expander and a polarizer. US Patent 6, 236, 461 B1, 2001.)

approach, for a ~100% laser beam polarized parallel to the plane of incidence, there is almost total transmission when the Glan–Thompson prism pair is oriented as in Figure 5.9a. As the prism pair is rotated about the axis of propagation, the intensity of the transmission decreases until it becomes zero once the angular displacement has reached $\pi/2$. With precision rotation of the prism pair, a scale of well-determined intensities can be easily obtained (Duarte 2001). This polarization attenuation technique has a number of applications including the generation of precise laser intensity scales for exposing instrumentation and laser printers used in imaging as discussed in Chapter 10 (Duarte 2001). Also, this technique has been successfully applied to *laser cooling* experiments to independently vary the intensity of the cooling and repumping lasers (Olivares et al. 2009).

5.8 POLARIZATION ROTATORS

Maximum transmission efficiency is always a goal in optical systems. If the polarization of a laser is mismatched to the polarization preference of the optics, then transmission efficiency will be poor. Furthermore, efficiency can be significantly improved if the polarization of a pump laser is matched to the polarization preference

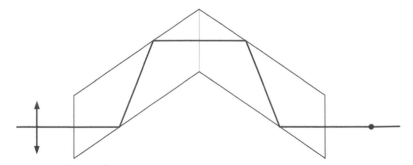

FIGURE 5.10 Side view of double Fresnel rhomb. Linearly polarized light is rotated by $\pi/2$ and exits with polarized orthogonally to the original polarization.

of the laser being excited (Duarte 1990). Although sometimes the efficiency can be improved, or even optimized, by the simple rotation of a laser, it is highly desirable and practical to have optical elements to perform this function. In this section, we shall consider two alternatives to perform such rotation: birefringent polarization rotators and prismatic rotators.

An additional alternative to rotate polarization are rhomboid configurations. For instance, double Fresnel rhomb (Figure 5.10), comprising two quarter-wave parallelepipeds, becomes a half-wave rhomb that rotates linearly polarized light by $\pi/2$. A commercially available Fresnel rhomb of this class, 10 mm wide and 53 mm long, offers a transmission efficiency close to 96% with broadband antireflection coatings. As indicated by Bennett and Bennett (1978), these achromatic rotators tend to offer a small useful aperture-to-length ratio and its achromaticity can be compromised by residual birefringence in the quartz. A description of the inner workings of Fresnel rhombs is offered by Born and Wolf (1999).

5.8.1 BIREFRINGENT POLARIZATION ROTATORS

In birefringent uniaxial crystalline materials, the ordinary and extraordinary rays propagate at different velocities. The generalized matrix for birefringent rotators is given by Equation 5.47:

$$\begin{pmatrix} a_{11} & a_{12} \\ a_{21} & a_{22} \end{pmatrix} = \begin{pmatrix} \cos\theta & \sin\theta \\ -e^{i\delta}\sin\theta & e^{i\delta}\cos\theta \end{pmatrix}$$

For a quarter-wave plate, $\delta = \pi/2$, the phase term is $e^{i\pi/2} = +i$, and the rotation matrix becomes

$$\begin{pmatrix} a_{11} & a_{12} \\ a_{21} & a_{22} \end{pmatrix} = \begin{pmatrix} \cos\theta & \sin\theta \\ -i\sin\theta & i\cos\theta \end{pmatrix} \qquad (5.55)$$

For a half-wave plate, $\delta = \pi$, and the phase term is $e^{i\pi} = -1$. Thus, the rotation matrix becomes

$$\begin{pmatrix} a_{11} & a_{12} \\ a_{21} & a_{22} \end{pmatrix} = \begin{pmatrix} \cos\theta & \sin\theta \\ \sin\theta & -\cos\theta \end{pmatrix} \qquad (5.56)$$

From the experiment, we know that a half-wave plate causes a rotation of a linearly polarized beam by $\theta = \pi/2$ so that Equation 5.56 reduces to

$$\begin{pmatrix} a_{11} & a_{12} \\ a_{21} & a_{22} \end{pmatrix} = \begin{pmatrix} 0 & 1 \\ 1 & 0 \end{pmatrix} \tag{5.57}$$

5.8.1.1 Example

Thus, if we send a beam polarized in the x-direction through a half-wave plate, the emerging beam polarization will be

$$\begin{pmatrix} 0 & 1 \\ 1 & 0 \end{pmatrix}\begin{pmatrix} 1 \\ 0 \end{pmatrix} = \begin{pmatrix} 0 \\ 1 \end{pmatrix} \tag{5.58}$$

which corresponds to a beam polarized in the y-direction as observed experimentally. In other words, polarization rotation by $\pi/2$ radians takes place.

5.8.2 BROADBAND PRISMATIC POLARIZATION ROTATORS

An alternative to frequency-selective polarization rotators are prismatic rotators (Duarte 1989). These devices work at normal incidence and apply the principle of total internal reflection. The basic operation of polarization rotation, by $\pi/2$, due to total internal reflection is shown in Figure 5.11. This operation, however, reflects the beam into a direction that is orthogonal to the original propagation. Furthermore, the beam is not in the same plane. In order to achieve collinear polarization rotation, by $\pi/2$, the beam must be displaced upward and then be brought into alignment with the incident beam while conserving the polarization rotation achieved by the initial double reflection operation. A collinear prismatic polarization rotator, which performs this task using seven total internal reflections, is depicted in Figure 5.12. For high-power laser applications, this rotator is best assembled using a high-precision mechanical mount that allows air interfaces between the individual prisms. For a particular rotator, the useful aperture is about 10 mm and its physical length is 30 mm.

It should be noted that despite the apparent complexity of this collinear polarization rotator, the transmission efficiency is relatively high using antireflection coatings. In fact, using broadband (425–675 nm) antireflection coatings with a nominal loss of 0.5% per surface, the measured transmission efficiency becomes 94.7% at $\lambda = 632.8$ nm. The predicted transmission losses using

$$L_r = 1 - (1 - L)^r \tag{5.59}$$

are 4.9%, with $L = 0.5\%$, compared to a measured value of 5.3%. Equation 5.59 is derived combining Equations 5.50 and 5.51 for the special case of identical reflection losses. Here, r is the total number of reflection surfaces. For this particular collinear rotator, $r = 10$. A further parameter of interest is the transmission fidelity of the rotator since it is also important to keep spatial distortions of the rotated beam to a minimum. The integrity of the beam due to transmission and rotation is quantified in Figure 5.13 where a very slight beam expansion of ~3.2% at full width at half maximum (FWHM) is evident (Duarte 1992).

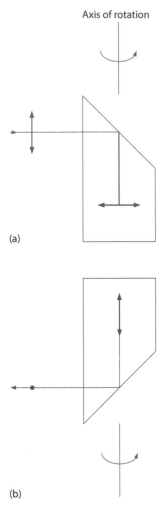

(a)

(b)

FIGURE 5.11 Basic prism operator for polarization rotation using two reflections. This can be composed of two 45° prisms adjoined $\pi/2$ to each other (note that it is also manufactured as one piece). (a) Side view of the rotator illustrating the basic rotation operation due to one reflection. The beam with the rotated polarization exists the prism into the plane of the figure. (b) The prism rotator is itself rotated anticlockwise by $\pi/2$ about the rotation axis (as indicated), thus providing an alternative perspective of the operation: the beam is now incident into the plane of the figure and it is reflected downward with its polarization rotated by $\pi/2$ relative to the original orientation. (Data from Duarte, F.J., Optical device for rotating the polarization of a light beam. US Patent 4, 822, 150, 1989.)

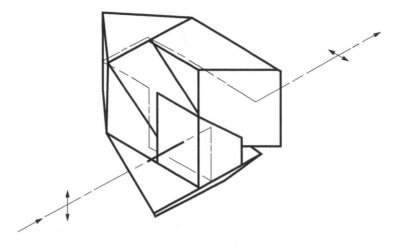

FIGURE 5.12 Broadband collinear prism polarization rotator. (Data from Duarte, F.J., Optical device for rotating the polarization of a light beam. US Patent 4, 822, 150, 1989.)

5.8.2.1 Example

The $\pi/2$ prismatic polarization rotator just described rotates linearly x-polarized radiation into linearly y-polarized radiation and vice versa. Considering first the case of $x \rightarrow y$ and using the Jones matrix formalism, we can write

$$\begin{pmatrix} a_{11} & a_{12} \\ a_{21} & a_{22} \end{pmatrix} \begin{pmatrix} 1 \\ 0 \end{pmatrix} = \begin{pmatrix} 1a_{11} + 0a_{12} \\ 1a_{21} + 0a_{22} \end{pmatrix} = \begin{pmatrix} 0 \\ 1 \end{pmatrix} \tag{5.60}$$

which means that $a_{11} = 0$ and $a_{21} = 1$. To find the other two components, we use the complementary rotation $y \rightarrow x$ that can be described as

$$\begin{pmatrix} a_{11} & a_{12} \\ a_{21} & a_{22} \end{pmatrix} \begin{pmatrix} 0 \\ 1 \end{pmatrix} = \begin{pmatrix} 0a_{11} + 1a_{12} \\ 0a_{21} + 1a_{22} \end{pmatrix} = \begin{pmatrix} 1 \\ 0 \end{pmatrix} \tag{5.61}$$

which implies that $a_{12} = 1$ and $a_{22} = 0$. Thus, the Jones matrix for $\pi/2$ rotation, and that applies directly to the prismatic rotator described in Figure 5.12, becomes

$$R = \begin{pmatrix} a_{11} & a_{12} \\ a_{21} & a_{22} \end{pmatrix} = \begin{pmatrix} 0 & 1 \\ 1 & 0 \end{pmatrix} \tag{5.62}$$

Thus, we have again arrived to the $\pi/2$ rotation matrix (Equation 5.57) using simple linear algebra.

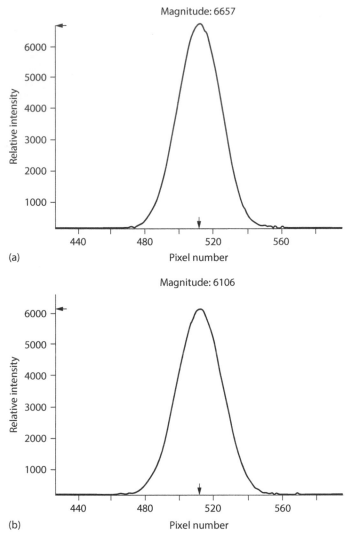

FIGURE 5.13 Transmission fidelity of the broadband collinear polarization rotator: (a) intensity profile of incident beam, prior to rotation, and (b) intensity profile of transmitted beam with rotated polarization. (Reproduced from Duarte, F.J., *Appl. Opt.*, 31, 3377–3378, 1992. With permission from the Optical Society.)

PROBLEMS

5.1. Design a single right-angle prism, made of fused silica, to expand a laser beam by a factor of 2 with orthogonal beam exit. Calculate $\mathcal{R}_{\parallel}$ and $\mathcal{R}_{\perp}$. (Use $n = 1.4583$ at $\lambda \approx 590$ nm.)

5.2. For a four-prism beam expander, with *orthogonal beam exit*, using fused silica prisms, calculate the overall beam magnification factor M for $\phi_{1,1} = \phi_{1,2} = \phi_{1,3} = \phi_{1,3} = 70°$. Also, calculate the overall transmission efficiency for a laser beam polarized parallel to the plane of incidence. (Use $n = 1.4583$ at $\lambda \approx 590$ nm.)

5.3. Use Maxwell equations in the Gaussian system, for $j = 0$ and $\rho = 0$ case, to derive the wave equations:

$$\nabla^2 E - (\varepsilon\mu)c^{-2}\frac{\partial^2 E}{\partial t^2} = 0$$

$$\nabla^2 H - (\varepsilon\mu)c^{-2}\frac{\partial^2 H}{\partial t^2} = 0$$

5.4. If a linearly polarized beam in the x-direction is sent through a rotator plate represented by the matrix

$$\begin{pmatrix} a_{11} & a_{12} \\ a_{21} & a_{22} \end{pmatrix} = \begin{pmatrix} 0 & 1 \\ -i & 0 \end{pmatrix}$$

What will be the polarization of the transmitted beam? What kind of plate would that be?

6 Laser Beam Propagation Matrices

6.1 INTRODUCTION

A powerful approach to characterize and design laser optics systems is the use of beam propagation matrices also known as *ray transfer matrices*. This is a practical method that applies to the propagation of laser beams with a Gaussian profile. Since most lasers can be designed to yield beams with Gaussian or near-Gaussian profiles, this is a widely applicable method. Laser beams with Gaussian or near-Gaussian spatial distributions result when the laser emission is single transverse mode or TEM_{00}, as explained in Chapter 7.

From a historical perspective, it should be mentioned that propagation matrices in optics have been known for a while. For early references in the subject, the reader is referred to Brouwer (1964), Kogelnik (1979), Siegman (1986), and Wollnik (1987). In this chapter, the basic principles of propagation matrices are outlined and a survey of matrices for various widely applicable optical elements is given. The emphasis is on the application of the method to practical optical systems. In addition, examples of some useful single-pass and multiple-pass calculations are given. Higher order matrices are also considered.

This chapter follows fairly closely the original version, on this subject matter, published in the work of Duarte (2003) while incorporating some improvements and corrections.

6.2 *ABCD* PROPAGATION MATRICES

The basic idea of propagation matrices is that one vector at a given plane is related to a second vector at a different plane via a linear transformation. This transformation is represented by a propagation matrix. This concept is applicable to the characterization of the deviation of a ray or beam of light through either free space or any linear optical media. The rays of light are assumed to be paraxial rays that propagate in proximity and almost parallel to the optical axis (Kogelnik 1979).

To illustrate this idea further consider the propagation of a paraxial ray of light from an original plane to a secondary plane in free space as depicted in Figure 6.1. In this geometry, the displacement distance l lies on the z-axis, which is perpendicular to x. In moving from the original plane to the secondary plane, the ray of light experiences a linear displacement in the x-direction and a very small angular deviation, that is,

$$x_2 = x_1 + l\theta_1 \tag{6.1}$$

$$\theta_2 = \theta_1 \tag{6.2}$$

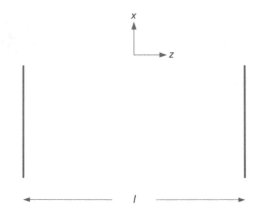

FIGURE 6.1 Geometry for propagation through distance l in free space. The displacement l is along the z-axis that is perpendicular to x.

which in matrix form can be stated as

$$\begin{pmatrix} x_2 \\ \theta_2 \end{pmatrix} = \begin{pmatrix} 1 & l \\ 0 & 1 \end{pmatrix} \begin{pmatrix} x_1 \\ \theta_1 \end{pmatrix} \tag{6.3}$$

The resulting 2×2 matrix is known as a *ray transfer matrix*. Here, it should be noted that some authors (Kogelnik 1979; Siegman 1986) use derivatives instead of the angular quantities, that is, $dx_1/dz = \theta_1$ and $dx_2/dz = \theta_2$.

For a thin lens, the geometry of propagation is illustrated in Figure 6.2. In this case, there is no displacement in the x-direction and the ray is concentrated, or focused, toward the optical axis so that

$$x_2 = x_1 \tag{6.4}$$

$$\theta_2 = -\left(\frac{1}{f}\right) x_1 + \theta_1 \tag{6.5}$$

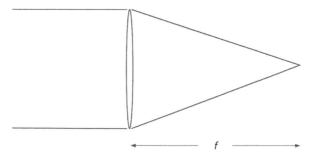

FIGURE 6.2 Thin convex lens.

which in matrix form can be expressed as

$$
\begin{pmatrix} x_2 \\ \theta_2 \end{pmatrix} = \begin{pmatrix} 1 & 0 \\ 1/f & 1 \end{pmatrix} \begin{pmatrix} x_1 \\ \theta_1 \end{pmatrix}
\tag{6.6}
$$

In more general terms, the X_2 vector is related to the X_1 vector by a transfer matrix T known as the $ABCD$ matrix so that

$$
X_2 = T X_1
\tag{6.7}
$$

where:

$$
T = \begin{pmatrix} A & B \\ C & D \end{pmatrix}
\tag{6.8}
$$

At this stage, it is useful to consider the dimensions of the components involved in these ray transfer matrices: A is a ratio of spatial dimensions, B is an optical length, while C is the reciprocal of an optical length. Consideration of various imaging systems leads to the conclusion that the spatial ratio represented by A is a beam magnification factor (M), whereas D is the reciprocal of such magnification ($1/M$). More explicitly, the $ABCD$ takes the dimensional format:

$$
T = \begin{pmatrix} M & B \\ C & 1/M \end{pmatrix}
\tag{6.9}
$$

These simple observations are very useful to verify the physical validity of newly derived matrices.

6.2.1 PROPERTIES OF *ABCD* MATRICES

A very useful property of $ABCD$ matrices is that they can be cascaded via matrix multiplication to produce a single overall matrix describing the propagation properties of an optical system. For example, if a linear optical system is composed of N optical elements deployed from left to right, as depicted in Figure 6.3, then the overall transfer matrix is given by the multiplication of the individual matrices in the reverse order, that is,

$$
\prod_{m=1}^{N} T_m = T_N \ldots T_3 T_2 T_1
\tag{6.10}
$$

It is easy to see that the complexity in the form of these product matrices can increase rather rapidly. Thus, it is always useful to remember that any resulting matrix must have the dimensions of Equation 6.9 and a determinant equal to unity, that is,

$$
AD - BC = 1
\tag{6.11}
$$

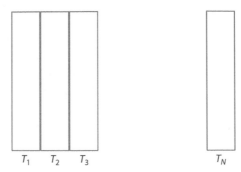

T_1 T_2 T_3 T_N

FIGURE 6.3 N optical elements in series.

6.2.2   SURVEY OF *ABCD* MATRICES

In Table 6.1, a number of representative and widely used optical components are represented in ray transfer matrix form. This is done without derivation and using the published literature as reference.

TABLE 6.1
***ABCD* Ray Transfer Matrices**

Optical Element or System	*ABCD* Matrix	Reference
Distance l in free space (Figure 6.1)	$\begin{pmatrix} 1 & l \\ 0 & 1 \end{pmatrix}$	Kogelnik (1979)
Distance l in a medium with refractive index n (Figure 6.4)	$\begin{pmatrix} 1 & l/n \\ 0 & 1 \end{pmatrix}$	Kogelnik (1979)
Slab of material with refractive index n (Figure 6.5)	$\begin{pmatrix} 1 & (l/n)(\cos\phi/\cos\psi)^2 \\ 0 & 1 \end{pmatrix}$	Duarte (1991)
Thin convex (positive) lens with focal length f (Figure 6.2)	$\begin{pmatrix} 1 & 0 \\ -1/f & 1 \end{pmatrix}$	Kogelnik (1979)
Thin concave (negative) lens (Figure 6.6)	$\begin{pmatrix} 1 & 0 \\ 1/|f| & 1 \end{pmatrix}$	Siegman (1986)

(Continued)

TABLE 6.1
(Continued) *ABCD* Ray Transfer Matrices

Optical Element or System	*ABCD* Matrix	Reference
Galilean telescope (Figure 6.7)	$\begin{pmatrix} f_2/\lvert f_1 \rvert & f_2 - \lvert f_1 \rvert \\ 0 & \lvert f_1 \rvert / f_2 \end{pmatrix}$	Siegman (1986)
Astronomical telescope (Figure 6.8)	$\begin{pmatrix} -f_2/f_1 & f_2 + f_1 \\ 0 & -f_1/f_2 \end{pmatrix}$	Siegman (1986)
Flat mirror (Figure 6.9)	$\begin{pmatrix} 1 & 0 \\ 0 & 1 \end{pmatrix}$	
Curved mirror (Figure 6.10)	$\begin{pmatrix} 1 & 0 \\ -2/R & 1 \end{pmatrix}$	Siegman (1986)
Double pass in Cassegrainian telescope (Figure 6.11)	$\begin{pmatrix} M & (M+1)L/M \\ 0 & 1/M \end{pmatrix}$	Siegman (1986)
Flat grating (Figure 6.12)	$\begin{pmatrix} \cos\Theta/\cos\Phi & 0 \\ 0 & \cos\Phi/\cos\Theta \end{pmatrix}$	Siegman (1986)
Flat grating in Littrow configuration (Figure 6.13)	$\begin{pmatrix} 1 & 0 \\ 0 & 1 \end{pmatrix}$	Duarte (1991)
Single right-angle prism (Figure 6.14)	$\begin{pmatrix} \cos\psi/\cos\phi & (l/n)\cos\phi/\cos\psi \\ 0 & \cos\phi/\cos\psi \end{pmatrix}$	Duarte (1989)
Multiple-prism beam expander (Figure 6.15)	$\begin{pmatrix} M_1 M_2 & B \\ 0 & (M_1 M_2)^{-1} \end{pmatrix}$	Duarte (1991)
Multiple-prism beam expander (return pass)	$\begin{pmatrix} (M_1 M_2)^{-1} & B \\ 0 & M_1 M_2 \end{pmatrix}$	Duarte (1991)

FIGURE 6.4 Geometry for propagation through distance *l* in region with refractive index *n*.

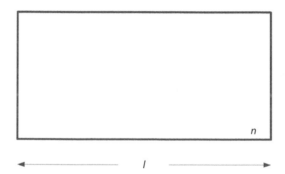

FIGURE 6.5 Slab of material with refractive index *n* such as an optical plate.

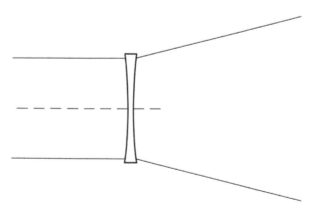

FIGURE 6.6 Concave lens.

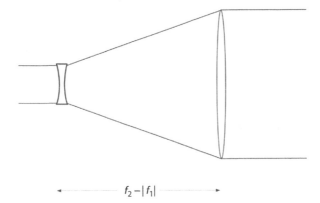

$$f_2 - |f_1|$$

FIGURE 6.7 Galilean telescope.

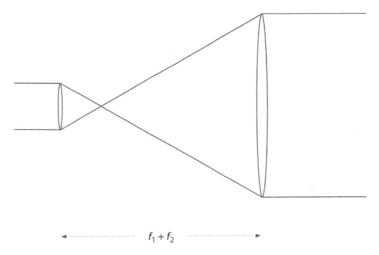

$$f_1 + f_2$$

FIGURE 6.8 Astronomical telescope.

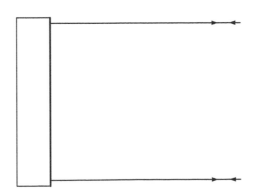

FIGURE 6.9 Flat mirror.

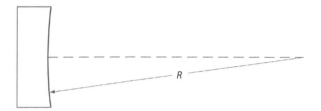

FIGURE 6.10   Curved mirror.

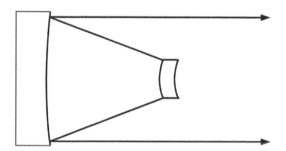

FIGURE 6.11   Reflective telescope of the Cassegrainian class. These reflective configurations are widely applied to unstable resonators, which in the far field can yield a near-TEM$_{00}$ laser beam profile.

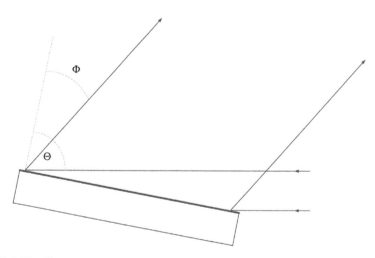

FIGURE 6.12   Generalized flat reflection grating.

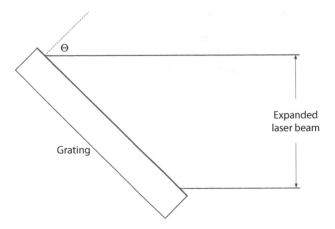

FIGURE 6.13   Flat reflection grating in Littrow configuration.

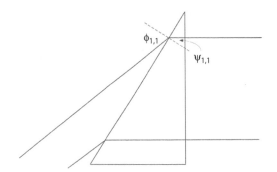

FIGURE 6.14   Single prism.

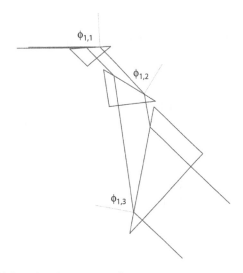

FIGURE 6.15   Multiple-prism beam expander.

6.2.3   THE ASTRONOMICAL TELESCOPE

The astronomical telescope (Figure 6.8) is composed of an input lens with focal length f_1, an intra lens distance L, and an output lens with focal lens f_2. Following Equation 6.10, the matrix multiplication proceeds as follows:

$$\begin{pmatrix} 1 & 0 \\ -1/f_2 & 1 \end{pmatrix} \begin{pmatrix} 1 & L \\ 0 & 1 \end{pmatrix} \begin{pmatrix} 1 & 0 \\ -1/f_1 & 1 \end{pmatrix} \tag{6.12}$$

Note that here and further, for notational succinctness, the left-hand side of the equation, that is, the $ABCD$ matrix, is abstracted. For a well-adjusted telescope, where

$$L = f_2 + f_1 \tag{6.13}$$

the transfer matrix becomes

$$\begin{pmatrix} A & B \\ C & D \end{pmatrix} = \begin{pmatrix} -f_2/f_1 & f_2 + f_1 \\ 0 & -f_1/f_2 \end{pmatrix} \tag{6.14}$$

which is the matrix given in Table 6.1. Defining

$$-M = -\frac{f_2}{f_1} \tag{6.15}$$

this matrix can be restated as

$$\begin{pmatrix} A & B \\ C & D \end{pmatrix} = \begin{pmatrix} -M & L \\ 0 & -1/M \end{pmatrix} \tag{6.16}$$

For this matrix, it can be easily verified that the condition $|AD - BC| = 1$ holds.

6.2.4   A SINGLE PRISM IN SPACE

For a right-angle prism, with orthogonal beam exit, preceded by a distance L_1 and followed by a distance L_2, as shown in Figure 6.16, the matrix multiplication becomes

$$\begin{pmatrix} 1 & L_2 \\ 0 & 1 \end{pmatrix} \begin{pmatrix} k & l/nk \\ 0 & 1/k \end{pmatrix} \begin{pmatrix} 1 & L_1 \\ 0 & 1 \end{pmatrix} \tag{6.17}$$

where:

$$k = \frac{\cos \psi}{\cos \phi} \tag{6.18}$$

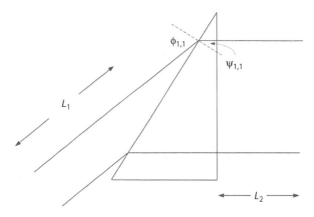

FIGURE 6.16   Single prism preceded by a distance L_1 and followed by a distance L_2.

Thus, the transfer matrix becomes

$$
\begin{pmatrix} A & B \\ C & D \end{pmatrix} = \begin{pmatrix} k & L_1 k + (L_2/k) + l/(nk) \\ 0 & 1/k \end{pmatrix}
\tag{6.19}
$$

which can be restated as

$$
\begin{pmatrix} A & B \\ C & D \end{pmatrix} = \begin{pmatrix} k & B \\ 0 & 1/k \end{pmatrix}
\tag{6.20}
$$

where:

$$
B = L_1 k + \frac{L_2}{k} + \frac{l}{nk}
\tag{6.21}
$$

which is the optical length of the system. Notice that both L_1 and L_2 are modified by the dimensionless beam magnification factor k, whereas l is divided by the dimensionless quantity nk.

6.2.5   MULTIPLE-PRISM BEAM EXPANDERS

For a generalized multiple-prism array, as illustrated in Figure 6.17, the ray transfer matrix is given by (Duarte 1989, 1991)

$$
\begin{pmatrix} A & B \\ C & D \end{pmatrix} = \begin{pmatrix} M_1 M_2 & B \\ 0 & (M_1 M_2)^{-1} \end{pmatrix}
\tag{6.22}
$$

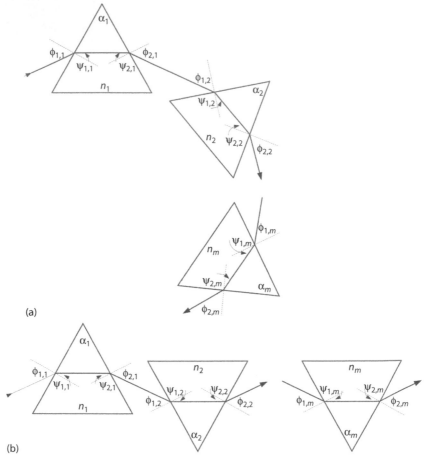

(a)

(b)

FIGURE 6.17   Generalized multiple-prism array (a) Describes an additive configuration and (b) a compensating configuration.

where:

$$M_1 = \prod_{m=1}^{r} k_{1,m} \tag{6.23}$$

$$M_2 = \prod_{m=1}^{r} k_{2,m} \tag{6.24}$$

and

$$B = M_1 M_2 \sum_{m=1}^{r-1} L_m \left(\prod_{j=1}^{m} k_{1,j} \prod_{j=1}^{m} k_{2,j} \right)^{-2} + \frac{M_1}{M_2} \sum_{m=1}^{r} \frac{l_m}{n_m} \left(\prod_{j=1}^{m} k_{1,j} \right)^{-2} \left(\prod_{j=m}^{r} k_{2,j} \right)^{2} \tag{6.25}$$

For a straightforward multiple-prism beam expander with orthogonal beam exit, $\cos\psi_{2,j} = 0$ and $k_{2,j} = 1$ so that the equations reduce to

$$\begin{pmatrix} A & B \\ C & D \end{pmatrix} = \begin{pmatrix} M_1 & B \\ 0 & (M_1)^{-1} \end{pmatrix} \tag{6.26}$$

where:

$$B = M_1 \sum_{m=1}^{r-1} L_m \left(\prod_{j=1}^{m} k_{1,j} \right)^{-2} + M_1 \sum_{m=1}^{r} \frac{l_m}{n_m} \left(\prod_{j=1}^{m} k_{1,j} \right)^{-2} \tag{6.27}$$

For a single prism, these equations reduce further to $M_1 = k_{1,1}$ and $B = l/(k_{1,1}n)$, which is the result for the single prism given in Table 6.1.

6.2.6 TELESCOPES IN SERIES

For some applications, it is necessary to propagate TEM_{00} laser beams through optical systems including telescopes in series as illustrated in Figure 6.18. For a series comprising a telescope followed by a free-space distance, followed by a second telescope, and so on, the single-pass cumulative matrix is given by

$$A = M^r \tag{6.28}$$

$$B = \sum_{m=1}^{r} M^{r-2m+2} L_m + M^{r-2m+1} B_{T_m} \tag{6.29}$$

$$C = 0 \tag{6.30}$$

$$D = M^{-r} \tag{6.31}$$

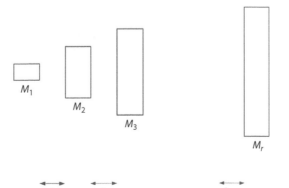

FIGURE 6.18 Series of telescopes separated by a distance L_m.

where:
  r is the total number of telescopes
  B_{T_m} is the B term of the mth telescope

This result applies to a series of well-adjusted Galilean or astronomical telescopes or a series of prismatic telescopes.

6.2.7 SINGLE RETURN-PASS BEAM DIVERGENCE

It can be shown (Duarte 1990) that the double-pass or single-return pass divergence in a dispersive laser cavity can be expressed as

$$\Delta\theta = \frac{\lambda}{\pi w}\left[1+\left(\frac{L_\mathfrak{R}}{B}\right)^2+\left(\frac{AL_\mathfrak{R}}{B}\right)^2\right]^{1/2} \tag{6.32}$$

where:

$$L_\mathfrak{R} = \left(\frac{\pi w^2}{\lambda}\right) \tag{6.33}$$

which is the Rayleigh length and A and B are the corresponding elements of the ray transfer matrix. The double-pass, or single-return pass, calculation for a narrow-linewidth multiple-prism grating oscillator, illustrated in Figure 6.19, can be performed using the reflection surface of the output coupler mirror as the reference point. For this purpose, the cavity is unfolded about the reflective surface of the output coupler as shown in Figure 6.20. Thus, from the output coupler to the grating and then proceeding from the grating to the output coupler, the matrix multiplication becomes

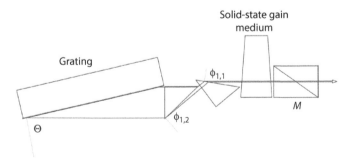

FIGURE 6.19 MPL grating laser oscillator. (Reproduced from Duarte, F.J., *Appl. Opt.*, 38, 6347–6349, 1999. With permission from the Optical Society.)

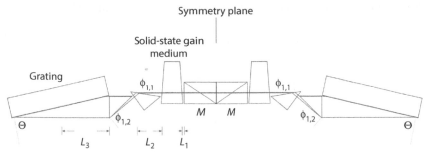

FIGURE 6.20 Unfolded laser cavity for multiple-return-pass analysis. L_1 is the intracavity distance between the polarizer output coupler and the gain medium, L_2 is the intracavity distance between the gain medium and the multiple-prism expander, and L_3 is the intracavity distance between the multiple-prism expander and the diffraction grating.

$$\begin{pmatrix} 1 & L_P/n_p \\ 0 & 1 \end{pmatrix} \begin{pmatrix} 1 & L_1 \\ 0 & 1 \end{pmatrix} \begin{pmatrix} \alpha & \beta \\ \chi & \delta \end{pmatrix} \begin{pmatrix} 1 & L_2 \\ 0 & 1 \end{pmatrix} \begin{pmatrix} M & B_{MP} \\ 0 & (1/M) \end{pmatrix}$$

$$\begin{pmatrix} 1 & L_3 \\ 0 & 1 \end{pmatrix} \begin{pmatrix} 1 & 0 \\ 0 & 1 \end{pmatrix} \begin{pmatrix} 1 & L_3 \\ 0 & 1 \end{pmatrix} \begin{pmatrix} (1/M) & B_{MP} \\ 0 & M \end{pmatrix} \begin{pmatrix} 1 & L_2 \\ 0 & 1 \end{pmatrix} \tag{6.34}$$

$$\begin{pmatrix} \alpha & \beta \\ \chi & \delta \end{pmatrix} \begin{pmatrix} 1 & L_1 \\ 0 & 1 \end{pmatrix} \begin{pmatrix} 1 & L_P/n_P \\ 0 & 1 \end{pmatrix} \begin{pmatrix} 1 & 0 \\ 0 & 1 \end{pmatrix}$$

which reduces to

$$\begin{pmatrix} 1 & \Lambda \\ 0 & 1 \end{pmatrix} \begin{pmatrix} \alpha & \beta \\ \chi & \delta \end{pmatrix} \begin{pmatrix} 1 & L_2 \\ 0 & 1 \end{pmatrix} \begin{pmatrix} M & B_{MP} \\ 0 & (1/M) \end{pmatrix}$$

$$\begin{pmatrix} 1 & 2L_3 \\ 0 & 1 \end{pmatrix} \begin{pmatrix} (1/M) & B_{MP} \\ 0 & M \end{pmatrix} \begin{pmatrix} 1 & L_2 \\ 0 & 1 \end{pmatrix} \tag{6.35}$$

$$\begin{pmatrix} \alpha & \beta \\ \chi & \delta \end{pmatrix} \begin{pmatrix} 1 & \Lambda \\ 0 & 1 \end{pmatrix}$$

where:

$$\Lambda = \frac{L_P}{n_P} + L_1 \tag{6.36}$$

Multiplication of these matrices *in reverse* leads to (Duarte et al. 1997)

$$A = \alpha^2 + \left[(\Lambda + \Xi)\alpha + \delta\Lambda + \beta \right]\chi + \Xi\Lambda\chi^2 \tag{6.37}$$

$$B = (\alpha^2 + \delta^2)\Lambda + (\alpha + \delta)\beta + \Xi\alpha\delta + \left[(\Xi + \Lambda)(\alpha + \delta) + 2\beta \right]\Lambda\chi + \Xi\Lambda^2\chi^2 \tag{6.38}$$

$$C = (\alpha + \delta)\chi + \Xi\chi^2 \tag{6.39}$$

$$D = \delta^2 + [(\Lambda + \Xi)\delta + \alpha\Lambda + \beta]\chi + \Xi\Lambda\chi^2 \tag{6.40}$$

where:

$$\Xi = 2L_2 + \frac{2B_{MP}}{M} + \frac{2L_3}{M^2} \tag{6.41}$$

For an ideal laser gain medium with $\alpha \approx \delta \approx 1$, $\chi \approx 0$, and $\beta \approx \beta$ so that

$$A \approx 1 \tag{6.42}$$

$$B \approx 2\Lambda + 2\beta + \Xi \tag{6.43}$$

$$C \approx 0 \tag{6.44}$$

$$D \approx 1 \tag{6.45}$$

which imply that the beam divergence will approach its diffraction limit as the optical length of the cavity B increases. This is in accordance with experimental observations.

6.2.8   MULTIPLE RETURN-PASS BEAM DIVERGENCE

The multiple-return pass laser linewidth in a dispersive tunable laser oscillator is given by (Duarte 2001)

$$\Delta\lambda = \Delta\theta_R (RM \; \nabla_\lambda\Theta_G + R \; \nabla_\lambda\Phi_P)^{-1} \tag{6.46}$$

where:
  M is the overall intracavity beam magnification
  R is the number of return passes
  $\nabla_\lambda\Theta_G$ is the grating dispersion
  $\nabla_\lambda\Phi_P$ is the return-pass multiple-prism dispersion

Here, the multiple return-pass beam divergence is given by

$$\Delta\theta_R = \frac{\lambda}{\pi w}\left[1+\left(\frac{L_\mathcal{R}}{B_R}\right)^2+\left(\frac{A_R L_\mathcal{R}}{B_R}\right)^2\right]^{1/2} \tag{6.47}$$

where:

A_R and B_R correspond to cumulative multiple-return pass transfer matrix coefficients

For a multiple-return pass analysis, the cavity is unfolded multiple times and the multiplication described for the single-return pass is performed multiple times. This leads to the following matrix components (Duarte 2001):

$$A_R = (\alpha A_{R-1}+\chi \mathcal{L}_{R-1})\left[\alpha+\chi(\Xi-L_2)\right]+\chi \mathcal{L}_{R-1}(\chi L_2+\delta)$$
$$+\chi A_{R-1}(\alpha L_2+\beta) \tag{6.48}$$

and

$$B_R = A_R\Lambda+(\alpha A_{R-1}+\chi \mathcal{L}_{R-1})\left[\beta+\delta(\Xi-L_2)\right]$$
$$+\delta \mathcal{L}_{R-1}(\chi L_2+\delta)+\delta A_{R-1}(\alpha L_2+\beta) \tag{6.49}$$

where:

$$\mathcal{L}_{R-1} = \Lambda A_{R-1}+B_{R-1} \tag{6.50}$$

For a single-return pass,

$$A_1 = (\alpha+\chi\Lambda)\left[\alpha+\chi(\Xi-L_2)\right]+\chi\Lambda(\chi L_2+\delta)+\chi(\alpha L_2+\beta) \tag{6.51}$$

and

$$B_1 = A_1\Lambda+(\alpha+\chi\Lambda)\left[\beta+\delta(\Xi-L_2)\right]+\delta\Lambda(\chi L_2+\delta)+\delta(\alpha L_2+\beta) \tag{6.52}$$

which reduce to Equations 6.37 and 6.38, respectively. Note that by definition $A_0=1$ and $B_0 = 0$. For an ideal gain medium, with little or no thermal lensing, $\alpha \approx \delta \approx 1$, $\chi \approx 0$, and $\beta \approx \beta$, so that

$$A_R \approx 1 \tag{6.53}$$

$$B_R \approx R(2\Lambda+2\beta+\Xi) \tag{6.54}$$

which are the multiple-pass versions of Equations 6.42 and 6.43, respectively. These results mean that, in the absence of thermal lensing, the beam divergence described by Equation 6.47 will decrease toward its diffraction limit as the number of intracavity passes increases. A discussion on the application of these equations to low-divergence narrow-linewidth tunable lasers has been given by Duarte (2001).

6.2.9 UNSTABLE RESONATORS

The subject of unstable resonators is a vast subject and has been treated in detail by Siegman (1986). Unstable resonators are cavities that are configured with curved mirrors in order to provide intracavity magnification, which in turn provides good transverse-mode discrimination and good far-field beam profiles. In addition to their application as intrinsic resonators, unstable resonators are also widely applied to configure the cavities of forced oscillators that amplify the emission from a master oscillator (see Chapter 7). Here, the discussion is limited to resonators incorporating Cassegrain or Cassegrainian telescopes. These are reflective telescopes, as illustrated in Figure 6.11, which are widely used in the field of astronomy. The double-pass, or single-return-pass ray transfer matrix for a telescope comprising a concave back reflector and a convex exit mirror is given in Table 6.1. The radius of curvature of the mirrors is considered positive if concave toward the resonator. In addition to the transfer matrix, the following relations are useful (Siegman 1986):

$$M = -\frac{R_2}{R_1} \tag{6.55}$$

and

$$L = \frac{R_1 + R_2}{2} \tag{6.56}$$

The condition for lasing in the unstable regime is determined by the inequality:

$$\left| \frac{A+D}{2} \right| > 1 \tag{6.57}$$

In addition to traditional unstable resonators of the Cassegrainian type, it should also be mentioned that multiple-prism grating oscillators can also meet the conditions of an unstable resonator when incorporating a gain medium that exhibits thermal lensing (Duarte et al. 1997). However, in the absence of thermal lensing, for an ideal gain medium, $\alpha = \delta = 1$, $\chi = 0$, and $\beta = \beta$, the A and D terms can have a value of unity.

6.3 HIGHER ORDER MATRICES

The description of optical systems using 3×3, 4×4, and 6×6 matrices has been considered by several authors (Brouwer 1964; Siegman 1986; Wollnik 1987). A more recent description of 4×4 matrices, by Kostenbauer (1990), uses the notation:

$$\begin{pmatrix} \partial x_2/\partial x_1 & \partial x_2/\partial \theta_1 & \partial x_2/\partial t_1 & \partial x_2/\partial v_1 \\ \partial \theta_2/\partial x_1 & \partial \theta_2/\partial \theta_1 & \partial \theta_2/\partial t_1 & \partial \theta_2/\partial v_1 \\ \partial t_2/\partial x_1 & \partial t_2/\partial \theta_1 & \partial t_2/\partial t_1 & \partial t_2/\partial v_1 \\ \partial v_2/\partial x_1 & \partial v_2/\partial \theta_1 & \partial v_2/\partial t_1 & \partial v_2/\partial v_1 \end{pmatrix} \tag{6.58}$$

which can be written as

$$
\begin{pmatrix}
A & B & 0 & E \\
C & D & 0 & F \\
G & H & 1 & I \\
0 & 0 & 0 & 1
\end{pmatrix}
\tag{6.59}
$$

where the *ABCD* terms have their usual meaning. For a plane mirror, this matrix becomes

$$
\begin{pmatrix}
1 & 0 & 0 & 0 \\
0 & 1 & 0 & 0 \\
0 & 0 & 1 & 0 \\
0 & 0 & 0 & 1
\end{pmatrix}
\tag{6.60}
$$

and for a thin concave lens,

$$
\begin{pmatrix}
1 & 0 & 0 & 0 \\
\left|f^{-1}\right| & 1 & 0 & 0 \\
0 & 0 & 1 & 0 \\
0 & 0 & 0 & 1
\end{pmatrix}
\tag{6.61}
$$

For the *r*th prism, some interesting components of the matrix are given by Duarte (1992):

$$
A_r = M_1 M_2
\tag{6.62}
$$

$$
B_r = M_1 M_2 \sum_{m=1}^{r-1} L_m \left(\prod_{j=1}^{m} k_{1,j} \prod_{j=1}^{m} k_{2,j} \right)^{-2} + \frac{M_1}{M_2} \sum_{m=1}^{r} \frac{l_m}{n_m} \left(\prod_{j=1}^{m} k_{1,j} \right)^{-2} \left(\prod_{j=m}^{r} k_{2,j} \right)^{2}
\tag{6.63}
$$

$$
C_r = 0
\tag{6.64}
$$

$$
D_r = (M_1 M_2)^{-1}
\tag{6.65}
$$

$$
F_r = \sum_{m=1}^{r} (\pm 1) \mathcal{H}_{1,m} \left(\prod_{j=m}^{r} k_{1,j} \prod_{j=m}^{r} k_{2,j} \right)^{-1} \nabla_v n_m
$$

$$
+ (M_1 M_2)^{-1} \sum_{m=1}^{r} (\pm 1) \mathcal{H}_{2,m} \left(\prod_{j=1}^{m} k_{1,j} \prod_{j=1}^{m} k_{2,j} \right) \nabla_v n_m
\tag{6.66}
$$

$$
G_r = \sum_{m=1}^{r} F_m \left(\prod_{j=1}^{m} k_{1,j} \prod_{j=1}^{m} k_{2,j} \right) \lambda^{-1}
\tag{6.67}
$$

where:

$$M_1 = \prod_{m=1}^{r} k_{1,m} \tag{6.68}$$

$$M_2 = \prod_{m=1}^{r} k_{2,m} \tag{6.69}$$

$$\mathcal{H}_{1,m} = \frac{\tan \phi_{1,m}}{n_m} \tag{6.70}$$

$$\mathcal{H}_{2,m} = \frac{\tan \phi_{2,m}}{n_m} \tag{6.71}$$

$$\nabla_v n_m = \frac{\partial n_m}{\partial v} \tag{6.72}$$

Going back to the generalized multiple-prism dispersion in its explicit form (Duarte and Piper 1982; Duarte 1988),

$$\nabla_\lambda \phi_{2,r} = \sum_{m=1}^{r} (\pm 1) \mathcal{H}_{1,m} \left(\prod_{j=m}^{r} k_{1,j} \prod_{j=m}^{r} k_{2,j} \right)^{-1} \nabla_\lambda n_m$$

$$+ (M_1 M_2)^{-1} \sum_{m=1}^{r} (\pm 1) \mathcal{H}_{2,m} \left(\prod_{j=1}^{m} k_{1,j} \prod_{j=1}^{m} k_{2,j} \right) \nabla_\lambda n_m \tag{6.73}$$

and comparing Equation 6.73 with 6.66, we can express the F_r term in shorthand notation as

$$F_r = \nabla_\lambda \phi_{2,r} \left(\frac{\partial \lambda}{\partial v} \right) \tag{6.74}$$

$$G_r = \sum_{m=1}^{r} F_m \left(\prod_{j=1}^{m} k_{1,j} \prod_{j=1}^{m} k_{2,j} \right) \lambda^{-1} \tag{6.75}$$

The explicit E_r, H_r, and I_r terms, for a generalized multiple-prism array, are a function of B_r. These terms are rather extensive and thus are not included in the text.

PROBLEMS

6.1 Derive the ray transfer matrix for the Galilean telescope given in Table 6.1.

6.2 Derive the single-pass ray transfer matrix corresponding to a multiple-prism beam expander, deployed in an additive configuration, comprising four prisms designed for orthogonal beam exit.

6.3 Perform the matrix multiplication for the intracavity single-return pass of the multiple-prism grating tunable laser oscillator depicted in Figure 6.19 that results in Equations 6.37 through 6.40.

6.4 Show that, for a single-pass analysis, Equations 6.51 and 6.52 reduce to the corresponding single-pass A and B terms given by Equations 6.37 and 6.38.

6.5 Use Equations 6.66 and 6.73 to show that the F_r term can be expressed in shorthand as $F_r = \nabla_\lambda \phi_{2,r}(\partial\lambda/\partial v)$.

7 Narrow-Linewidth Tunable Laser Oscillators

7.1 INTRODUCTION

In this chapter, the basics of interference, the uncertainty principle, polarization, and beam propagation are applied to the design, architecture, and engineering of narrow-linewidth tunable lasers. The principles discussed here apply in general to tunable lasers and are not limited by the type or class of gain media. Thus, the gain media assumed here are generic broadly tunable media in the gas, liquid, or solid state.

A narrow-linewidth tunable laser oscillator is defined as a source of highly coherent continuously tunable laser emission. That is, a laser source that emits highly directional radiation of an extremely pure color. Pure emission in the visible spectrum is defined as having a linewidth narrower than $\Delta\nu \approx 3$ GHz, which translates approximately into $\Delta\lambda \approx 0.0017$ nm at $\lambda \approx 510$ nm. This criterion is provided by the laser linewidth requirements to excite single vibro-rotational levels in the $B^3\Pi_{ou}^+ - X^1\Sigma_g^+$ electronic transition of the iodine molecule at room temperature. Although most of the discussion is oriented toward high-power pulsed tunable lasers, mention of continuous wave (CW) lasers is also made when appropriate.

7.2 TRANSVERSE AND LONGITUDINAL MODES

A broadband unrefined laser can emit in beams spatially integrating many transverse modes and each of these transverse modes can include a multitude of longitudinal modes. A refined, truly coherent, laser should emit in a single transverse mode (TEM$_{00}$) and, ideally, in a single longitudinal mode (SLM). That orderly, low-entropy emission reaches toward the ideal of monochromatic emission and certainly provides a population of indistinguishable photons.

7.2.1 Transverse Mode Structure

The most straightforward laser cavity comprises a gain medium and two mirrors, as illustrated in Figure 7.1. The physical dimension of the intracavity aperture relative to the separation of mirrors, or cavity length, determines the number of transverse electromagnetic modes. *The narrower the width of the intracavity aperture and the longer the cavity length, the lower the number of transverse modes.* The single-pass transverse mode structure in one dimension can be characterized using the generalized interferometric equation introduced in Chapter 2 (Duarte 1991a, 1993b)

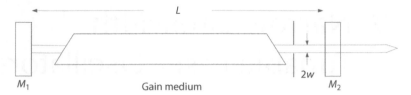

FIGURE 7.1 Mirror–mirror laser cavity. The physical dimensions of the intracavity aperture relative to the cavity length determine the number of transverse modes.

$$\left| \langle x | s \rangle \right|^2 = \sum_{j=1}^{N} \Psi(r_j)^2 + 2 \sum_{j=1}^{N} \Psi(r_j) \left[\sum_{m=j+1}^{N} \Psi(r_m) \cos(\Omega_m - \Omega_j) \right] \qquad (7.1)$$

and in two dimensions using (Duarte 1995b)

$$\left| \langle x | s \rangle \right|^2 = \sum_{z=1}^{N} \sum_{y=1}^{N} \Psi(r_{zy}) \sum_{q=1}^{N} \sum_{p=1}^{N} \Psi(r_{pq}) e^{i(\Omega_{qp} - \Omega_{zy})} \qquad (7.2)$$

In addition, a useful tool to determine the number of transverse modes is the *Fresnel number* (Siegman 1986):

$$N_F = \frac{w^2}{L\lambda} \qquad (7.3)$$

The single-pass approximation to estimate the transverse mode structure assumes that in a laser with a given cavity length most of the emission generated next to the output coupler mirror is in the form of spontaneous emission and thus highly divergent. Thus, only the emission generated at the opposite end of the cavity and that propagates via an intracavity length L contributes to the initial transverse mode structure.

In order to illustrate the use of these equations, let us consider a hypothetical laser with a 10 cm cavity emitting at $\lambda \approx 590\,nm$ incorporating a $2w = 2\,mm$-wide one-dimensional aperture. Using Equation 7.1, the intensity distribution of the emission is calculated as shown in Figure 7.2. Each ripple represents a transverse mode. An estimate of this number can be obtained by counting the ripples in Figure 7.2, which yields an approximate number of 17. This number of ripples should be compared with the Fresnel number, which is $N_F = 16.94$.

For the same wavelength, $\lambda \approx 590\,nm$ and a cavity length of $L = 10$ cm, reducing the dimensions of the intracavity aperture to $2w = 250\,\mu m$, yields $N_F \approx 0.26$. For such dimensions, the calculated intensity distribution, using Equation 7.1, is given in Figure 7.3. The distribution in Figure 7.3 indicates that most of the emission intensity is contained in a central near-Gaussian distribution, which is the case in practice since the secondary modes can be absent due to cavity losses.

In practice, the spatial distribution of the emission for tunable lasers with the above cavity length-to-intracavity aperture ratio (100,000:125), at visible wavelengths

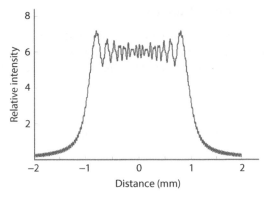

FIGURE 7.2   Cross section of diffraction distribution corresponding to a large number of transverse modes. Here, $w = 1.5$ mm, $L = 10$ cm, $\lambda = 632.82$ nm, and the Fresnel number becomes $N_F \approx 35.56$.

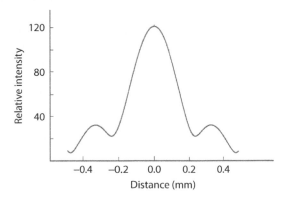

FIGURE 7.3   Cross section of diffraction distribution corresponding to a near-TEM$_{00}$ corresponding to $w = 250$ μm, $L = 40$ cm, and $\lambda = 632.82$ nm, so that $N_F \approx 0.25$.

$\lambda \approx 590$ nm, is a near-Gaussian distribution characteristic of what is known as single transverse mode, or TEM$_{00}$, emission. Additional examples of the beneficial effect derived from the combination of long cavity lengths and narrow intracavity apertures are found in CW lasers, emitting via atomic transitions, that typically yield TEM$_{00}$ emission. Examples of such coherent sources are the He–Ne, He–Cd, and He–Zn lasers.

The concept of *mode matching* in optically pumped laser systems refers to adjusting the dimensions of the excitation laser beam, via optical focusing, to sustain TEM$_{00}$ emission in the secondary or optically pumped laser. As just explained, these dimensions depend on the necessary cavity length-to-intracavity aperture ratio dictated by the interferometric equation. Reducing the transverse mode distribution to TEM$_{00}$ emission is the first step in the design of narrow-linewidth tunable lasers. The task of the designer consists in achieving TEM$_{00}$ emission in the shortest possible cavity length.

7.2.2   Longitudinal Mode Emission

Once TEM$_{00}$ emission has been established, the task consists in controlling the number of longitudinal modes in the cavity. In a laser with cavity length L, the longitudinal mode spacing, in the frequency domain, is given by (see Chapter 3)

$$\delta v = \frac{c}{2L} \tag{7.4}$$

and the number of longitudinal modes N_{LM} is given by

$$N_{LM} = \frac{\Delta v}{\delta v} \tag{7.5}$$

where:
Δv is the measured laser linewidth

Thus, for a laser with a 30 cm cavity length and a measured linewidth of $\Delta v = 3$ GHz, the number of longitudinal modes becomes $N_{LM} \approx 6$. If the cavity length is reduced to 10 cm, then the number of longitudinal modes is reduced to $N_{LM} \approx 2$ and the emission would be called double-longitudinal-mode (DLM) emission. If the cavity length is reduced to 5 cm, then $N_{LM} \approx 1$ and the laser is said to be undergoing SLM oscillation. These simple examples highlight the advantages of compact cavity designs, provided the active medium can sustain the gain to overcome threshold.

An alternative to reducing the cavity, and still achieving SLM emission, is to optimize the beam divergence and to increase the intracavity dispersion to yield a narrower cavity linewidth that would restrict oscillation to the SLM regime. In this context, the linewidth

$$\Delta \lambda \approx \Delta \theta (\nabla_\lambda \theta)^{-1} \tag{7.6}$$

is converted to Δv using the identity

$$\Delta v = \Delta \lambda \left(\frac{c}{\lambda^2} \right) \tag{7.7}$$

and applying the criterion

$$\Delta v \leq \delta v \tag{7.8}$$

to guide the design of the dispersive oscillator.

Multiple-longitudinal-mode emission appears complex and chaotic in both the frequency and temporal domains. DLM and SLM emission can be characterized in the frequency domain using Fabry–Pérot interferometry or in the temporal domain by observing the shape of the temporal pulse. In the case of DLM emission, the interferometric rings appear to be double. In the temporal domain, *mode beating* is still observed when the intensity ratio of the primary to the secondary mode is 100:1 or even higher. Mode beating of two longitudinal modes, as illustrated in Figure 7.4, can be characterized using a simple wave representation (Pacala et al. 1984) where

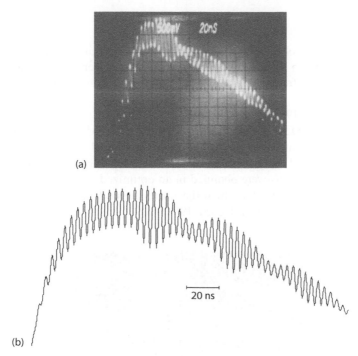

(a)

(b)

FIGURE 7.4 Mode beating resulting from DLM oscillation. (a) Measured temporal pulse. (b) Calculated temporal pulse assuming interference between the two longitudinal modes. (Reproduced from Duarte, F.J., et al., *Appl. Opt.*, 27, 843–846, 1988. With permission from the Optical Society.)

each mode of amplitudes E_1 and E_2, with frequencies ω_1 and ω_2, combine to produce a resulting field

$$E = E_1 \cos(\omega_1 t - k_1 z) + E_2 \cos(\omega_2 t - k_2 z) \tag{7.9}$$

For incidence at $z = 0$ on a square-law temporal detector, the intensity can be expressed as

$$E^2 = \frac{1}{2}\left(E_1^2 + E_2^2\right) + \frac{1}{2}\left(E_1^2 \cos 2\omega_1 t + E_2^2 \cos \omega_2 t\right)$$
$$+ E_1 E_2 \cos(\omega_1 + \omega_2)t + E_1 E_2 \cos(\omega_1 - \omega_2)t \tag{7.10}$$

Detectors in the nanosecond regime respond only to the first and last terms of this equation so that the equation can be approximated by

$$E^2 \approx \frac{1}{2}\left(E_1^2 + E_1^2\right) + E_1 E_2 \cos(\omega_1 - \omega_2)t \tag{7.11}$$

Using this approximation and a non-Gaussian temporal representation, derived from experimental data, for the amplitudes of the form

$$E_1(t) = \left(a_2 t^2 + a_1 t + a_0\right)\left(b_1 t + b_0\right)^{-1} \qquad (7.12)$$

a calculated version of the experimental waveform exhibiting mode beating can be obtained as shown in Figure 7.4b. For this particular dispersive oscillator lasing in a DLM, the ratio of frequency jitter $\delta\omega$ to cavity mode spacing $\Delta\omega \approx (\omega_1 - \omega_2)$ was represented by a sinusoidal function at 20 MHz. The initial mode intensity ratio is 200:1 (Duarte et al. 1988; Duarte 1990a).

In the case of SLM emission, the Fabry–Pérot interferometric rings appear singular and well defined (see Figure 7.5). In this case, mode beating in the temporal domain is absent and the pulses assume a smooth near-Gaussian distribution (see Figure 7.6). These results were obtained in an optimized solid-state multiple-prism grating dye laser oscillator for which the limit derived from the uncertainty principle, $\Delta\nu\, \Delta t \approx 1$, is approached (Duarte 1999).

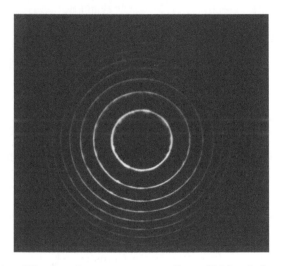

FIGURE 7.5   Fabry–Pérot interferogram corresponding to SLM emission at $\Delta\nu \approx 350$ MHz. (Reproduced from Duarte, F.J., *Appl. Opt.*, 38, 6347–6349, 1999. With permission from the Optical Society.)

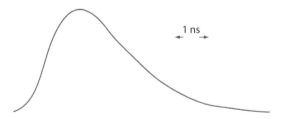

FIGURE 7.6   Near-Gaussian temporal pulse corresponding to SLM emission. The temporal scale is 1 ns/div. (Reproduced from Duarte, F.J., *Appl. Opt.*, 38, 6347–6349, 1999. With permission from the Optical Society.)

7.3 TUNABLE LASER OSCILLATOR ARCHITECTURES

Tunable laser oscillators can be configured in a variety of cavity designs. Here, a brief survey of these alternatives is presented. In general, these cavity architectures can be classified into tunable laser oscillators *without intracavity beam expansion* and tunable laser oscillators *with intracavity beam expansion* (Duarte 1991a). Further, each of these classes can be divided into *open* and *closed cavity* designs (Duarte and Piper 1980). It is assumed that oscillators considered in this section are designed to yield narrow-linewidth emission. The architectures are relevant to a variety of tunable gain media and are applicable to both CW and pulsed lasers.

7.3.1 TUNABLE LASER OSCILLATORS *WITHOUT* INTRACAVITY BEAM EXPANSION

Tunable laser oscillators *without* intracavity beam expansion are those laser resonators in which the intrinsic narrow beam waist at the gain region is not expanded using intracavity optics. The most basic of tunable laser designs is that incorporating an output coupler mirror and a tuning grating in Littrow configuration as illustrated in Figure 7.7. Tuning is accomplished by slight rotation of the grating. This cavity configuration will yield relatively broad tunable emission in a short pulsed laser, such a high-power pulsed dye laser, but could emit fairly narrow emission if applied to a CW semiconductor laser. A refinement of this cavity consists in inserting one or more intracavity etalons to further narrow the emission wavelength as illustrated in Figure 7.8. Such multiple-etalon grating cavity could yield very

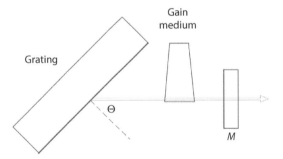

FIGURE 7.7   Grating-mirror tunable laser cavity.

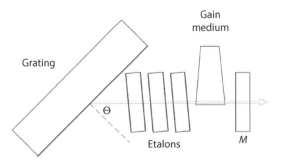

FIGURE 7.8   Grating-mirror laser cavity incorporating intracavity etalons.

narrow pulsed emission, down to the SLM regime, in high-power pulsed lasers. The introduction of the etalons provides further avenues of wavelength tuning. The main disadvantage of this class of resonators, in the pulsed regime is the very high intracavity power flux that can induce optical damage in the grating and the coating of the etalons.

An important configuration in the tunable laser oscillators *without* intracavity beam expansion class, which employs only the natural divergence of the intracavity beam for total illumination of the diffractive element, is the grazing-incidence grating design (Shoshan et al. 1977; Littman and Metcalf 1978). In these lasers, the grating is deployed at a high angle of incidence. The diffracted beam is subsequently reflected back toward the grating by the tuning mirror. A variation on this design is the replacement of the tuning mirror by a grating deployed in Littrow configuration (Littman 1978). This cavity has been configured as a open cavity, as shown in Figure 7.9a, or as a closed cavity, as shown in Figure 7.9b. A further alternative is the inclusion of an intracavity etalon (Saikan 1978) for further linewidth narrowing. Grazing-incidence grating cavities have the advantage of being fairly compact and are widely used in both the pulsed regime and the CW regime. A limitation of these cavities is the relatively high losses associated with the deployment of the diffraction grating at a high angle of incidence as illustrated in Figure 7.10 (Duarte and Piper 1981).

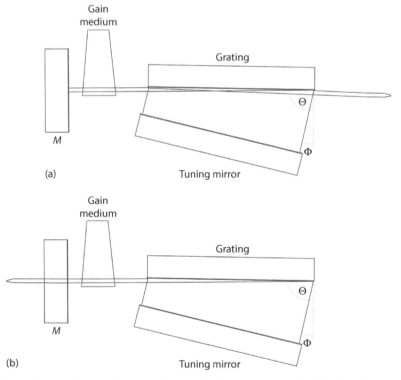

FIGURE 7.9   Grazing-incidence grating cavities: (a) open cavity; (b) closed cavity.

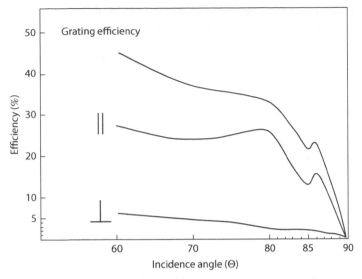

FIGURE 7.10 Grating efficiency curve as a function of angle of incidence at $\lambda = 632.82$ nm. (Reproduced from Duarte, F.J., and Piper, J.A., *Appl. Opt.*, 20, 2113–2116, 1981. With permission from the Optical Society.)

For tunable laser oscillators without intracavity beam expansion, and in the absence of intracavity etalons, the cavity linewidth equation takes the fairly simple form

$$\Delta\lambda \approx \Delta\theta_R (R\nabla_\lambda\Theta_G)^{-1} \tag{7.13}$$

where:

$\nabla_\lambda\Theta_G$ is the grating dispersion either in Littrow configuration

$$\nabla_\lambda\Theta_G = \frac{2\tan\Theta}{\lambda} \tag{7.14}$$

$$\nabla_\lambda\Theta_G = \frac{m}{d\cos\Theta} \tag{7.15}$$

or in grazing-incidence configuration

$$\nabla_\lambda\Theta_G = \frac{2(\sin\Theta + \sin\Phi)}{\lambda\cos\Theta} \tag{7.16}$$

$$\nabla_\lambda\Theta_G = \frac{2m}{d\cos\Theta} \tag{7.17}$$

where the multiple-return-pass beam divergence $\Delta\theta_R$ is given by

$$\Delta\theta_R = \frac{\lambda}{\pi w}\left[1 + \left(\frac{L_\mathcal{R}}{B_R}\right)^2 + \left(\frac{A_R L_\mathcal{R}}{B_R}\right)^2\right]^{1/2} \tag{7.18}$$

and the appropriate propagation terms can be calculated according to the optical architecture of the cavity as described in Chapter 6. In Equations 7.13 and 7.18, R refers to the number of intracavity return passes.

For a simple mirror-grating cavity, as illustrated on Figure 7.7, and a single return pass ($R = 1$), Equation 7.13 reduces to

$$\Delta\lambda \approx \Delta\theta(\nabla_\lambda\Theta_G)^{-1}$$

which provides an estimate of the dispersive linewidth derived from the dispersion of the grating. For a tunable oscillator incorporating a diffraction grating, and an intra-cavity etalon as outlined in Figure 7.8, the dispersive linewidth $\Delta\lambda$ can be used via its frequency domain version given by Equation 7.7, that is, Δv, to guide the design of the intracavity etalon with a free spectral range (FSR) is given by (Born and Wolf 1999)

$$\text{FSR} = \frac{c}{2nl_e} = \delta v_{\text{FSR}} \tag{7.19}$$

where:

n is the refractive index of the etalon's material
l_e is the distance between the reflective surfaces

The expression for the FSR has its origin in $\Delta v = c/\Delta x$ as discussed in Chapter 3.

The minimum resolvable linewidth, or resulting laser linewidth, obtainable from the etalon is given by

$$\Delta v_{\text{FSR}} = \frac{\delta v_{\text{FSR}}}{\mathcal{F}} \tag{7.20}$$

where:

$\mathcal{F}$ is the *effective finesse* of the etalon

The finesse of the etalon is a function of the flatness of the surfaces (often in the $\lambda/100$–$\lambda/50$ range), the dimensions of the aperture, and the reflectivity of the surfaces. The effective finesse is given by (Meaburn 1976)

$$\mathcal{F}^{-2} = \mathcal{F}_R^{-2} + \mathcal{F}_F^{-2} + \mathcal{F}_A^{-2} \tag{7.21}$$

where:

$\mathcal{F}_R$, $\mathcal{F}_F$, and $\mathcal{F}_A$ are the reflective, flatness, and aperture finesses, respectively

The reflective finesse is given by (Born and Wolf 1999)

$$\mathcal{F}_R = \frac{\pi\sqrt{\mathcal{R}}}{(1-\mathcal{R})} \tag{7.22}$$

where:

$\mathcal{R}$ is the reflectivity of the surface

Further details on Fabry–Pérot etalons are given in Chapter 11.

Multiple-etalon systems are described in detail in the literature (Maeda et al. 1975; Pacala et al. 1984). As implied earlier, these multiple-etalon assemblies are designed so that the FSR of the etalon to be introduced is compatible with the measured laser linewidth attained with the previous etalon or etalons. Although this is a very effective avenue to achieve fairly narrow linewidths, the issue of optical damage, due to high intracavity power densities, does introduce limitations. The performance of representative tunable laser oscillators *without* intracavity beam expansion is summarized in Table 7.1.

7.3.2 Tunable Laser Oscillators *with* Intracavity Beam Expansion

Equation 7.6 indicates that a key principle in achieving narrow-linewidth emission consists in augmenting the intrinsic intracavity dispersion provided by the diffraction grating. This is accomplished by the total illumination of the diffraction surface of the tuning element. In the case of the grazing-incidence grating cavities, this is done by deploying the grating at a high angle of incidence; however, this can be associated with very low diffraction efficiencies. An alternative method is to illuminate the diffractive element via intracavity beam expansion. Tunable laser oscillators with intracavity

---

TABLE 7.1
Performance of Tunable Laser Oscillators *without* Intracavity Beam Expansion

Gain Medium	Cavity	λ (nm)	Δv	Energy[a]	η (%)	Reference
Gas Lasers						
XeCl	GI	308	~1 GHz	4 mJ		Sugii et al. (1987)
XeCl	3 etalons	308	~150 MHz	2–5 μJ		Pacala et al. (1984)
CO_2	GI	10,591	117 MHz	140 mJ		Duarte (1985a)
Liquid Lasers						
Rh 590	GI	600	300 MHz		2	Littman (1978)
Solid-State Lasers						
Ti:Al$_2$O$_3$	GI[b]		1.5 GHz	2 mJ	5	Kangas et al. (1989)
Semiconductor Lasers						
GaAlAs	Etalon[c]		4 kHz			Harrison and Mooradian (1989)
GaAlAs	GI[c,d]	780	10 kHz			Harvey and Myatt (1991)

[a] Output energy per pulse.
[b] Tuning range: $746 \leq \lambda \leq 918$ nm
[c] CW regime.
[d] Approximate tuning range: $770 \leq \lambda \leq 790$ nm
Rh, rhodamine.

beam expansion are divided into two subclasses: those using *two-dimensional* beam expansion and those using *one-dimensional* beam expansion.

Initially, intracavity beam expansion was accomplished utilizing two-dimensional beam expansion and a diffraction grating deployed in Littrow configuration as illustrated in Figure 7.11 (Hänsch 1972). Two-dimensional beam expansion has also been demonstrated using reflection telescopes (Beiting and Smith 1979). Transmission telescopes can be either of the Galilean or astronomical class, whereas the reflection telescope can be of the Cassegrainian type (see Chapter 6). The main advantage of this approach is the significant reduction of intracavity energy incident on the tuning grating, thus vastly reducing the risk of optical damage. The main disadvantages of the two-dimensional intracavity beam expansion is the requirement of expensive circular diffraction gratings, a relatively difficult alignment process, and the need for long cavities that incorporate thermal dependence consequences. Since the telescopes mentioned here can provide low dispersion, the cavity linewidth equation reduces to

$$\Delta\lambda \approx \Delta\theta_R (RM\nabla_\lambda\Theta_G)^{-1} \tag{7.23}$$

where:

$\nabla_\lambda\Theta_G$ is the grating dispersion in either Littrow or grazing-incidence configuration, as given in Equations 7.14 through 7.17, and the multiple-return-pass beam divergence $\Delta\theta_R$ is given by Equation 7.18.

Certainly, it should be indicated that for a single return pass ($R = 1$), Equation 7.23 assumes the form of expression introduced by Hänsch (1972):

$$\Delta\lambda \approx \Delta\theta \ (M\nabla_\lambda\Theta_G)^{-1} \tag{7.24}$$

One-dimensional intracavity beam expansion uses multiple-prism beam expanders rather than conventional telescopes to perform the beam expansion. Multiple-prism grating tunable laser oscillators are classified as multiple-prism Littrow (MPL)

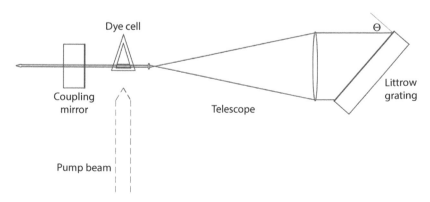

FIGURE 7.11 Two-dimensional transmission telescope Littrow grating laser cavity. This class of telescopic cavity was first introduced by Hänsch.

grating laser oscillators (Kasuya et al. 1978; Klauminzer 1978; Wyatt 1978; Duarte and Piper 1980) and hybrid multiple-prism grazing-incidence (HMPGI) grating laser oscillators (Duarte and Piper 1981, 1984a). MPL and HMPGI grating laser oscillators are depicted in Figures 7.12 through 7.14. Both of these oscillator subclasses belong to the closed cavity class.

In these laser oscillators, the intracavity beam expansion is one dimensional, thus facilitating the alignment process significantly. In addition, the requirements of the dimensions of the diffraction grating, perpendicular to the plane of incidence, are reduced significantly. Another advantage is compactness since these high-power tunable laser oscillators can be configured in architectures requiring cavity lengths in the 50–100 mm range.

The solid-state organic MPL grating oscillator depicted in Figure 7.12 incorporates a compensating $(+, +, +, -)$ multiple-prism configurations with $M \approx 90$ yielding near-zero prismatic dispersion at $\lambda \approx 590$ nm. The laser linewidth was measured

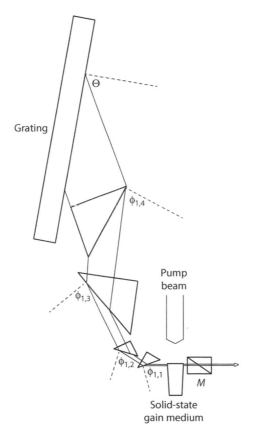

FIGURE 7.12   Long pulse MPL grating solid-state organic dye laser oscillator incorporating a $(+, +, +, -)$ compensating multiple-prism configuration. Laser linewidth is $\Delta v \approx 650$ MHz at a pulse length of $\Delta t \approx 105$ ns. (Reproduced from Duarte, F.J., et al., *Appl. Opt.*, 37, 3987–3989, 1998. With permission from the Optical Society.)

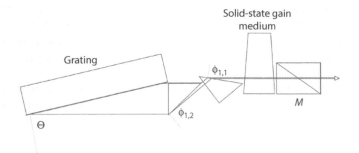

FIGURE 7.13 Optimized compact MPL grating solid-state organic dye laser oscillator. Laser linewidth is $\Delta v \approx 350$ MHz at a pulse length of $\Delta t \approx 3$ ns. (Reproduced from Duarte, F.J., *Appl. Opt.*, 38, 6347–6349, 1999. With permission from the Optical Society.)

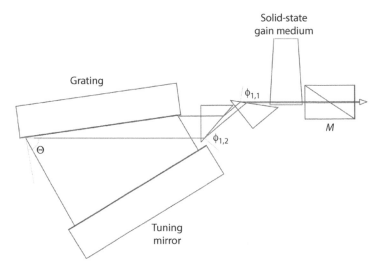

FIGURE 7.14 Solid-state HMPGI grating organic dye laser oscillator. Laser linewidth is $\Delta v \approx 375$ MHz at a pulse length of $\Delta t \approx 7$ ns. (Reprinted from *Opt. Laser Technol.*, 29, Duarte, F.J., Multiple-prism near-grazing-incidence grating solid-state dye laser oscillator, 512–516, Copyright 1997, with permission from Elsevier.)

at $\Delta v \approx 650$ MHz for a pulse length of $\Delta t \approx 105$ ns (Duarte et al. 1998). This type of oscillator configuration often uses rather large intracavity beam expansion factors, which in practice can be in the $100 \leq M \leq 200$ range. Oscillator designs with multiple-prism beam magnification factors as high as $M \approx 990$ have been described (Shay and Duarte 2009). Further configurational alternatives are discussed in Chapter 4.

An optimized dispersive oscillator architecture where a high-density diffraction grating (3300 lines/mm) allows relatively high incidence angles, at $\Theta \approx 77°$, in a Littrow configuration is depicted in Figure 7.13 (Duarte 1999). Thus, the required illumination of the diffraction element can proceed with rather modest intracavity beam expansion factors of $M \approx 44$. Consequently, the compactness of the cavity is significantly improved. This optimized dispersive solid-state tunable laser oscillator demonstrated laser linewidths down to $\Delta v \approx 350$ MHz, for $\Delta t \approx 3$ ns, which is close to the limit allowed by Heisenberg's uncertainty principle at a pulse power of ~33 kW (Duarte 1999).

The HMPGI grating oscillator depicted in Figure 7.14 is inherently compact since the grating is deployed at a *near*-grazing-incidence angle and the required intracavity beam expansion can be provided by a double-prism expander deployed in a compensating configuration to yield $M \approx 30$. The difference in efficiency performance between a grating deployed at grazing incidence and a grating deployed at near-grazing incidence can be significant as demonstrated by Duarte and Piper (1981, 1984b) with a clear advantage for the latter. To illustrate this point explicitly, an efficiency curve, as a function of angle of incidence, is provided for a typical diffraction grating in Figure 7.10. Using HMPGI grating oscillator configurations, SLM oscillation at linewidths in the $400 \le \Delta v \le 650$ MHz have been demonstrated in CVL-pumped tunable liquid organic lasers (Duarte and Piper 1984b) and $\Delta v \approx 375$ MHz in solid-state dye-doped organic polymer gain media (Duarte 1997).

It is interesting to note that most of external semiconductor lasers use the near-grazing-incidence configuration rather than the pure grazing-incidence scheme given the highly divergent beams available from the narrow gain regions of semiconductor lasers. MPL and HMPGI grating oscillator configurations are directly applicable to tunable high-power gas lasers such as excimer lasers and CO_2 lasers (Duarte 1985a, 1985b, 1985c) as depicted in Figure 7.15. Using MPL grating configurations, incorporating ZnSe prisms, variable linewidth emission in the $250 \le \Delta v \le 650$ MHz range was accomplished as a function of M (Duarte 1985b), while an SLM linewidth corresponding to $\Delta v \approx 107$ MHz was measured in a high-power TEA CO_2 HMPGI grating oscillator at $\lambda \approx 10{,}591$ nm (Duarte 1985c).

MPL and HMPGI grating oscillators have been shown to inherently yield extremely low levels of amplified spontaneous emission (ASE), which is a very desirable feature in many applications including high-resolution spectroscopy (Duarte

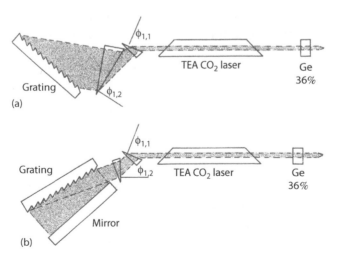

FIGURE 7.15 (a) MPL and (b) HMPGI grating high-power pulsed CO_2 laser oscillators. The prisms are made of ZnSe and the output couplers are made of Ge. (Reproduced from Duarte, F.J., *Appl. Opt.*, 24, 1244–1245, 1985b. With permission from the Optical Society.)

and Piper 1980, 1981). The MPL and HMPGI grating oscillators described here (Figures 7.12 through 7.14) incorporate a polarizer output coupler mirror, which is antireflection coated toward the gain medium, whereas its output surface is broadband coated with ~20% reflectivity. The function of this polarizer output coupler mirror is to further suppress the single-pass ASE since the laser emission from these oscillators is intrinsically highly polarized parallel to the plane of incidence. Using these oscillator architectures, ASE levels, as determined by the spectral density ratio ρ_{ASE}/ρ_l (Duarte 1990b), are in the $10^{-7} - 10^{-6}$ range (Duarte 1990b, 1997, 1999).

The multiple-return-pass linewidth for a multiple-prism grating oscillator is given by (Duarte and Piper 1984b; Duarte 2001)

$$\Delta\lambda = \Delta\theta_R (RM \nabla_\lambda \Theta_G + R \nabla_\lambda \Phi_P)^{-1} \tag{7.25}$$

where:
   $\nabla_\lambda \Theta_G$ is the grating dispersion in either Littrow or grazing-incidence configuration, as given in Equations 7.14 through 7.17, and the multiple-return-pass beam divergence $\Delta\theta_R$ is given by Equation 7.18 (see Chapter 4).

In addition to the dispersion of the grating, the multiple-prism must also be considered and is given by (Duarte and Piper 1982a; Duarte 1985c, 1989)

$$\nabla_\lambda \Phi_P = 2M_1 M_2 \sum_{m=1}^{r} (\pm 1)\mathcal{H}_{1,m} \left(\prod_{j=m}^{r} k_{1,j} \prod_{j=m}^{r} k_{2,j} \right)^{-1} \nabla_\lambda n_m$$

$$+2 \sum_{m=1}^{r} (\pm 1)\mathcal{H}_{2,m} \left(\prod_{j=1}^{m} k_{1,j} \prod_{j=1}^{m} k_{2,j} \right) \nabla_\lambda n_m \tag{7.26}$$

As indicated in Chapter 4, the above equation can be either used to evaluate the return-pass multiple-prism dispersion or applied to design zero-dispersion beam expanders at any given wavelength. In the latter case, $\nabla_\lambda \Phi_P \approx 0$ and Equation 7.25 reduces to Equation 7.23. As described in Chapter 5, the transmission efficiency of the multiple-prism beam expander can be quantified using

$$L_{1,m} = L_{2,(m-1)} + [1 - L_{2,(m-1)}] \mathcal{R}_{1,m} \tag{7.27}$$

and

$$L_{2,m} = L_{1,m} + (1 - L_{1,m}) \mathcal{R}_{2,m} \tag{7.28}$$

where:
   $L_{1,m}$ and $L_{2,m}$ represent the losses at the incidence and exit surfaces of the mth prism, respectively
   $\mathcal{R}_{1,m}$ and $\mathcal{R}_{2,m}$ are the respective Fresnel reflection factors given in Chapter 5

The performance of tunable laser oscillators with intracavity beam expansion is summarized in Table 7.2.

TABLE 7.2
Performance of Tunable Laser Oscillators *with* Intracavity Beam Expansion

Gain Medium	Cavity	λ (nm)	Tuning (nm)	Δv	Energy[a] (mJ)	η(%)	Reference
Gas Lasers							
XeCl	HMPGI[b]	308		1.8 GHz			Duarte (1991b)
CO_2	MPL	10,591		140 MHz	200		Duarte (1985b)
CO_2	HMPGI	10,591		107 MHz	85		Duarte (1985b)
Liquid Dye Lasers							
Rh 590	Telescopic	600		2.5 GHz		20	Hänsch (1972)
Rh 590	Telescopic[c]	600		300 MHz		2–4	Hänsch (1972)
C 500	MPL	510	$490 \leq \lambda \leq 530$	1.61 GHz		14	Duarte and Piper (1980)
Rh 590	MPL[c]	572		60 MHz		5	Bernhardt and Rasmussen (1981)
Rh 590	MPL	575	$565 \leq \lambda \leq 605$	1.4 GHz		5	Duarte and Piper (1984b)
C 500	HMPGI	510	$490 \leq \lambda \leq 530$	1.15 GHz		4	Duarte and Piper (1981)
Rh 590	HMPGI	575	$565 \leq \lambda \leq 603$	650 MHz		4	Duarte and Piper (1984b)
Solid-State Dye Lasers							
Rh 590	HMPGI	580	$565 \leq \lambda \leq 610$	375 MHz		4	Duarte (1997)
Rh 590	MPL	590	$550 \leq \lambda \leq 603$	350 MHz		5	Duarte (1999)
External Cavity Semiconductor Lasers[d]							
InGaAsP/InP	MPG		$1255 \leq \lambda \leq 1335$	100 kHz			Zorabedian (1992)

(Continued)

TABLE 7.2

(Continued) Performance of Tunable Laser Oscillators *with* Intracavity Beam Expansion

Gain Medium	Cavity	λ (nm)	Tuning (nm)	Δv	Energy[a] (mJ)	η(%)	Reference
GaInP–AlGaInP	Littrow[e]		$640 \leq \lambda \leq 652$	4 MHz			Laurila et al. (2002)
InAs (QD)[f]	Littrow[g]		$1125 \leq \lambda \leq 1288$	200 kHz			Nevsky et al. (2008)
	Littrow[g]		$1420 \leq \lambda \leq 1620$	8.7 kHz			Bennetts et al. (2014)

[a] Output energy per pulse.
[b] Calculated return-pass dispersive linewidth.
[c] Includes intracavity etalon.
[d] CW regime.
[e] Transmission grating.
[f] Quantum dot.
[g] Uses intracavity lens to collimate beam illuminating the grating.
Rh, rhodamine.

7.3.3   WIDELY TUNABLE NARROW-LINEWIDTH EXTERNAL CAVITY SEMICONDUCTOR LASERS

The external cavity semiconductor laser (ECSL) has become widely used in a number of contemporaneous applications including laser cooling and Bose–Einstein condensation. Central to the attractiveness of ECSLs is their tunability, narrow-linewidth characteristics, stability, compactness, and low cost. Two of the most popular external cavity configurations belong to the *open cavity* class in the form of a Littrow grating cavity design (Wieman and Hollberg 1991) and a grazing-incidence grating design (Harvey and Myatt 1991). The adoption of open cavity designs appears to have resulted from a need to adapt to the availability of commercial semiconductor lasers with one end of the cavity sealed from access.

The advantages of closed cavity tunable laser oscillators, over open cavity alternatives, in the pulsed high-power domain, were outlined by Duarte and Piper (1980, 1981). These advantages include significantly reduced optical noise levels and independence from external feedback effects. Albeit the first phenomenon is less pronounced, in semiconductor lasers, the second observation still applies. Moreover, in ECSLs incorporating open Littrow grating configurations, the diffracted beam is directed back to the gain region, while the output is coupled via the reflection losses of the grating. This means that as the wavelength is tuned, the direction of the output beam changes. Thus, additional external optics is required for beam correction (Hawthorn et al. 2001). In ECSLs incorporating open grazing-incidence configuration, tighter boundaries for thermal stability are required (Laurila et al. 2002).

This is interesting given the fact that the closed cavity ECSLs preceded the open cavity alternatives (Voumard 1977; Fleming and Mooradian 1981; Belenov et al. 1983). The main difference is that closed cavity ECSLs require access to both ends of the gain region. Ideally, both ends of the gain region should be antireflection coated as indicated by Fleming and Mooradian (1981) so that the characteristics of the emission are entirely controlled by the external cavity as illustrated in Figure 7.16 (Duarte 1993a). Alternatively, at least one of the cavity ends should be antireflection coated. When this is the case, the frequency-selective elements are deployed next to the antireflection end of the gain region, while the laser output is coupled from the partially reflective extreme (Notomi et al. 1990).

Perhaps, one of the most thoroughly engineered ECSLs to date is that reported by Zorabedian (1992), which was inspired on a closed cavity MPL grating design. Zorabedian reports on a laser linewidth of $\Delta v = 100$ kHz sustainable over a 60 nm tuning range, with a side-mode suppression ratio of 70 dB (Zorabedian 1992).

The closed cavity approach has also been demonstrated in the form of a cavity incorporating a transmission Littrow grating (Laurila et al. 2002). These authors report laser linewidths of $\Delta v \approx 4$ MHz over a tuning range of 12 nm centered at $\lambda = 646$ nm. The laser beam has an elliptical profile of 8 mm × 1.5 mm (Laurila et al. 2002).

Elliptical TEM_{00} laser beam profiles are typical in semiconductor lasers and they are a result of the rectangular cross section of the gain region. For instance, rectangular dimensions of 4 μm × 1 μm are not uncommon (Fox et al. 1997). One way to produce, intracavity, a circular beam with tunable narrow-linewidth characteristics

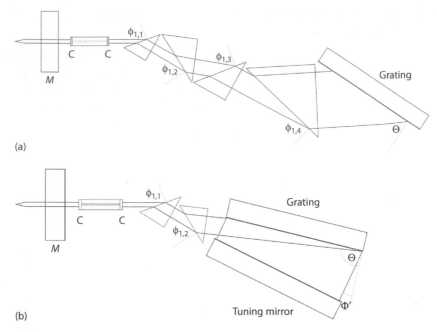

(a)

(b)

Tuning mirror

FIGURE 7.16 MPL (a) and HMPGI (b) grating semiconductor laser oscillators. (From Duarte, F. J. (ed.), Broadly tunable external-cavity semiconductor lasers. In *Tunable Laser Applications*, 2nd edn., CRC Press, New York, Chapter 5, 2009.)

is to introduce multiple-prism beam expansion prior to illumination of the transmission grating as illustrated in Figure 7.17a. In this design, the gain region is oriented to emit its elliptical beam with the long axis perpendicular to the plane of incidence. The multiple-prism beam expander then elongates the narrow dimension of the beam to yield a circular profile incident on the grating. As in previous oscillator configurations, the beam expansion, which contributes to the line narrowing, takes place at a plane parallel to the plane of incidence (Duarte 2003). A better solution is to use two antireflection ends to the gain region, deploy the gain region so that the long axis of the beam is perpendicular to the plane of incidence, and provide a large multiple-prism beam expansion prior to the illumination of the diffraction grating deployed in Littrow configuration as shown in Figure 7.17b. At the other end of the cavity, only the necessary beam expansion, parallel to the plane of incidence, is provided to yield a circular beam (Duarte 2009).

The multiple-prism grating configurations incorporating ZnSe prisms, demonstrated in CO_2 lasers (Duarte 1985a, 1985b, 1985c), should be applicable to the design and construction of closed cavity quantum cascade lasers oscillating near $\lambda \approx 10.6$ µm. The performance of tunable external cavity quantum cascade lasers is considered in Chapter 9.

A quantum dot semiconductor laser yielding a linewidth of $\Delta v = 200$ KHz, over a tuning range of $1125 \leq \lambda \leq 1288$ nm was reported by Nevsky et al. (2008). In this cavity, the initially divergent beam illuminates a collimating lens so that the Littrow grating is illuminated by an expanded intracavity beam. The performance of various

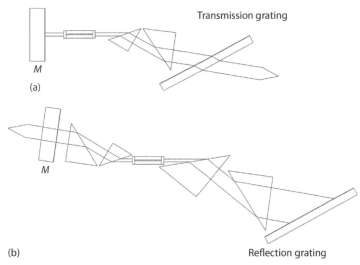

FIGURE 7.17 (a) MPL transmission grating semiconductor laser oscillator designed to produce circular TEM_{00} emission. The semiconductor is oriented to emit its elongated beam with the long axis perpendicular to the plane of propagation. Intracavity prismatic beam expansion (parallel to the plane of propagation) renders a nearly circular output beam. (b) MPL grating semiconductor laser oscillator configuration designed to produce circular TEM_{00} emission. The same expansion strategy is used as in (a) while using a reflection grating in Littrow configuration and beam expansion at both ends of the cavity. The multiple-prism beam expansion illuminating the grating can be as large as necessary to yield very narrow linewidths. (From Duarte, F.J., *Tunable Laser Applications*, CRC Press, New York, 2009. With permission.)

tunable semiconductor lasers incorporating intracavity beam expansion is summarized in Table 7.2. Comprehensive reviews on the principles of ECSLs are given by Zorabedian (1995), Duarte (1995a), and Fox et al. (1997).

7.3.4 DISTRIBUTED FEEDBACK LASERS

A laser architecture that does not belong to either class of resonators previously described, since it does not incorporate mirrors in the optical axis of the emission, is that of the distributed feedback (DFB) laser introduced by Kogelnik and Shank (1971). In these laser configurations, depicted in Figure 7.18, the excitation laser beam is divided into two sub-beams that are recombined at the gain medium where they produce an interference signal. In DFB lasers, feedback is provided by backward Bragg scattering induced by the periodic perturbations of the refractive index at the gain medium. Kogelnik and Shank (1972) have modeled basic gain and spectral features of these lasers using a representation of the scalar wave equation in the form of

$$\frac{\partial^2 E}{\partial z^2} + \kappa^2 E = 0 \tag{7.29}$$

and a two counter-propagating wave approach to describe the electric field in the form of

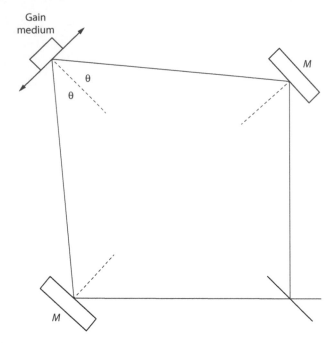

FIGURE 7.18   Generic DFB laser configuration.

$$E(z) = R(z)e^{-i\kappa z/2} + S(z)e^{i\kappa z/2} \tag{7.30}$$

where:

  z is in the direction of propagation

These authors also assume a periodic spatial variation in the refractive index represented by

$$n(z) = n + n_1 \cos \kappa z \tag{7.31}$$

where:

  κ is related to the fringe spacing Λ by

$$\kappa = \frac{2\pi}{\Lambda} \tag{7.32}$$

and the wavelength of oscillation is defined by the Bragg condition (see Section 7.4.4)

$$\lambda_l = 2n\Lambda \tag{7.33}$$

The passive linewidth of a DFB laser is given by Bor (1979):

$$\Delta\lambda_l = \left(\frac{\lambda_l}{\lambda_p}\right) \Delta\lambda_p \tag{7.34}$$

where:

λ_l is the wavelength of the DBF laser

λ_p is the wavelength of the pump laser

$\Delta\lambda_p$ is the linewidth of the pump laser

This equation indicates that the linewidth of the DFB laser is determined by the spectral properties of the pump laser. Thus, narrow-linewidth emission in DBF lasers requires a narrow-linewidth pump laser. This is quite different from the lasers using dispersive oscillators whose emission linewidth is independent of the bandwidth of the excitation source.

Tuning DFB lasers can be accomplished by either varying the refractive index of the gain medium or changing Λ. The refractive index can be changed by thermal means or by varying the composition of the gain medium. In the case of DFB dye lasers, it is done by altering the composition of the solvents. For a DFB laser configuration allowing rotation of the mirrors, as shown in Figure 7.18, the fringe separation is given by the first-order grating equation

$$\Lambda = \frac{\lambda_p}{2\sin\theta} \tag{7.35}$$

so that (Shank et al. 1971)

$$\lambda_l = \frac{n\lambda_p}{\sin\theta} \tag{7.36}$$

which allows the laser to be tuned by geometrical means.

Given their typical linewidth characteristics, $\Delta v \approx 45$ GHz (Bor 1979), DFB lasers are well-known sources of pulses in the picosecond regime. DFB laser configurations originally applied to liquid dye lasers have also been demonstrated with solid-state dye laser gain media (Wadsworth et al. 1999; Zhu et al. 2000).

In the CW regime, DFB semiconductor lasers have been demonstrated to yield linewidths as low as 2 MHz in the 1.30–1.55 μm wavelength range (Wolf et al. 1991).

7.4 WAVELENGTH TUNING TECHNIQUES

Laser oscillators incorporating dispersive, diffractive, and interferometric elements such as prisms, gratings, and etalons are intrinsically tunable. Here, several wavelength tuning techniques based on these elements are described.

7.4.1 PRISMATIC TUNING TECHNIQUES

For a single prism, the exit angle is a function of wavelength as given by (Duarte 1990b)

$$\phi_{2,1} = \arcsin\left(n(\lambda,T)\sin\left\{ \alpha_1 - \arcsin\left[\frac{\zeta}{n(\lambda,T)} \right] \right\} \right) \tag{7.37}$$

where:

$\zeta = \sin \phi_{1,1}$ is a constant for a stationary prism

α_1 is its apex angle

$n(\lambda, T)$ is the appropriate material dispersion function

Similarly, the exit angle at the mth prism can be expressed as

$$\phi_{2,m} = \arcsin\left(n_m(\lambda,T)\sin\left\{ \alpha_m - \arcsin\left[\frac{\sin\phi_{1,m}}{n_m(\lambda,T)} \right] \right\} \right) \qquad (7.38)$$

where:

$\phi_{1,m}$ is related geometrically to the exit angle of the preceding prism in the array $\phi_{2,(m-1)}$

Equation 7.37 highlights the fact that the exit angle at a given prism can be slightly altered by changing the refractive index of the prism, which can be done, for instance, by changing the temperature that exploits the $\partial n/\partial T$ of the prism's material. Also, these equations illustrate the fact that the rays of different wavelengths follow a slightly different geometrical path through the prism sequence.

A purely geometrical tuning approach can be described considering an array of r identical prisms, deployed in a symmetric configuration, and a mirror as shown in Figure 7.19. The cumulative angular displacement at the exit surface of the last prism is augmented according to (Duarte 1990a)

$$\nabla_\lambda \phi_{2,r} = r\nabla_\lambda \phi_{2,1} \qquad (7.39)$$

Again this means that a wave front of different wavelengths emerges at a slightly different angle at the exit prism of the array. Thus, depending on the cumulative dispersion, the tuning mirror can only reflect back light within a narrow bandwidth. At a slightly different angular position, the light of a slightly different wavelength is reflected back to the gain region. Using this approach, Strome and Webb (1971) report a tuning range from 571 to 615 nm using a four-prism sequence in a pulsed dye laser.

7.4.2 Diffractive Tuning Techniques

Diffraction gratings are the main tuning elements in narrow-linewidth dispersive laser oscillators. In a cavity incorporating a grating deployed in a grazing-incidence, or near-grazing-incidence, configuration (as illustrated in Figure 7.20), the wavelength is changed by rotating the tuning mirror. Wavelength tuning in this configuration is described by the grating equation:

$$m\lambda = d(\sin\Theta + \sin\Phi) \qquad (7.40)$$

where:

Θ is the angle of incidence

Φ is the angle of refraction

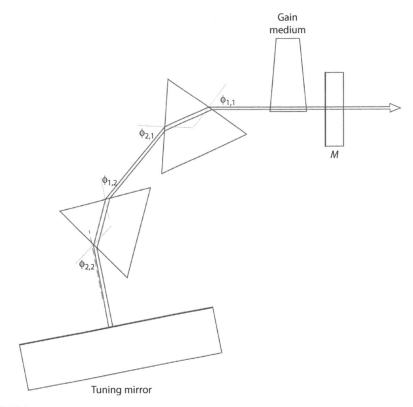

FIGURE 7.19 Multiple-prism tuning.

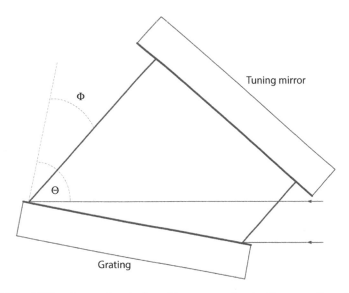

FIGURE 7.20 Diffraction grating deployed in near-grazing-incidence configuration.

For a fixed angle of incidence, different wavelengths diffract at different angles. Hence, the precise angular position of the tuning mirror determines the wavelength of the radiation that will return to the gain region for further amplification. Geometrically, the tuning is described by Equation 7.40, where m, d, and Θ are kept constant so that λ becomes a function of $sin\ \Phi$.

For a grating in Littrow configuration, $\Theta = \Phi$ (see Figure 7.21), and the grating equation becomes

$$m\lambda = 2d\sin\Theta \qquad (7.41)$$

and with m and d being fixed, λ becomes solely a function of $sin\ \Theta$. Using these simple diffraction techniques, narrow-linewidth dispersive laser oscillators have been smoothly tuned over 50 nm or more (Duarte 1990a, 1999).

Accurate angular displacement of tuning gratings requires the use of high-quality kinematic mounts with 0.1 s of arc or better. For an MPL grating oscillator, a frequency shift of $\delta v \approx 250$ MHz, comparable to the Δv of the dispersive oscillator, requires an angular rotation of $\delta\Theta \approx 10^{-6}$ radians at the grating (Duarte et al. 1988).

7.4.2.1 Example

In an optimized multiple-prism grating tunable laser oscillator, as illustrated in Figure 7.13, the multiple-prism expander is designed to yield zero dispersion at the central wavelength of emission, in this case $\lambda \approx 590$ nm, and the tuning characteristics of the laser are entirely dominated by the 3300 lines/mm diffraction grating deployed in a Littrow configuration at $m = 1$. Thus, to tune from $\lambda_1 \approx 590.0000$ to $\lambda_2 \approx 590.0004$, it is necessary to rotate the diffraction grating by a minute angle determined by $\lambda_1 = 2d\sin\Theta_1$ and $\lambda_2 = 2d\sin\Theta_2$, since $m = 1$ in Equation 7.41. Substitution by the given quantities, and computation of the angles related to the given wavelengths, yields an angular difference of $\delta\Theta \approx 2.8860 \times 10^{-6}$ radians. This

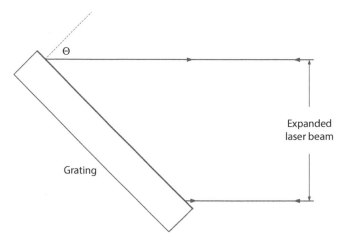

FIGURE 7.21 Diffraction grating deployed in Littrow configuration. Here, the angle of incidence equals the angle of diffraction ($\Theta = \Phi$).

is the angular displacement necessary to tune approximately by $\Delta v \approx 350$ MHz, which is the measured linewidth of this multiple-prism grating laser oscillator (Duarte 1999). This example illustrates the functional elegance of this class of high-power pulsed tunable oscillator.

7.4.3 SYNCHRONOUS TUNING TECHNIQUES

As the wavelength of the oscillator is being tuned using a dispersive element such as a diffraction grating, in either grazing-incidence or Littrow configuration, the FSR of the cavity, or the spacing of the longitudinal modes, varies according to

$$\text{FSR} = \frac{\lambda^2}{2L} \tag{7.42}$$

This can lead to abrupt jumps in the longitudinal mode selection, which is also known as *mode hopping*. One way to suppress this effect is to adjust the cavity length accordingly so that the FSR of the cavity is not altered. This is known as *synchronous tuning*. In cavities tuned by a grating deployed in Littrow configuration, synchronous tuning can be achieved by synchronizing the position of the output coupler mirror with the rotation of the grating. Although this is a simple principle, its successful practical implementation requires the application of accurate wavelength monitoring and high-precision servomechanisms.

An alternative approach to synchronous wavelength tuning, applicable to SLM lasers, was introduced by Littman (1981) who realized that the cavity length (see Figure 7.22) is an integral number of half wavelengths so that

$$\lambda = \frac{2}{N}(L_f + L_p \sin \Phi) \tag{7.43}$$

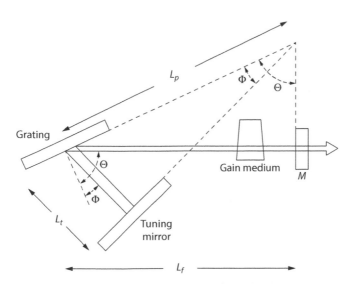

FIGURE 7.22   A synchronous wavelength tuning configuration.

where:

N is the number of longitudinal modes

Comparison of this equation with

$$\lambda = \frac{d}{m}(\sin \Theta + \sin \Phi)$$

indicates that equivalence is achieved if (Littman 1981)

$$\frac{2L_f}{N} = \frac{d}{m}\sin \Theta \qquad (7.44)$$

and

$$\frac{2L_p}{N} = \frac{d}{m} \qquad (7.45)$$

Careful selection of the cavity parameters involved in these equations can lead to the establishment of a fairly wide single-mode scanning range. This scheme requires high-precision rotation of the tuning mirror since the mechanical displacement equivalent to half a wavelength can result in mode hopping (Littman 1981).

7.4.4 BRAGG GRATINGS

One further method of wavelength tuning, applicable to ECSLs and fiber lasers, is the use of Bragg gratings (Kogelnik and Shank 1972). In a fiber laser, for instance, a grating can be engraved in the fiber with the grooves perpendicular to the plane of propagation using suitable laser radiation. A Bragg grating can be visualized as a wavelength-selective mirror satisfying the Bragg condition:

$$\lambda = 2n\Lambda \qquad (7.46)$$

where:

n is the refractive index
Λ is the grating period

The linewidth selectivity can be estimated using (see Chapter 3)

$$\Delta\lambda \approx \frac{\lambda^2}{\Delta x} \qquad (7.47)$$

with $\Delta x = 2nd$, for propagation in a bulk material of refractive index n, so that

$$\Delta\lambda = \frac{\lambda^2}{2nd} \qquad (7.48)$$

where:

d is the thickness of the grating

The gating period can also be defined as $\Lambda = d/N$, where N is the number of planes in the grating. Thus, in terms of explicit grating parameters, the wavelength can expressed as

$$\lambda \approx \frac{2nd}{N} \tag{7.49}$$

and the linewidth as

$$\Delta\lambda \approx \frac{2nd}{N^2} \tag{7.50}$$

For assessment and comparison purposes, it is useful to restate the interferometric identity in frequency units

$$\Delta\nu \approx \frac{c}{\Delta x} \tag{7.51}$$

or

$$\Delta\nu \approx \frac{c}{2n\Lambda N} \tag{7.52}$$

7.4.5 INTERFEROMETRIC TUNING TECHNIQUES

In addition to prismatic and diffractive tuning methods, intracavity etalons provide a further alternative for fine wavelength tuning. According to Born and Wolf (1999), the etalon can be considered as a periodic wavelength filter that satisfies the following condition for maxima:

$$m_e\lambda = 2n(\lambda,T)d_e \cos\psi_e \tag{7.53}$$

where:
m_e is an integer
d_e is the distance between the reflective surfaces
ψ_e is the refraction angle that is related to the tilt angle by

$$\sin\phi_e = n(\lambda,T)\sin\psi_e \tag{7.54}$$

$n(\lambda,T)$ is the refractive index of the substrate of the etalon (see Figure 7.23)

The angular dispersion of the etalon can be written as (Duarte 1990b)

$$\nabla_\lambda\phi_e = \frac{\sin\psi_e}{\cos\phi_e}\nabla_\lambda n + n\left(\frac{\cos\psi_e}{\cos\phi_e}\right)\nabla_\lambda\psi_e \tag{7.55}$$

where:

$$\nabla_\lambda\psi_e = \left(n^{-1}\nabla_\lambda n - \lambda^{-1}\right)(\tan\psi_e)^{-1} \tag{7.56}$$

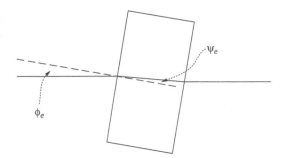

FIGURE 7.23   Solid etalon depicting incidence and refraction angles.

For $n = 1$, the condition $\nabla_\lambda \phi_e = \nabla_\lambda \psi_e$ arises so that

$$\nabla_\lambda \phi_e = -(\lambda \tan \phi_e)^{-1} \tag{7.57}$$

as given by Schäfer (1990). For the special case of $n = 1$, it can be shown that the wavelength shift resulting from a displacement in ϕ_e (from $\phi_e = 0$ at λ_1) is given by (Schäfer 1990)

$$\delta\lambda = (1 - \cos \phi_e)\lambda_1 \tag{7.58}$$

In addition to angular rotation, an alternative fine-tuning technique exploits the thermal dependence of the refractive index of the etalon's substrate. Using Equation 7.53, the wavelength difference from λ_1 to λ_2 corresponding to a change in temperature from T_1 to T_2 can be written as

$$\delta\lambda = \frac{2l_e}{m_e}\left[n(T_1)\cos\psi_{e_1} - n(T_2)\cos\psi_{e2}\right] \tag{7.59}$$

which for $\psi_{e_1} = 0$ at λ_1 reduces to

$$\delta\lambda = \left\{1 - \left[\frac{n(T_2)}{n(T_1)}\right]\cos\psi_{e2}\right\}\lambda_1 \tag{7.60}$$

For the special case of $n(T_1) = n(T_2) = 1$, this equation takes the form of Equation 7.58.

In addition to solid etalons, gas-spaced Fabry–Pérot interferometers can also be used for fine frequency tuning. Meaburn (1976) indicates that the refractive index of the gas can be varied, as a function of pressure, according to

$$K\Delta P \approx (n - 1) \tag{7.61}$$

where:
  K is a constant related to the intrinsic properties of the gas
  ΔP is the pressure gradient applied over the plates of the interferometer

The wavelength shift thus obtained can be expressed as (Meaburn 1976)

$$\delta\lambda \approx \lambda K \Delta P \tag{7.62}$$

This principle was applied by Wallenstein and Hänsch (1974) to vary the frequency of a telescopic dye laser oscillator, including a tilted etalon and a grating mounted in Littrow configuration. According to these authors, the laser frequency changes almost linearly, as a function of pressure, according to

$$\delta v \approx -\left(\frac{c}{\lambda}\right) K \Delta P \tag{7.63}$$

7.4.6 LONGITUDINAL TUNING TECHNIQUES FOR LASER MICROCAVITIES

Longitudinal tuning is the result of slight changes in cavity length as illustrated in Figure 7.24. This section is based on the original discussion on longitudinal tuning techniques by Duarte (2003) and refinements made in the work of Duarte (2014).

The principle of longitudinal tuning is based on the interferometric identity

$$\delta \lambda = \frac{\lambda^2}{\delta x} \tag{7.64}$$

which can also be stated as

$$\delta v = \frac{c}{\delta x} \tag{7.65}$$

Assume that the cavity linewidth $\Delta \lambda$ remains constant at two relatively close but different wavelengths as illustrated in Figure 7.25. This is a reasonable assumption for a laser emitting a diffraction-limited beam. Under these circumstances, the spacing of the intracavity modes will change within the transmission window so that

$$\delta \lambda_1 = \frac{\lambda_1^2}{2L} \tag{7.66}$$

$$\delta \lambda_2 = \frac{\lambda_2^2}{2(L \pm \Delta L)} \tag{7.67}$$

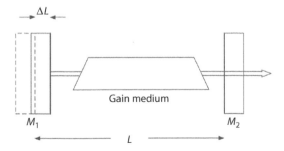

FIGURE 7.24 Longitudinal tuning applicable to laser microcavities. The cavity length L is changed by a minute amount ΔL (see text).

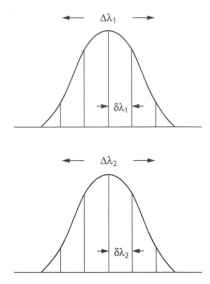

FIGURE 7.25 Dispersive cavity linewidth at two slightly different frequencies. It is assumed that $\Delta\lambda_1 \approx \Delta\lambda_2$, while the intracavity FSR changes.

or

$$N_1 = \frac{\Delta\lambda_1}{\delta\lambda_1} \tag{7.68}$$

$$N_2 = \frac{\Delta\lambda_2}{\delta\lambda_2} \tag{7.69}$$

for $\Delta\lambda_1 \approx \Delta\lambda_2$,

$$\lambda_2 \approx \lambda_1 \left(\frac{\delta\lambda_2}{\delta\lambda_1} \right)^{1/2} \left(1 \pm \frac{\Delta L}{L} \right)^{1/2} \tag{7.70}$$

or

$$\lambda_2 \approx \lambda_1 \left(\frac{N_1}{N_2} \right)^{1/2} \left(1 \pm \frac{\Delta L}{L} \right)^{1/2} \tag{7.71}$$

for $\Delta L = 0$, $\delta\lambda_1 \approx \delta\lambda_2$, and $\lambda_1 \approx \lambda_2$. For multilongitudinal mode oscillation, this approach requires counting the number of modes at the two wavelengths. For SLM oscillation, $N_1 = N_2 = 1$ and

$$\lambda_2 \approx \lambda_1 \left(1 \pm \frac{\Delta L}{L} \right)^{1/2} \tag{7.72}$$

This equation indicates that tuning, by changing the cavity length, depends directly on the $(\Delta L/L)$ ratio. For instance, a tunable dispersive oscillator yielding SLM emission ($\Delta v \approx 350$ MHz) with a cavity length of $L = 75$ mm (Duarte 1999) can tune its emission wavelength from 590.0000 to 590.0197 nm by increasing its cavity length just by $\Delta L = 0.005$ mm. This wavelength change translates to a frequency shift of $\delta v \approx 17$ GHz. Given a laser linewidth of $\Delta v \approx 350$ MHz, such change in the cavity length translates into an enormous frequency shift. This implies that the cavity length, and thus the thermal conditions, of such oscillators must be carefully controlled to ensure stability in the frequency domain. For the case of a miniature laser cavity, with a length of 1 mm, tuned with MEMS methods and lasing at 650 nm, the frequency can be tuned in steps of 2.3 GHz by changing the cavity length every 10 nm.

7.4.6.1 Example

Uenishi et al. (1996) reported on experiments using the $\Delta L/L$ method to perform wavelength tuning in a MEMS-driven semiconductor laser cavity. In such experiments, they observed wavelength tuning, in the absence of mode hopping, as long as the change in wavelength did not exceed $\lambda_2 - \lambda_1 \approx 1$ nm. Using their graphical data for the scan initiated at $\lambda_1 \approx 1547$ nm, it is established that $\Delta L \approx 0.4$ μm, and using $L \approx 305$ μm, Equation 7.72 yields $\lambda_2 \approx 1548$ nm, which approximately agrees with their observations (Uenishi et al. 1996). In this regard, it should be mentioned that Equation 7.72 was implicitly derived with the assumption of a wavelength scan obeying the condition $\delta\lambda_1 \approx \delta\lambda_2$. Albeit here, we use the term *microcavity*, this approach should also apply to cavities in the submicrometer regime or nanocavities.

7.4.7 BIREFRINGENT FILTERS

Birefringence in intracavity wavelength selectivity occurs via the use of birefringent filters (see also Chapter 5). The spatial phase component in the wave equation

$$\psi = \psi_0 e^{-i(\omega t - kx)} \tag{7.73}$$

for a uniaxial crystal plate can be expressed in more detail as (Born and Wolf 1999)

$$\delta = k(n_e - n_o)x \tag{7.74}$$

which can be written as

$$\delta = \frac{2\pi x \Delta n}{\lambda} \tag{7.75}$$

Maximum transmission, for a single birefringent plate, will occur for values of $\delta = 2\pi m$ (where $m = 1, 2, 3,...$) so that, at a transmission maximum designated by λ_{max},

$$m\lambda_{max} = x\Delta n \tag{7.76}$$

so that using $\Delta\lambda \approx \lambda^2/2x\Delta n$ the linewidth becomes

$$\Delta\lambda \approx \frac{\lambda_{max}}{2m} \tag{7.77}$$

Birefringent filters for wavelength tuning applications are often used in stacks and are known as *Lyot filters*. The use of a three-plate birefringent filter as an intracavity wavelength selectivity element was described by Johnston and Duarte (2002).

7.5   POLARIZATION MATCHING

The polarization characteristics of a given tunable laser oscillator depend on the intrinsic polarization of the gain medium, the angle of the laser windows (or emission exit surfaces), the configuration of the dispersive elements, and the diffraction grating. For optimum laser conversion efficiency, it is important to perform a *polarization matching* of the gain region to that of the optical elements integrating the laser cavity. For the purpose of this discussion, the plane of incidence is as defined in Chapter 5.

The polarization characteristics of a given gain medium depend on the atomic or molecular composition of such medium. A gain medium such as a laser dye responds differently to different orientations of the polarization of the pump laser (Schäfer 1990). As illustrated in Figure 7.26 (Duarte 1990b), excitation

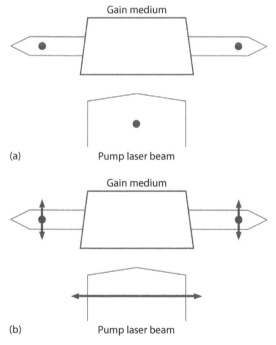

FIGURE 7.26   Polarization preference as explained by Duarte (1990b): (a) An excitation beam polarized perpendicular to the plane of propagation yields emission also polarized perpendicular to the plane of propagation; (b) the same excitation emission polarized parallel to the plane of propagation yields emission, in the same molecular active medium, partially polarized in both directions (see text).

with the pump laser beam polarized perpendicular to the plane of propagation yields single-pass emission, which is also polarized perpendicular to the plane of incidence when using rhodamine 590 molecules, as the gain medium, and the $^2P_{3/2} - {}^2D_{5/2}$ transition of the copper laser ($\lambda = 510.554$ nm) as the excitation radiation. However, for excitation parallel to the plane of incidence, the emission is almost unpolarized.

The first opportunity to induce a given polarization is presented in the selection of the angle of the laser windows. Windows deployed at an angle, thus creating a gain region in form of a trapezoid, greatly reduce internal reflections, contribute significantly to reduce optical noise, and facilitate the control of the spectral characteristics by the dispersive optics. The reflection losses for the given components of polarization are given by (Born and Wolf 1999)

$$\mathcal{R}_{\parallel} = \left[\frac{\tan^2(\phi - \psi)}{\tan^2(\phi + \psi)} \right] \tag{7.78}$$

$$\mathcal{R}_{\perp} = \left[\frac{\sin^2(\phi - \psi)}{\sin^2(\phi + \psi)} \right] \tag{7.79}$$

For a multiple-prism grating oscillator, the reflection losses at the multiple-prism beam expander are calculated using Equations 7.27 and 7.28. The polarization efficiency response of a typical holographic grating suitable for linewidth narrowing and tuning in dispersive tunable laser oscillators is illustrated in Figure 7.10.

Given that the polarization efficiency response of the grating clearly favors radiation polarized parallel to the plane of incidence, and the fact that the natural deployment of the multiple-prism beam expander (see Figure 7.13) also induces polarization parallel to the plane of incidence, it is logical to deploy the gain medium as illustrated in Figure 7.26, that is, with the windows at an angle, relative to the orthogonal to the optical axis, and the trapezoid parallel to the plane of incidence. As described in Chapter 5, the closer this angle gets to the Brewster angle, the greater the preference for radiation polarized parallel to the plane of incidence. Finally, the orientation of the polarization of the excitation laser should be selected so that the single-pass emission is compatible with the polarization preference of the architecture of the oscillator. In this particular case, the pump laser radiation should be selected to yield unpolarized single-pass emission as shown in Figure 7.26b.

Using the approach described here, Duarte and Piper (1984a) demonstrated SLM emission ~100% polarized parallel to the plane of incidence in a high-pulse-repetition-frequency copper laser-pumped dye laser. Polarization matching of the excitation laser to the polarization preference of the multiple-prism grating oscillator yields an increase of 15% in the laser conversion efficiency (Duarte and Piper 1982b). It should be indicated that although a specific example was used to describe the idea of polarization matching, the simple principles described are applicable to any laser system and even to extracavity optics for the efficient transmission of laser radiation.

7.6 DESIGN OF EFFICIENT NARROW-LINEWIDTH TUNABLE LASER OSCILLATORS

The design of efficient optically pumped high-power pulsed tunable laser oscillators follows a well-defined series of stages that are outlined in chronological order.

1. Select the most efficient pump laser for the given medium. In the case of molecular gain media, the efficiency follows approximately the ratio (Shank 1975):

$$\eta \approx \frac{\lambda_e}{\lambda_p} \qquad (7.80)$$

where:
   λ_e is the emission wavelength
   λ_p is the wavelength of the pump source
In addition to better efficiency, in the case of visible dye lasers, an excitation source with a wavelength close to the emission wavelength enhances significantly the lifetime of the gain medium. Specifically, the lifetime of rhodamine dye solutions using copper vapor laser excitation is vastly superior to the lifetime of the same molecular medium under the excitation of ultraviolet lasers.

2. Determine the energy density threshold for optical damage of the gain medium.

3. Select the geometry of the gain medium. A trapezoid geometry is recommended to eliminate internal reflections and thus facilitate the frequency control with the intracavity optics.

4. Once the approximate cavity length has been determined, select the dimensions of the beam waist (w) at the gain region adequate to yield TEM$_{00}$ emission. This could be done applying Equations 7.1 through 7.3 or experimentally. An important issue here is not to exceed the energy density threshold for optical damage of the gain medium. For a given cavity length, verify experimentally that TEM$_{00}$ emission is present.

5. Select the cavity architecture that best matches the application. Here, issues of compactness, simplicity, and efficiency play an important role.

6. Select an efficient high-density diffraction grating suitable for the required tuning range. Use Equations 7.14 through 7.17 to determine the dispersion and Equations 7.40 and 7.41 to calculate the tuning range of a given grating.

7. If intracavity beam expansion is required, select the appropriate method of expansion. For telescopic beam expansion, use the matrix equations in Chapter 6 to design an appropriate two-dimensional beam expander. If a multiple-prism beam expander is selected, use Equation 7.26 to design the multiple-prism array and Equations 7.27 and 7.28 to determine its transmission efficiency. The beam expander selected should have polarization characteristics compatible with those of the diffraction grating.

8. Perform a return-pass linewidth calculation using Equation 7.25. Adjust the cavity length to satisfy the criterion established in the inequality $\Delta v \leq \delta v$.

9. If the dispersion of the cavity is not sufficient to satisfy the criterion in the inequality $\Delta v \leq \delta v$, either redesign the oscillator to attain a shorter cavity or insert an intracavity etalon. The etalon can be designed using Equations 7.19 through 7.22.
10. Depending on the optical components integrating the cavity, select the method of wavelength tuning.
11. Determine the polarization preference of the cavity and compare it to the polarization of the excitation source. Perform polarization matching if necessary.
12. Determine the ρ_{ASE}/ρ_l ratio. If necessary, replace the output coupler mirror by a polarizer output coupler mirror as illustrated in Figure 7.13.
13. Optimize alignment and measure the conversion efficiency, tuning range, $\Delta\theta$, and $\Delta\lambda$. If SLM oscillation is not observed, either decrease the cavity length or increase the intracavity dispersion.

7.6.1 Useful Axioms for the Design of Narrow-Linewidth Tunable Laser Oscillators

The design of efficient dispersive narrow-linewidth tunable laser oscillators does benefit from the careful application of a number of well-defined, and specific, rules of physics that can be classified as follows:

1. The longer the cavity and the narrower the beam waist, the better the beam quality of the laser emission, or

$$\left|\langle x|s\rangle\right|^2 = \sum_{j=1}^{N} \Psi(r_j)^2 + 2\sum_{j=1}^{N} \Psi(r_j)\left[\sum_{m=j+1}^{N} \Psi(r_m)\cos(\Omega_m - \Omega_j)\right]$$

2. The larger the optical length of the cavity, the lower the beam divergence, or

$$\Delta\theta_R = \frac{\lambda}{\pi w}\left[1 + \left(\frac{L_\mathcal{R}}{B_R}\right)^2 + \left(\frac{A_R L_\mathcal{R}}{B_R}\right)^2\right]^{1/2}$$

3. For diffraction-limited TEM_{00} emission, the narrower the beam waist w, the larger the beam divergence, or

$$\Delta\theta = \frac{\lambda}{\pi w}$$

4. The larger the beam magnification, the larger the intracavity dispersion, and the narrower the linewidth, or

$$\Delta\lambda = \Delta\theta_R\left(RM\ \nabla_\lambda\Theta_G + R\ \nabla_\lambda\Phi_P\right)^{-1}$$

5. Also, from the previous axiom, the lower the beam divergence, the narrower the linewidth.

6. From the second and fourth axioms, the larger the number of intracavity passes, the lower the beam divergence and the narrower the linewidth.
7. The shorter the cavity, the longer the longitudinal mode spacing, or

$$\delta v = \frac{c}{\delta x}$$

These axioms clearly illustrate that some of the design parameters have a competing effect on the overall physics. The task of the designer is to apply these principles in a balanced approach to optimize the beam divergence and linewidth performance in a compact cavity architecture.

7.7   NARROW-LINEWIDTH OSCILLATOR-AMPLIFIERS

The dispersive tunable laser oscillators so far described yield SLM emission at very low levels of ASE. However, for many applications, high energies or high-average powers are required. For that purpose, the exquisite emission from the oscillator must be amplified by one or several amplification stages. A review on this subject, and its literature, is given by Duarte (1990b). Here, the focus will be on the fundamentals and the performance of various representative systems.

7.7.1   Laser-Pumped Narrow-Linewidth Oscillator-Amplifiers

Amplification of coherent optical radiation in laser-excited systems is well illustrated by the configurations developed for dye lasers. Two of the most interesting features of these systems are that amplification is performed in a *single pass* and that several stages of amplification are often employed. As such, given that lasing in these systems occurs in the nanosecond regime, it is important to synchronize the arrival of the oscillator pulse with the excitation of the amplifier. This is arranged by allowing the excitation geometry to delay the pump pulse as illustrated in Figure 7.27. An additional aspect important to the design of multistage oscillator-amplifier systems is the geometrical matching of the oscillator, or preamplified, beam with the focused excitation laser at the corresponding amplifier stage. This helps to maintain the cumulative ASE at low levels. For optimum efficiency, proper distribution of the pump energy is required with only a fraction (often less than a 5%) of it being used to excite the oscillator. Correct polarization matching is also important.

The performances of illustrative multistage oscillator-amplifier laser systems are listed in Table 7.3. Bos (1981) reports 6% efficiency at the oscillator, 20% at the pre-amplifier, and 60% ant the amplifiers. The overall gain factor is about 229. The copper vapor laser-pumped dye laser reported from Lawrence Livermore (Bass et al. 1992) operates at a pulse repetition frequency of 13.2 kHz and comprises several master oscillator (MO) power amplifier (MOPA) chains performing at a 50%–60% overall conversion efficiency. The MOs are of the MPL grating class and incorporate an intracavity etalon, and each MOPA chain includes three to four amplifiers in series.

Although most oscillator-amplifiers systems considered here utilize high-performance pulsed MOs, the alternative of semiconductor laser oscillators, lasing in the CW regime, is also available as demonstrated by Farkas and Eden (1993).

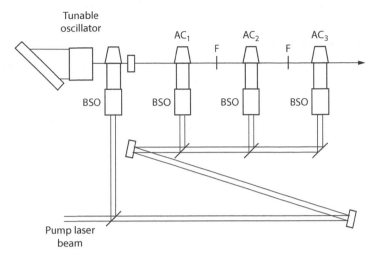

FIGURE 7.27   Multiple-stage single-pass laser amplification.

TABLE 7.3
Laser-Pumped Narrow-Linewidth Oscillator-Amplifiers

Oscillator	λ (nm)	Δv	Amplification Stages	Gain	Energy[a] (mJ)	Power[b] (kW)	η (%)	Reference
Telescopic[c]	590	320 MHz	3	229	165		55	Bos (1981)
HMPGI	440	650 MHz	2	~700	3.5		~9	Dupre (1987)
MPL[c]	~590	0.5–5 GHz	3–4[d]		190	2.5[e]	~60	Bass et al. (1992)

[a]   Output energy per pulse.
[b]   Output average power.
[c]   Includes intracavity etalon.
[d]   At each of four amplification chains.
[e]   At a prf of 13.2 kHz.

These authors used a five-stage dye laser amplification system to produce pulses of 1.2 mJ at 786 nm with $\Delta v = 118$ MHz.

7.7.2   NARROW-LINEWIDTH MO FORCED OSCILLATORS

The MO is composed of narrow-linewidth dispersive laser oscillators already discussed. The forced oscillator (FO), however, is an amplifier stage comprising a gain region within a resonator as depicted in Figure 7.28. The resonator of the amplifier stage can be an unstable resonator. Several aspects are rather critical to the efficient frequency locking of these configurations. First, the alignment of the MO relative to the FO must be concentric. Second, there are stringent requirements for the timing of the excitation that impose arrival of the MO pulse at the onset of the FO

TABLE 7.4
Narrow-Linewidth MOFOs

MO[a]	λ (nm)	Δv (MHz)	FO[b]	Gain	Energy[c] (mJ)	Reference
Two etalons	589	346	Flat-mirror cavity		300	Flamant and Maillard (1984)
MPL	590	≤375	Unstable resonator	~51	600	Duarte and Conrad (1987)
CWDL[d]		80	Ring cavity		50	Blit et al. (1977)

[a] MO configuration.
[b] FO configuration.
[c] Output energy per pulse.
[d] CW dye laser.

pulse buildup. Efficient frequency locking also occurs. Optimum lasing is achieved when the emission wavelength of the MO is tuned to the central wavelength of the gain spectrum of the FO. The performance of some representative MOFO systems is described in Table 7.4.

The FO cavity depicted in Figure 7.28 is configured after a Cassegrainian telescope and is cataloged as a confocal *unstable resonator* of the positive branch (Siegman 1986). Here, the radius of curvature of the large concave mirror is R_2, whereas the radius of the small mirror is R_1 and has a negative value. The magnification of the resonator is given by (Siegman 1986)

$$M = -\frac{R_2}{R_1} \tag{7.81}$$

and its length is

$$L = \frac{(R_1 + R_2)}{2} \tag{7.82}$$

As discussed by Siegman (1986), the condition for oscillation in the unstable regime is satisfied by

$$\left| \frac{A + D}{2} \right| > 1 \tag{7.83}$$

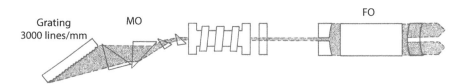

FIGURE 7.28 Master oscillator Forced oscillator laser configuration. (Reproduced from Duarte, F.J., and Conrad, R.W., *Appl. Opt.*, 26, 2567–2571, 1987. With permission from the Optical Society.)

where:

A and D are the matrix elements introduced in Chapter 6

For the resonator depicted in Figure 7.27, $R_1 = -2$ m and $R_2 = 4$ m, so that $M = 2$, $L = 1$ m, and the condition for lasing in the unstable regime is satisfied.

In Chapter 6, it is mentioned that narrow-linewidth multiple-prism grating oscillators can lase in the unstable regime in the presence of thermal lensing at the gain medium (Duarte et al. 1997). Finally, although the focus given here is on high-power pulsed lasers in the liquid phase, the principles illustrated apply to lasers in general.

7.8 DISCUSSION

In this section, a concept previously introduced in Section 7.2 is discussed further. Besides the obvious practical significance and advantages of tunable narrow-linewidth laser emission, there is an important physical dimension to this topic. Raw unrefined broadband high-power multiple-transverse-mode emission, with each of those transverse modes containing a multitude of longitudinal modes, is a manifestation of entropy in directed energy. Refining that broadband emission to a TEM_{00} and then to a SLM is a transparent example of entropy reduction in nature, which is beautifully captured by the cavity linewidth equation:

$$\Delta\lambda = \Delta\theta \left(\frac{\partial\theta}{\partial\lambda} \right)^{-1}$$

which tells the designer that to produce very pure narrow-linewidth emission, with a minimum of entropy, the beam divergence must be minimized and the dispersion of the cavity must be augmented. The word *dispersion* here means the continuous orderly arrangement of wavelengths in the spectrum. High intracavity dispersion leads to the emission of indistinguishable photons in a narrow-linewidth distribution. In other words, intracavity dispersion reduces entropy.

PROBLEMS

7.1 For a laser, emitting at $\lambda = 590$ nm, with a beam waist of $w = 100\ \mu$m and a cavity length of 10 cm, calculate the Fresnel number. Comment on the likely beam profile of this laser.

7.2 For a dispersive oscillator lasing with a TEM_{00} beam profile at $\lambda = 590$ nm and a cavity length of 10 cm, calculate the longitudinal mode spacing. Assuming that the calculated single-return-pass dispersive linewidth is $\Delta\nu = 500$ MHz, determine whether this oscillator design is likely to yield an SLM emission, given that the temporal pulse is known to be 5 ns long at full width at half maximum.

7.3 For a dispersive laser oscillator yielding DLM emission with $\delta\nu = 1$ GHz, design a suitable etalon to restrict oscillation to an SLM. Assume a fairly high-surface finesse and neglect the aperture finesse. Select a suitable reflectivity to allow maximum transmission while restricting oscillation to an SLM.

7.4 For a multiple-prism grating laser oscillator with $w = 117$ μm and $M = 100$, calculate the single-return-pass linewidth for a 5 cm grating with 3000 lines/mm. Assume $\lambda = 590$ nm, a diffraction-limited laser beam, and that the multiple-prism expander was designed to yield zero dispersion at this wavelength.

7.5 For an optimized tunable laser oscillator with a cavity length of 50 mm and lasing in an SLM at $\lambda = 590$ nm, calculate the wavelength shift due to a decrease of 50 μm in cavity length.

7.6 Design, step by step, a multiple-prism grating oscillator capable of yielding an SLM lasing at $\lambda = 590$ nm. Assume that a beam waist of $w = 100$ μm can be attained at the gain region and that a 5 cm grating is available which has 3300 lines/mm. Use a double-prism beam expander and configure it to yield zero dispersion at the given wavelength. Select a cavity length providing a longitudinal mode spacing approximately equal to the return-pass linewidth.

8 Nonlinear Optics

8.1 INTRODUCTION

The subject of polarization as related to reflection and transmission in isotropic homogeneous optical media, such as optical glass, was considered via Maxwell's equations in Chapter 5. Here, we consider the subject of propagation and polarization in crystalline media that gives origin to the subject of nonlinear optics. The brief treatment given here is at an introductory level and designed only to highlight the main features relevant to frequency conversion. For a detailed treatment on the subject of nonlinear optics, the reader is referred to a collection of books on nonlinear optics including Bloembergen (1965), Baldwin (1969), Shen (1984), Yariv (1985), Mills (1991), and Boyd (1992). Besides some revision and update on the subject of optical clockworks, this chapter remains fairly much as in its original version published in 2003.

8.1.1 INTRODUCTION TO NONLINEAR POLARIZATION

For propagation in an isotropic media, the polarization P is related to the electric field by the following identity:

$$P = \chi^{(1)} E \tag{8.1}$$

where:
$\chi^{(1)}$ is known as the *electric susceptibility*

In a crystal, the propagating field induces a polarization that depends on the direction and magnitude of this field, and the simple definition given in Equation 8.1 must be extended to include the second- and third-order susceptibilities so that

$$P = \chi^{(1)} E + \chi^{(2)} E^2 + \chi^{(3)} E^3 + \cdots \tag{8.2}$$

Second harmonic generation, sum-frequency generation, and optical parametric oscillation depend on $\chi^{(2)}$, whereas third-harmonic generation depends on $\chi^{(3)}$.

The second-order nonlinear polarization $P^{(2)} = \chi^{(2)} E^2$ can be expressed in more detail using

$$E(t) = E_1 e^{-i\omega_1 t} + E_2 e^{-i\omega_2 t} + \cdots \tag{8.3}$$

so that (Boyd 1992)

$$P^{(2)} = \chi^{(2)} \left(\begin{array}{l} E_1^2 e^{-2i\omega_1 t} + E_2^2 e^{-2i\omega_2 t} + 2E_1 E_2 e^{-i(\omega_1+\omega_2)t} \\ + 2E_1 E_2^* e^{-i(\omega_1-\omega_2)t} + 2E_1^* E_2 e^{-i(\omega_1+\omega_2)t} + \cdots \end{array} \right)$$
$$+2\chi^{(2)} \left(E_1 E_1^* + E_2 E_2^* \right)$$

(8.4)

The first two terms of this equation relate to second harmonic generation, the third term to sum-frequency generation, and the fourth term to difference-frequency generation.

Nonlinear susceptibility is described using tensors, which for the second order take the form of $\chi_{ijk}^{(2)}$. In shorthand notation, these are described by

$$d_{ijk} = \tfrac{1}{2} \chi_{ijk}^{(2)}$$

(8.5)

In Table 8.1, second-order nonlinear susceptibilities are listed for some well-known crystals.

Identities useful in this chapter are

$$k_m = \frac{n_m \omega_m}{c}$$

(8.6)

$$k_m = \frac{2\pi n_m}{\lambda_m}$$

(8.7)

and

$$n_m = (\varepsilon \omega_m)^{1/2}$$

(8.8)

TABLE 8.1
Second-Order Nonlinear Optical Susceptibilities

Crystal	$d_{il} = \tfrac{1}{2}\chi^{(2)}$	Reference
ADP	$d_{36} = 0.53$	Barnes (1995)
KDP	$d_{36} = 0.44$	Barnes (1995)
LiNbO$_3$	$d_{22} = 2.76$, $d_{31} = -5.44$	Barnes (1995)
BBO	$d_{22} = 2.22$, $d_{31} = 0.16$	Barnes (1995)
KTP	$d_{31} = 2.5$, $d_{32} = 4.4$, $d_{33} = 16.9$, $d_{24} = 3.6$, $d_{15} = 1.9$	Orr et al. (1995)
AgGaS$_2$	$d_{36} = 13.5$	Orr et al. (1995)
AgGaSe$_2$	$d_{36} = 33$	Orr et al. (1995)

Note: Units of d_{il} are in 10^{-12} m V^{-1}. The d_{il} matrix element is a contracted notation for d_{ijk}; see, for example, Boyd, R.W., *Nonlinear Optics*, Academic Press, New York, 1992.

8.2   GENERATION OF FREQUENCY HARMONICS

In this section, a basic description of second harmonic, sum-frequency, and difference-frequency generation is given. The difference-frequency generation section is designed to describe some of the salient aspects of optical parametric oscillation.

8.2.1   SECOND HARMONIC AND SUM-FREQUENCY GENERATION

Previously, Maxwell's equations were applied to describe propagation in isotropic linear optical media. Here, the propagation of electromagnetic radiation in crystals is considered from a practical perspective consistent with the previous material on polarization.

Maxwell's equations in the Gaussian system of units are given by (Born and Wolf 1999)

$$\nabla \cdot \boldsymbol{B} = 0 \tag{8.9}$$

$$\nabla \cdot \boldsymbol{E} = 4\pi\rho \tag{8.10}$$

$$\nabla \times \boldsymbol{H} = \frac{1}{c}\left(\frac{\partial \boldsymbol{D}}{\partial t} + 4\pi \boldsymbol{j}\right) \tag{8.11}$$

$$\nabla \times \boldsymbol{E} = -\frac{1}{c}\frac{\partial \boldsymbol{B}}{\partial t} \tag{8.12}$$

For a description of propagation in a crystal, we adopt the approach of Boyd (1992) and further consider a propagation medium characterized by $\rho = 0, \boldsymbol{j} = 0$, and $\boldsymbol{B} = \boldsymbol{H}$. The nonlinearity of the medium introduces

$$\boldsymbol{D} = \boldsymbol{E} + 4\pi \boldsymbol{P} \tag{8.13}$$

As done in Chapter 5, taking the curl of both sides of Equation 8.12 and using Equation 8.13 lead to

$$\nabla \times \nabla \times \boldsymbol{E} = -c^{-2}\left(\nabla_t^2 \boldsymbol{E} + 4\pi \nabla_t^2 \boldsymbol{P}\right) \tag{8.14}$$

which is the generalized wave equation for nonlinear optics. Here, $\nabla_t^2 = (\partial^2/\partial t^2)$.

Following Boyd (1992), it is useful to provide a number of definitions starting by separating the polarization into its linear and nonlinear components so that

$$\boldsymbol{P} = \boldsymbol{P}_L + \boldsymbol{P}_{NL} \tag{8.15}$$

followed by the separation of the displacement into

$$\boldsymbol{D} = \boldsymbol{D}_L + 4\pi \boldsymbol{P}_{NL} \tag{8.16}$$

where:

$$\boldsymbol{D}_L = \boldsymbol{E} + 4\pi \boldsymbol{P}_L \tag{8.17}$$

Using this definition, the nonlinear wave equation can be rewritten as

$$\nabla \times \nabla \times E = -c^{-2}\left(\nabla_t^2 D_L + 4\pi\nabla_t^2 P_{NL}\right) \tag{8.18}$$

In Chapter 5, for an isotropic material, we have seen that

$$D_L = \varepsilon E \tag{8.19}$$

For the case of a crystal, this definition can be modified to

$$D_L(r,t) = \varepsilon(\omega) \cdot E(r,t) \tag{8.20}$$

which includes a real frequency-dependent dielectric tensor. Using Equation 8.20, the nonlinear wave equation can be restated as (Armstrong et al. 1962)

$$\nabla \times \nabla \times E(r,t) = -c^{-2}\left[\varepsilon(\omega) \cdot \nabla_t^2 E(r,t) + 4\pi\nabla_t^2 P_{NL}(r,t)\right] \tag{8.21}$$

where:

$$E(r,t) = E(r)e^{-i\omega t} + \cdots \tag{8.22}$$

and

$$P_{NL}(r,t) = P_{NL}(r)e^{-i\omega t} + \cdots \tag{8.23}$$

Now, with the nonlinear wave equation established, we proceed to describe the process of second harmonic generation or frequency doubling. This is illustrated schematically in Figure 8.1 and consists in the basic process of radiation of ω_1 incident on a nonlinear crystal to yield collinear output radiation of frequency $\omega_2 = 2\omega_1$. We proceed as done in Chapter 5 using the identity

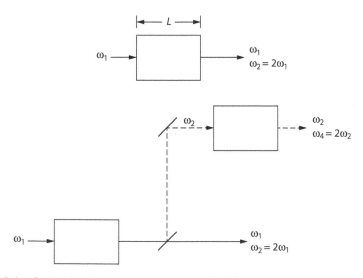

FIGURE 8.1 Optical configuration for frequency doubling generation.

$$\nabla \times \nabla \times E = \nabla \ \nabla \cdot E - \nabla^2 E \qquad (8.24)$$

The wave equation can be restated in scalar form as

$$\nabla^2 E_m(z,t) = -c^{-2}\left[\varepsilon(\omega_m)\nabla_t^2 E_m(z,t) + 4\pi\nabla_t^2 P_m(z,t)\right] \qquad (8.25)$$

After Boyd (1992), we use the following expressions for $m = 2$:

$$E_m(z,t) = A_m(z)e^{ik_m z}e^{-i\omega_m t} +\cdots \qquad (8.26)$$

$$E(z,t) = E_1(z,t) + E_2(z,t) \qquad (8.27)$$

$$P_m(z,t) = P_m(z)e^{-i\omega_m t} +\cdots \qquad (8.28)$$

$$P_1(z) = 4dA_2 A_1^* e^{i(k_2 - k_1)z} \qquad (8.29)$$

$$P_2(z) = 2dA_1^2 e^{ik_1 z} \qquad (8.30)$$

$$P(z,t) = P_1(z,t) + P_2(z,t) \qquad (8.31)$$

Following differentiation and substitution into the wave equation, the $\partial^2 A_1 / \partial z^2$ and $\partial^2 A_2 / \partial z^2$ terms are neglected so that the coupled amplitude equations can be expressed as (Boyd 1992)

$$\frac{dA_1}{dz} = i\left(\frac{8\pi d\omega_1^2}{k_1 c^2}\right)A_1^* A_2 e^{-i\Delta kz} \qquad (8.32)$$

and

$$\frac{dA_2}{dz} = i\left(\frac{4\pi d\omega_2^2}{k_2 c^2}\right)A_1^2 \ e^{i\Delta kz} \qquad (8.33)$$

where:

$$\Delta k = 2k_1 - k_2 \qquad (8.34)$$

Integration of Equation 8.33 leads to

$$A_2 A_2^* = \left(\frac{4\pi d\omega_2^2}{kc^2}\right)^2 A_1^4 L^2 \left[\frac{\left(\sin^2 L\Delta k/2\right)}{\left(L\Delta k/2\right)^2}\right] \qquad (8.35)$$

The above equation illustrates the nonlinear dependence of the frequency-doubled output on the input signal and indicates its relation to the $(L\Delta k/2)$ parameter. This dependence implies that conversion efficiency decreases significantly as $(L\Delta k/2)$ increases. The distance

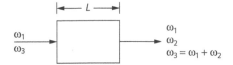

FIGURE 8.2 Optical configuration for sum-frequency generation.

$$L_c = \frac{2}{\Delta k} \tag{8.36}$$

is referred to as the *coherence length* of the crystal and provides a measure of the length of the crystal necessary for the efficient generation of second-harmonic radiation.

Sum-frequency generation is outlined in the third term of Equation 8.4 and involves the interaction of radiation at two different frequencies in a crystal to produce radiation at a third distinct frequency. This process is illustrated schematically in Figure 8.2 and consists in the normal incidence radiation of ω_1 and ω_2 onto a nonlinear crystal to yield collinear output radiation of frequency $\omega_3 = \omega_1 + \omega_2$. Using the appropriate expressions for $E_m(z,t)$ and $P_m(z,t)$, in the wave equation, it can be shown that (Boyd 1992)

$$\Delta k = k_1 + k_2 - k_3 \tag{8.37}$$

and the output intensity again depends on $\operatorname{sinc}^2(L\Delta k/2)$.

The ideal condition of *phase matching* is achieved when

$$\Delta k = 0 \tag{8.38}$$

and it offers the most favorable circumstances for a high conversion efficiency. When this condition is not satisfied, there is a strong decrease in the efficiency of sum-frequency generation.

8.2.2 DIFFERENCE-FREQUENCY GENERATION AND OPTICAL PARAMETRIC OSCILLATION

The process of difference-frequency generation is outlined in the fourth term of Equation 8.4 and involves the interaction of radiation at two different frequencies in a crystal to produce radiation at a third distinct frequency. This process is illustrated schematically in Figure 8.3 and consists in the normal incidence radiation of ω_1 and ω_3 onto a nonlinear crystal to yield collinear output radiation of frequency $\omega_2 = \omega_3 - \omega_1$.

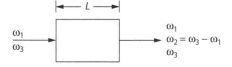

FIGURE 8.3 Optical configuration for difference-frequency generation.

Assuming that ω_3 is the frequency of a high-intensity pump laser beam, which remains undepleted during the excitation process, then A_3 can be considered a constant, and using an analogous approach to that adopted in Section 8.2.1, it is found that (Boyd 1992)

$$\frac{dA_1}{dz} = i\left(\frac{8\pi d\omega_1^2}{k_1 c^2}\right) A_3 A_2^* e^{i\Delta kz} \tag{8.39}$$

$$\frac{dA_2}{dz} = i\left(\frac{8\pi d\omega_2^2}{k_2 c^2}\right) A_3 A_1^* e^{i\Delta kz} \tag{8.40}$$

$$\frac{dA_3}{dz} = 0 \tag{8.41}$$

where:

$$\Delta k = k_3 - k_2 - k_1 \tag{8.42}$$

If the nonlinear crystal involved in the process of frequency difference is deployed, and properly aligned at the propagation axis of an optical resonator, as illustrated in Figure 8.4, then the intracavity intensity can build to very high values. This is the essence of an *optical parametric oscillator* (OPO). Early papers on OPOs are those of Giordmaine and Miller (1965), Akhmanov et al. (1966), Byer et al. (1968), and Harris (1969). Recent reviews are given by Barnes (1995) and Orr et al. (2009).

In the OPO literature, ω_3 is known as the *pump* frequency, ω_1 as the *idler* frequency, and ω_2 as the *signal* frequency. Thus, Equation 8.42 can be restated as

$$\Delta k = k_P - k_S - k_I \tag{8.43}$$

Equations 8.39 and 8.40 can be used to provide equations for the signal under various conditions of interest. For example, for the case when the initial idler intensity is zero, and $\Delta k \approx 0$, it can be shown that

$$A_S(L)A_S^*(L) \approx \tfrac{1}{4} A_S(0)A_S^*(0) \left(e^{\gamma L} + e^{-\gamma L}\right)^2 \tag{8.44}$$

where:
$A_S(0)$ is the initial amplitude of the signal

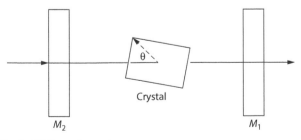

FIGURE 8.4   Basic OPO configuration.

Here, the parameter γ is defined as (Boyd 1992)

$$\gamma = \left(64\pi^2 d^2 \omega_I^2 \omega_S^2 k_I^{-1} k_S^{-1} c^{-4} |A_P|^2\right)^{1/2} \tag{8.45}$$

Equation 8.44 indicates that for the ideal condition of $\Delta k \approx 0$, the signal experiences an exponential gain as long as the pump intensity is not depleted.

Frequency selectivity in pulsed OPOs has been studied in detail by Brosnan and Byer (1979) and Barnes (1995). Wavelength tuning by angular and thermal means is discussed by Barnes (1995). Considering the frequency difference

$$\omega_S = \omega_P - \omega_I \tag{8.46}$$

and Equation 8.43, it can be shown that for the case of $\Delta k \approx 0$ (Orr et al. 1995)

$$\lambda_S \approx \frac{\lambda_P(n_S - n_I)}{(n_P - n_I)} \tag{8.47}$$

which illustrates the dependence of the signal wavelength on the refractive indices. An effective avenue to change the refractive index is to vary the angle of the optical axis of the crystal, relative to the optical axis of the cavity, as indicated in Figure 8.4. For instance, Brosnan and Byer (1979) report that changing this angle from 45° to 49°, in a Nd:YAG laser-pumped $LiNbO_3$ OPO, tunes the wavelength from ~2 μm to beyond 4 μm. The angular dependence of refractive indices in uniaxial birefringent crystals is discussed by Born and Wolf (1999).

It should be mentioned that the principles discussed in Chapters 4 and 7 can be applied toward the tuning and linewidth narrowing in OPOs. However, there are some unique features of nonlinear crystals that should be considered in some detail. Central to this discussion is the issue of phase matching or allowable mismatch. It is clear that a resonance condition exists around $\Delta k \approx 0$, and from Equation 8.44, it is seen that the output signal from an OPO can experience a large increase when this condition is satisfied. Thus, $\Delta k \approx 0$ is a desirable feature. Here, it should be mentioned that some authors define slightly differently what is known as *allowable mismatch*. For instance, Barnes (1995) defines it as

$$\Delta k = \frac{\pi}{L} \tag{8.48}$$

which is slightly broader than the definition given in Equation 8.36.

The discussion on frequency selectivity in OPOs benefits significantly by expanding Δk in a Taylor series (Barnes and Corcoran 1976) so that

$$\Delta k = \Delta k_0 + \left(\frac{\partial \Delta k}{\partial x}\right)\Delta x + \frac{1}{2!}\left(\frac{\partial^2 \Delta k}{\partial x^2}\right)\Delta x^2 + \cdots \tag{8.49}$$

Here, this process is repeated for other variables of interest:

$$\Delta k = \Delta k_0 + \left(\frac{\partial \Delta k}{\partial \theta}\right)\Delta \theta + \frac{1}{2!}\left(\frac{\partial^2 \Delta k}{\partial \theta^2}\right)\Delta \theta^2 + \cdots \tag{8.50}$$

$$\Delta k = \Delta k_0 + \left(\frac{\partial \Delta k}{\partial \lambda} \right) \Delta \lambda + \frac{1}{2!} \left(\frac{\partial^2 \Delta k}{\partial \lambda^2} \right) \Delta \lambda^2 + \cdots \qquad (8.51)$$

$$\Delta k = \Delta k_0 + \left(\frac{\partial \Delta k}{\partial T} \right) \Delta T + \frac{1}{2!} \left(\frac{\partial^2 \Delta k}{\partial T^2} \right) \Delta T^2 + \cdots \qquad (8.52)$$

Equating the first two series, and ignoring the second derivatives, it is found that (Barnes 1995)

$$\Delta \lambda = \Delta \theta \left(\frac{\partial \Delta k}{\partial \theta} \right) \left(\frac{\partial \Delta k}{\partial \lambda} \right)^{-1} \qquad (8.53)$$

This linewidth equation shows a dependence on the beam divergence, which is determined by the geometrical characteristics of the pump beam and the geometry of the cavity. It should be noted that this equation provides an estimate of the *intrinsic linewidth* available from an OPO in the absence of intracavity dispersive optics or injection seeding from external sources. Barnes (1995) reports that for a $AgGaSe_2$ OPO pumped by an Er:YLF laser, the linewidth is $\Delta \lambda = 0.0214$ µm at $\lambda = 3.82$ µm.

Introduction of the intracavity dispersive techniques described in Chapter 7 produces much narrower emission linewidths. A dispersive OPO is illustrated in Figure 8.5. For this oscillator, the multiple-return-pass linewidth is determined by

$$\Delta \lambda = \Delta \theta_R \left(RM \nabla_\lambda \Theta_G + R \nabla_\lambda \Phi_P \right)^{-1} \qquad (8.54)$$

where the various coefficients are defined in Chapter 7. It should be apparent that Equation 8.54 has its origin in

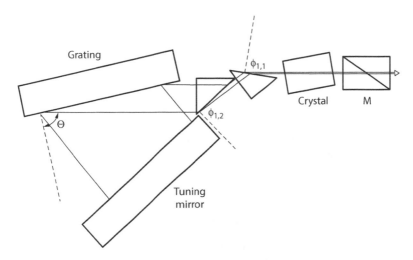

FIGURE 8.5 Dispersive OPO using a HMPGI grating configuration as described by Duarte. (Data from Duarte, F.J., *Tunable Laser Optics*, Elsevier Academic Press, New York, 2003.)

$$\Delta\lambda = \Delta\theta \left(\frac{\partial\theta}{\partial\lambda} \right)^{-1} \tag{8.55}$$

which is a simplified version of Equation 8.53. Hence, we have demonstrated a simple mathematical approach to arrive to the linewidth equation, which was previously derived using interferometric arguments in Chapter 2.

Using a dispersive cavity incorporating an intracavity etalon in a $LiNbO_3$ OPO excited by a Nd:YAG laser, Brosnan and Byer (1979) achieved a linewidth of $\Delta\nu = 2.25$ GHz. Also using a Nd:YAG-pumped $LiNbO_3$ OPO, and a similar interferometric technique, Milton et al. (1989) achieved single-longitudinal-mode emission at a linewidth of $\Delta\nu \approx 30$ MHz.

A further aspect illustrated by the Taylor series expansion is that equating the second and third series, it is found that (Duarte 2003)

$$\Delta\theta = \Delta T \left(\frac{\partial\Delta k}{\partial T} \right) \left(\frac{\partial\Delta k}{\partial\theta} \right)^{-1} \tag{8.56}$$

which indicates that the beam divergence is a function of temperature, which should be considered when contemplating thermal tuning techniques. Chapter 9 includes a section on the emission performance of various OPOs.

8.2.3 THE REFRACTIVE INDEX AS A FUNCTION OF INTENSITY

Using a Taylor series to expand an expression for the refractive index yields

$$n = n_0 + \left(\frac{\partial n}{\partial I} \right) I + \frac{1}{2!} \left(\frac{\partial^2 n}{\partial I^2} \right) I^2 + \cdots \tag{8.57}$$

Neglecting the second-order and higher terms, this expression reduces to

$$n = n_0 + \left(\frac{\partial n}{\partial I} \right) I \tag{8.58}$$

where:

n_0 is the normal weak-field refractive index defined in Chapter 13 for various materials

The quantity $(\partial n/\partial I)$ is not dimensionless and has units that are the inverse of the laser intensity or W^{-1} cm^2. Using polarization arguments, this derivative can be expressed as (Boyd 1992)

$$\frac{\partial n}{\partial I} = \frac{12\pi^2 \chi^3}{n_0^2(\omega)c} \tag{8.59}$$

This quantity is known as the second-order index of refraction and is traditionally referred to as n_2. Setting $(\partial n/\partial I) = n_2$, Equation 8.58 can be restated in its usual form as

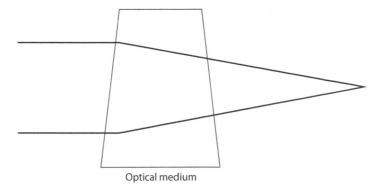

Optical medium

FIGURE 8.6 Simplified representation of self-focusing, due to $n = n_0 + n_2 I$, in an optical medium due to propagation of a laser beam with a near-Gaussian intensity profile.

$$n(\omega) = n_0(\omega) + n_2(\omega) I(\omega) \tag{8.60}$$

The change in refractive index, as a function of laser intensity, is known as the optical Kerr effect.

A well-known consequence of the optical Kerr effect is the phenomenon of *self-focusing*. This results from the propagation of a laser beam with a near-Gaussian spatial intensity profile since, according to Equation 8.60, the refractive index at the center of the beam is higher than the refractive index at the wings of the beam. This results in an intensity-dependent lensing effect as illustrated in Figure 8.6.

The phenomenon of self-focusing, or intensity-dependent lensing, is important in ultrafast lasers or femtosecond lasers (Diels 1990; Diels and Rudolph 2006), where it gives rise to what is known as Kerr lens mode locking (KLM). This is applied to spatially select the high-intensity mode-locked pulses from the background continuous-wave lasing. This can be simply accomplished by inserting an aperture near the gain medium to restrict lasing to the central, high-intensity, portion of the intracavity beam. This technique has become widely used in femtosecond laser cavities.

8.3 OPTICAL PHASE CONJUGATION

Optical phase conjugation is a technique that is applied to correct laser beam distortions, either intracavity or extracavity. A proof of the distortion correction properties of phase conjugation was provided by Yariv (1977) and is outlined here. Consider a propagating beam, in the $+z$ direction, represented by

$$E(r,t) = A_1(r)e^{-i(\omega t - kz)} + \cdots \tag{8.61}$$

and the scalar version of the nonlinear wave equation given in Equation 8.25, assuming that the spatial variations of ε are much larger than the optical wavelength. Neglecting the polarization term, one can write

$$\frac{\partial^2 A_1}{\partial z^2} + i2k\left(\frac{\partial A_1}{\partial z}\right) - \left(\frac{\varepsilon \omega^2}{c^2} + k^2\right)A_1 = 0 \tag{8.62}$$

The complex conjugate of this equation is

$$\frac{\partial^2 A_1^*}{\partial z^2} - i2k\left(\frac{\partial A_1^*}{\partial z}\right) - \left(\frac{\varepsilon\omega^2}{c^2} + k^2\right)A_1^* = 0 \qquad (8.63)$$

which is the same wave equation of a wave propagating in the $-z$ direction of the form:

$$E(r,t) = A_2(r)e^{-i(\omega t + kz)} + \cdots \qquad (8.64)$$

provided

$$A_2(r) = aA_1^*(r) \qquad (8.65)$$

where:

a is a constant

Here, the presence of a distorting medium is represented by the *real* quantity ε (Yariv 1977). This exercise illustrates that a wave propagating in the *reverse direction* of $A_1(r)$, and whose complex amplitude is *everywhere* the complex conjugate of $A_1(r)$, satisfies the same wave equation satisfied by $A_1(r)$. From a practical perspective, this implies that a phase-conjugate mirror (PCM) can generate a wave propagating in reverse, to the incident wave, whose amplitude is the complex conjugate of the incident wave. Thus, the wave fronts of the reverse wave coincide with those of the incident wave. This concept is illustrated in Figure 8.7.

A PCM, as depicted in Figure 8.8, is generated by a process called degenerate four-wave mixing (DFWM), which itself depends on $\chi^{(3)}$ (Yariv 1985). This process can be described considering plane wave equations of the form

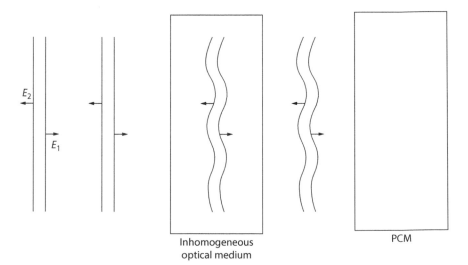

FIGURE 8.7   The concept of optical phase conjugation.

FIGURE 8.8   Basic phase-conjugated laser cavity.

$$E_m(r,t) = A_m(r)e^{-i(\omega t - k_m \cdot r)} + \cdots \qquad (8.66)$$

where:

$m = 1, 2, 3, 4$

k and r are vectors

Using these equations and the simplified equations for the four polarization terms (Boyd 1992),

$$P_1 = 3\chi^{(3)}\left(E_1^2 E_1^* + 2E_1 E_2 E_2^*\right) \qquad (8.67a)$$

$$P_2 = 3\chi^{(3)}\left(E_2^2 E_2^* + 2E_2 E_1 E_1^*\right) \qquad (8.67b)$$

$$P_3 = 3\chi^{(3)}\left(2E_3 E_1 E_1^* + 2E_3 E_2 E_2^* + 2E_1 E_2 E_4^*\right) \qquad (8.67c)$$

$$P_4 = 3\chi^{(3)}\left(2E_4 E_1 E_1^* + 2E_4 E_2 E_2^* + 2E_1 E_2 E_3^*\right) \qquad (8.67d)$$

into the generalized wave equation

$$\nabla^2 E_m(z,t) = -c^{-2}\left[\varepsilon(\omega_m)\nabla_t^2 E_m(z,t) + 4\pi\nabla_t^2 P_m(z,t)\right]$$

eventually lead to expressions for the amplitudes, which show that the generated field is driven only by the *complex conjugate* of the input amplitude.

An issue of practical interest is the representation of a PCM in transfer matrix notation, as introduced in Chapter 6. This problem was solved by Auyeung et al. (1979), who, using the argument that the reflected field is the conjugate replica of the incident field, showed that the *ABCD* matrix (see Chapter 6) is given by

$$\begin{pmatrix} A & B \\ C & D \end{pmatrix} = \begin{pmatrix} 1 & 0 \\ 0 & -1 \end{pmatrix} \qquad (8.68)$$

that should be compared to

$$\begin{pmatrix} A & B \\ C & D \end{pmatrix} = \begin{pmatrix} 1 & 0 \\ 0 & 1 \end{pmatrix} \qquad (8.69)$$

for a conventional optical mirror. A well-known nonlinear material suitable as a PCM is CS_2 (Yariv 1985). Fluctuations in the phase-conjugated signal generated by DFWM in sodium were investigated by Kumar et al. (1984).

8.4   RAMAN SHIFTING

Stimulated Raman scattering (SRS) is an additional, and very useful, tool to extend the frequency range of fixed frequency and tunable lasers. SRS, also known as Raman shifting, can be accomplished by focusing a TEM_{00} laser beam on to a nonlinear medium, such as H_2 (as illustrated in Figure 8.9), to generate emission at a series of wavelengths above and below the wavelength of the laser pump. The series of longer wavelength emissions are known as Stokes and are determined by (Hartig and Schmidt 1979)

$$\nu_{S_m} = \nu_P - m\nu_R \qquad (8.70)$$

where:

$\quad \nu_{S_m}$ is the frequency of a given Stokes
$\quad \nu_P$ is the frequency of the pump laser
$\quad \nu_R$ is the intrinsic Raman frequency
$\quad m = 1, \ 2, \ 3, \ 4 \ ...$ for successively higher Stokes

For the series of shorter anti-Stokes wavelengths,

$$\nu_{AS_m} = \nu_P + m\nu_R \qquad (8.71)$$

where:

$\quad \nu_{AS_m}$ is the frequency of a given anti-Stokes

It should be noted that ν_{S_1} and ν_{AS_1} are generated by the pump radiation, while these fields, in turn, generate ν_{S_2} and ν_{AS_2}. In other words, for $m = 2, \ 3, \ 4...$, ν_{S_m} and ν_{AS_m} are generated by $\nu_{S_{(m-1)}}$ and $\nu_{AS_{(m-1)}}$, respectively. Hence, the most intense radiation occurs for $m = 1$ with successively weaker emission for $m = 2, \ 3, \ 4...$ as depicted in Figure 8.10. For instance, efficiencies can decrease progressively from 37% (first Stokes), to 18% (second Stokes), to 3.5% (third Stokes) (Berik et al. 1985). For the H_2 molecule, $\nu_R \approx 124.5637663$ THz (or 4155 cm^{-1}) (Bloembergen 1967).

Using the wave equation and assuming solutions of the form

$$E_S(z,t) = A_s(z)e^{-i(\omega_S t - k_S z)} + \cdots \qquad (8.72)$$

and

$$E_P(z,t) = A_p(z)e^{-i(\omega_P t - k_P z)} + \cdots \qquad (8.73)$$

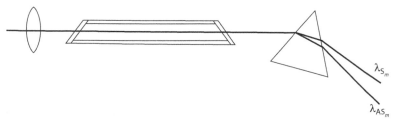

FIGURE 8.9   Optical configuration for H_2 Raman shifter. The output window and the dispersing prism are made of CaF_2.

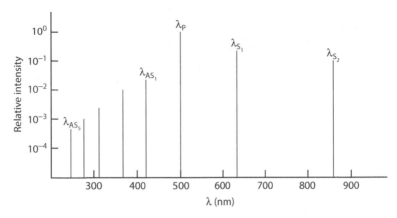

FIGURE 8.10   Stokes and anti-Stokes emission in H_2 for $\lambda_P = 500\,\text{nm}$.

it can be shown, using the fact that the Stokes polarization depends on $\chi^{(3)}E_P E_P^* E_S$, that the gain at the Stokes frequency depends on the intensity of the pump radiation, the population density, and the inverse of the Raman linewidth, among other factors (Trutna and Byer 1980). It is interesting to note that the Raman gain can be independent of the linewidth of the pump laser (Trutna et al. 1979). A detailed description on the mechanics of SRS is provided by Boyd (1992).

SRS in H_2 has been widely used to extend the frequency range of tunable lasers, such as dye lasers. This technique was first demonstrated by Schmidt and Appt (1972) using room-temperature hydrogen at a pressure of 200 atm. This is mentioned since, albeit simple, the use of pressurized hydrogen requires the use of stainless steel cells and detailed attention to safety procedures. Using a dye laser with an emission wavelength centered around 563 nm, Wilke and Schmidt (1978) generated SRS radiation in H_2 from the eight anti-Stokes (at 198 nm) to the third Stokes (at 2064 nm) at an overall conversion efficiency of up to 50%. Using the second harmonic of the dye laser, the same authors generated from the fourth anti-Stokes to the fifth Stokes, as illustrated in Table 8.2, at an overall conversion efficiency of up to 75%. Using a similar dye laser configuration, Hartig and Schmidt (1979) employed a capillary waveguide H_2 cell to generate tunable first, second, and third Stokes spanning in the 0.7–7 µm wavelength range.

Using a dye laser system incorporating an MPL grating oscillator and two stages of amplification, Schomburg et al. (1983) achieved generation up to the thirteenth anti-Stokes at 138 nm. Brink and Proch (1982) reported on a 70% conversion efficiency at the seventh anti-Stokes by lowering the H_2 temperature to 78°K. Hanna et al. (1985) reported on a 90% conversion efficiency to the first Stokes using an oscillator-amplifier configuration for SRS in H_2.

In addition to H_2, numerous materials have been characterized as SRS media (Bloembergen 1967; Yariv 1975). Other gaseous media include I_2 (Fouche and Chang 1972), Cs (Wyatt and Cotter 1980), Ba (Manners 1983), Sn and Ti (White and Henderson 1983; Ludewigt et al. 1984), and Pb (Marshall and Piper 1990). A review on SRS in optical fibers is given by Toulouse (2005).

TABLE 8.2
Tunable Raman Shifting in Hydrogen

Anti-Stokes λ Range (nm)	Tunable Laser[a] λ Range (nm)	Stokes λ Range (nm)[b]
$\lambda_4 \approx 192\ (\delta\lambda_4 \approx 5.8)^c$	$275 \leq \lambda \leq 287$	$309 \leq \lambda_1 \leq 326$
$\lambda_3 \approx 210\ (\delta\lambda_3 \approx 7.2)$		$355 \leq \lambda_2 \leq 378$
$\lambda_2 \approx 229\ (\delta\lambda_2 \approx 8.9)$		$418 \leq \lambda_3 \leq 450$
$\lambda_1 \approx 251\ (\delta\lambda_1 \approx 10.7)$		$505 \leq \lambda_4 \leq 550$
		$640 \leq \lambda_5 \leq 711$

Source: Wilke, V., and Schmidt, W., *Appl. Phys.*, 16, 151–154, 1978.
[a] Second-harmonic from a dye laser.
[b] Approximate values.
[c] Corresponds to a quoted range of 188.7 nm $\leq \lambda_1 \leq$ 194.5 nm. All other values are approximated.

8.5 OPTICAL CLOCKWORK

Perhaps the most well-known application of nonlinear optics in the laser optics field is in the generation of second, third, and fourth harmonics of some well-established laser sources, including the Nd:YAG laser. Table 8.3 lists the laser fundamental and its three harmonics. This frequency multiplication can be accomplished using nonlinear crystals such as KDP and ADP. Certainly, it should be apparent that the generation of frequency harmonics is not just limited to the Nd:YAG laser, but it is also practiced with a variety of laser sources including widely tunable lasers such as the Ti:Sappire laser and various fiber lasers (see Chapter 9).

One application that integrates various aspects of laser optics, including harmonic generation, is known as *optical clockwork* (Holzwarth et al. 2001). This involves the generation of a phase-locked white-light continuum for absolute frequency measurements. This is an idea originally outlined by Hänsch and colleagues in the mid- to late 1970s (Eckstein et al. 1978).

The basic tools in this technique are a stabilized femtosecond laser, a nonlinear crystal fiber capable of self-modulation, a stabilized narrow-linewidth laser, and a

TABLE 8.3
Harmonics of the $^4F_{3/2} - {}^4I_{11/2}$ Transition of the Nd:YAG Laser

Fundamental	Harmonics
$\nu \approx 2.82 \times 10^{14}$ Hz ($\lambda \approx 1064$ nm)	$2\nu \approx 5.64 \times 10^{14}$ Hz ($\lambda \approx 532$ nm)
	$3\nu \approx 8.46 \times 10^{14}$ Hz ($\lambda \approx 355$ nm)
	$4\nu \approx 1.13 \times 10^{15}$ Hz ($\lambda \approx 266$ nm)

frequency doubling crystal. Briefly, the concept consists in generating a periodic train of pulses, also known as a *comb* or *ruler*, with each pulse separated in the frequency domain by an interval Δ, for an entire optical octave. This is accomplished by focusing a high-intensity femtosecond laser beam onto a $\chi^{(3)}$ medium. This medium is a crystal fiber also known as a photonic crystal fiber whose refractive index behaves according to

$$n(t) = n_0 + n_2 I(t) \tag{8.74}$$

Propagation in such medium causes red spread at the leading edge of the pulse and a blue spread at the trailing edge of the pulse since the field experiences a time-dependent shift described by (Bellini and Hänsch 2000)

$$\Delta\omega(t) = -\frac{\omega_0 n_2 L}{c}\left[\frac{dI(t)}{dt}\right] \tag{8.75}$$

Thus, a high intensity of ~20 fs pulse focused on a $\chi^{(3)}$ medium, a few centimeters long, can give rise to a continuum (Holzwarth et al. 2001).

The frequency-stabilized, and broadened, pulse train is made collinear with the fundamental of the narrow-linewidth stabilized laser, to be measured, and its second harmonic (Diddams et al. 2000). The combined laser beam containing the pulse train, ν, and 2ν, is then dispersed by a grating and two detectors are combined to determine the frequency beating between the pulse train with ν and 2ν, thus determining the beat frequencies δ_1 and δ_2. From Figure 8.11, the frequency difference $(2\nu - \nu)$ is given by Diddams et al. (2000)

$$2\nu - \nu = n\Delta \pm (\delta_1 \pm \delta_2) \tag{8.76}$$

where:

$$\Delta = \frac{\upsilon_g}{2L} \tag{8.77}$$

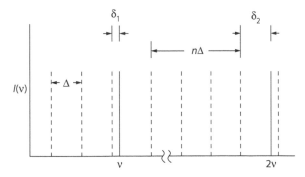

FIGURE 8.11 Schematics for determining the frequency difference $(2\nu - \nu)$ in the optical clockwork approach. This is a simplified version of the intensity versus frequency diagram considered by Diddams et al. (Adapted from Diddams, S.A., et al., *Phys. Rev. Lett.*, 84, 5102–5105, 2000.)

is determined by controlling L, which is the cavity length of the stabilized femtosecond laser (here, v_g is the group velocity). Using this method, Diddams et al. (2000) determined v for an $^{127}I_2$-stabilized Nd:YAG laser to be 281, 630, 111, and 740 kHz with an uncertainty of 50 Hz or 2×10^{-13}. More recently, further refinement in optical clockworks has resulted in optical frequency synthesizers capable of providing an upper limit for the measurement uncertainty of several parts in 10^{-18} (Diddams 2010). Applications for optical combs include next-generation optical clocks and precision spectroscopy. In astronomy, applications include calibration sources for the search of exoplanets and measurement of fundamental constants of the early universe (Diddams 2010).

PROBLEMS

8.1 Use Maxwell's equations to derive the generalized wave equation of nonlinear optics, that is, Equation 8.14.

8.2 Use Equations 8.39 and 8.40 to arrive at Equation 8.44 using the approximation $\Delta k \approx 0$.

8.3 Use the scalar form of the wave equation (Equation 8.25) to arrive at Equation 8.62.

8.4 Derive the linewidth equation for an OPO, that is, Equation 8.53.

8.5 Determine the wavelengths for the Stokes radiation at $m = 1$, 2, 3 and for the anti-Stokes radiation at $m = 1$, 2, 3, 4, 5 for H_2, given that the laser excitation is at $\lambda = 600$ nm.

9 Lasers and Their Emission Characteristics

9.1 INTRODUCTION

The aim of this chapter is to provide an utilitarian and succinct overview of the emission characteristics of lasers. Lasers included in this survey emit mostly in the ultraviolet–visible–near-infrared part of the spectrum and offer one, or more, of the following features:

1. Narrow linewidth emission
2. TEM_{00} beam profile
3. High pulsed powers, high average powers, or high continuous-wave (CW) powers

At this stage, it should be indicated that lasers that exhibit naturally a high degree of spectral coherence tend to be lasers that derive their emission from discrete atomic transitions. Examples of such coherent sources are found among the CW gas lasers, which also tend to emit laser beams with TEM_{00} characteristics. High-gain, high-power, lasers can be configured to yield both TEM_{00} beam characteristics and narrow-linewidth emission using a variety of optical architectures, as described in Chapter 7.

As discussed in a previous publication (Duarte 1995a), different types of lasers cover different spectral regions and offer different optimized modes of operation and emission. As such, different lasers should be considered from a perspective of *complementarity*. In this context, *it is the application itself that should determine the selection of a particular laser.* Among the characteristics that determine the suitability of a particular, pulsed or CW, laser for a given application are

1. Spectral region of emission
2. Wavelength tunability
3. Laser linewidth
4. Amplified spontaneous emission (ASE) level
5. Wavelength and linewidth stability
6. Beam profile
7. Pulse duration
8. Pulse energy
9. Average power
10. CW power

11. Physical and optical ruggedness
12. Physical dimensions
13. Cost and environment

An example of complementarity in the laser field can be provided considering the application of laser isotope separation. Here, the spectroscopic information can be obtained using either CW dye lasers or tunable external cavity diode lasers (see, e.g., Olivares et al. 2002). However, large-scale isotope separation requires high-power narrow-linewidth dye lasers efficiently excited by copper vapor lasers (CVLs) operating in the high-pulse repetition frequency (prf) regime (Broyer et al. 1984; Bass et al. 1992; Singh et al. 1994; Sugiyama et al. 1996). A further example of complementarity is provided by recent applications in laser cooling where the lasers of choice are external cavity tunable semiconductor lasers, with CW dye lasers being employed to provide highly coherent radiation in spectral regions unattainable with the compact diode lasers (Bradley et al. 2000).

For the purpose of this survey, high-power pulsed lasers are those lasers belonging to a class capable of generating pulsed powers starting in the tens of kilowatts regime. High-energy lasers are those lasers capable of generating at least tens of joules per pulse. Lasers classified as high average, or high CW, powers are those lasers capable of delivering powers starting at the several tens of watts level.

It should be noted that this survey should not be considered a historical introduction to the subject of lasers since references were mainly selected for the instructional value of their data. Furthermore, information about high-power gas lasers and high-power liquid lasers is included for three reasons: (1) the energy and power performance of some of these lasers has not yet been superseded by the output characteristics of recent semiconductor, or semiconductor-excited solid-state, laser alternatives; (2) it is important to emphasize the vast variety of laser materials from the gaseous to the solid state; and (3) technology can always surprise us with the revival of previous accomplishments.

9.2 GAS LASERS

The emission of high pulse energies, or high average powers, requires the rapid removal of heat. Gas lasers are ideally suited for the efficient removal of excess heat. In this section, representative gas lasers from the subclasses of molecular and atomic transitions are included.

9.2.1 PULSED MOLECULAR GAS LASERS

Among the most widely used high-power gas lasers are those that derive their radiation from molecular transitions. Table 9.1 includes these lasers and relevant emission characteristics. Excimer lasers can be excited either via e-beam technology or, more commonly, using electrical discharges. An excimer molecule, such as XeCl, does not readily exist in nature and is created by the interaction of energetic electrons with a gas mixture that includes a buffer gas that participates in the excitation process, such as He, and the components of the lasing molecules such as Xe and Cl.

TABLE 9.1

Molecular Ultraviolet and Visible Pulsed Gas Lasers

Laser	Transition	λ (nm)	prf[a] (Hz)	Bandwidth[b]	Reference[c]
KrF	$B^2\Sigma^+_{1/2}-X^2\Sigma^+_{1/2}$	248	200	10,500 GHz[d]	Loree et al. (1978)
		248		2,583 GHz[d]	Caro et al. (1982)
XeCl	$B-X$	308[e]	1,000	204 GHz	McKee (1985)
XeF	$B-X$	351[e]	200	187 GHz	Yang et al. (1988)
	$C-A$	482.5[e]		40 nm[d]	Tittel et al. (1986)
N_2	$C^3\Pi_u-B^3\Pi_g$	337.1	100	203 GHz	Woodward et al. (1973)
HgBr	$B^2\Sigma^+_{1/2}-X^2\Sigma^+_{1/2}$	502[e]	~100	918 GHz	Shay et al. (1981)

Source: Duarte, F.J., *Quantum Optics for Engineers*, CRC Press, New York, 2014.

[a] prf values do not represent absolute limits.

[b] Bandwidth, or tuning range, is given in either GHz units or nm (see Chapter 3).

[c] References relate to bandwidth, or tuning range, exclusively.

[d] Tuning range.

[e] Principal wavelength.

Discharge-excited excimer lasers are available in a variety of designs that provide an ample choice of performance parameters from joule-class pulse energies, at low repetition rates, to millijoule-type lasers operating at a prf of hundreds of hertz and approaching the kilohertz regime. Commercial excimer lasers offer average powers from a few watts to hundreds of watts. Pulse duration in excimer lasers is a function of the discharge configuration and the type of excimer. Most excimer lasers provide pulses in the 20–30 ns range; however, using inductance stabilized circuits, pulses as long as 180 ns in KrF and 250 ns in XeCl have been reported (Sze and Harris 1995). Excimer lasers tend to yield wide beams with multiple–transverse-mode structures. Adoption of unstable resonator techniques can be used to improve the beam quality and linewidth performance can be improved significantly using, for example, intracavity multiple-prism grating techniques (Ludewigt et al. 1987; Duarte 1991a). For a comprehensive review on excimer lasers, the reader should refer to Rhodes (1979). Sze and Harris (1995) provide an excellent discussion of tunable excimer lasers.

Nitrogen lasers are reliable, simple to build, and easy to operate. The N_2 laser emits via two electronic transitions: one is the $C^3\Pi_u - B^3\Pi_g$ transition, in the ultraviolet, and the other is the $B^3\Pi_g - A^3\Sigma^+_g$ transition, in the near infrared. In the ultraviolet system, three vibrational transitions, corresponding to $v' = 0 \rightarrow v'' = 0$, $v' = 0 \rightarrow v'' = 1$, and $v' = 1 \rightarrow v'' = 0$, are observed at 337.1, 357.7, and 315.9 nm, respectively (Willett 1974). Of these transitions, the one at 337.1 nm has the largest Franck–Condon factor. N_2 lasers emitting at this wavelength often deliver energies of about 10 mJ in the pulsed regime, which is approximately 10 ns wide (full width at half maximum [FWHM]) at prfs from a few hertz up to 100 Hz.

CO_2 lasers emit in the infrared mainly in the 10 μm region. These lasers can be configured into powerful sources of coherent emission in both the pulsed and the CW regime. Information about the transitions of CO_2 lasers was provided by Willett (1974). In the case of pulsed TEA CO_2 lasers, emission occurs via the $P20$ $(00°1 - 10°0)$ transition corresponding to $\lambda = 10,590$ nm and their emission bandwidth prior to narrowing can be in the 3–4 GHz range (Duarte 1985a, 1985b, 1985c).

Since molecular transitions give origin to emission bands, these lasers are tunable. Application of the line-narrowing techniques described in Chapter 7 can yield tunable narrow-linewidth emission. Narrow-linewidth tunable pulsed excimer lasers are summarized in Table 9.2 and narrow-linewidth tunable pulsed CO_2 lasers in Table 9.3.

TABLE 9.2
Narrow-Linewidth Tunable Pulsed Excimer Laser Oscillators

Laser	Oscillator Configuration	λ (nm)	Δv	Energy[a]	Reference
ArF	MPL	193	10 GHz	150 μJ	Ludewigt et al. (1987)
KrF	GI	248	9 GHz	15 μJ	Caro et al. (1982)
XeCl	GI[b]	308	31 GHz	50 mJ	Buffa et al. (1983)
XeCl	GI	308	1.5 GHz	1 mJ	Sze et al. (1986)
XeCl	GI	308	1 GHz	4 mJ	Sugii et al. (1987)
XeCl	Three etalons	308	150 MHz	2–5 μJ	Pacala et al. (1984)

Source: Duarte, F.J., *Quantum Optics for Engineers*, CRC Press, New York, 2014.
[a] Output energy per pulse.
[b] Open cavity configuration.

TABLE 9.3
Narrow-Linewidth Tunable Pulsed CO_2 Laser Oscillators

Laser	Oscillator Configuration	λ (nm)	Δv (MHz)	Energy[a] (mJ)	Reference
CO_2	GI[b]	10,591	117	140	Duarte (1985a)
CO_2	GI[b]	10,591	700	230	Bobrovskii et al. (1987)
CO_2	MPL	10,591	140	200	Duarte (1985b)
CO_2	HMPGI	10,591	107[c]	85	Duarte (1985b)

Source: Duarte, F.J., *Quantum Optics for Engineers*, CRC Press, New York, 2014.
[a] Output energy per pulse.
[b] Open cavity configuration.
[c] Corresponds to SLM oscillation.

9.2.2 PULSED ATOMIC METAL VAPOR LASERS

Emission characteristics of atomic pulsed metal vapor lasers are given in Table 9.4. Perhaps, the most well-known member of this subgroup is the copper laser, which is also referred to as the copper vapor laser. These lasers have found numerous applications due to their ability to emit large average powers in the green at $\lambda = 510.554$ nm and, at their secondary emission wavelength, at $\lambda = 578.21$ nm. CVLs use a buffer gas, such as He or Ne, and operate at a high prf in order to attain the necessary metal vapor pressure. Excitation of the upper ^2P state occurs mainly via direct electron excitation (Harstad 1983). Output parameters from individual Cu lasers have been reported to cover a wide range of values. At pulsed energies of up to several millijoules per pulse, these lasers can emit pulses in the 10–60 ns range, at prfs from 2 to 32 kHz (Webb 1991). For instance, a specific CVL can yield an average power of 100 W, at 20 mJ per pulse, and a prf of 5 kHz (Webb 1991). Integrated CVL systems have been reported to yield average powers of up to 7 kW at a prf of 26 kHz (Bass et al. 1992). Copper lasers can also be operated at low repetition rates using copper halides to attain the necessary vapor pressures at relatively low temperatures (Piper 1978; Brandt and Piper 1981).

CVLs have been reported in a variety of cavity arrangements including plane mirror resonators and various unstable resonator configurations (Webb 1991). These lasers are useful for a number of applications, including the excitation of tunable dye lasers at low and high prfs (Duarte and Piper 1982, 1984; Webb 1991).

9.2.3 CW GAS LASERS

Perhaps the most well-known gas laser is the helium–neon (He–Ne) laser. Its most well-known line corresponds to the $3s_2 - 2p_4$ transition at $\lambda = 632.82$ nm. This transition is made possible by excitation transfer from atoms at the helium metastable level $He^* (2^3 S_1)$ to ground-state Ne atoms (Willett 1974). The emission from this laser is characterized by a most beautiful TEM_{00} beam at a laser linewidth typically less than a few gigahertz in the absence of frequency-selective intracavity optics. At the referred wavelength, and depending on the discharge length, available powers vary

---

TABLE 9.4
Atomic Pulsed Gas Lasers

Laser	Transition	λ (nm)	prf[a] (kHz)	Bandwidth (GHz)	Reference[b]
Cu	$^2P_{3/2} - ^2D_{5/2}$	510.5	2–30	7	Tenenbaum et al. (1980)
	$^2P_{1/2} - ^2D_{3/2}$	578.2	2–30	11	Tenenbaum et al. (1980)
Au	$^2P_{1/2} - ^2D_{3/2}$	627.8	5–20		

Source: Duarte, F.J., *Quantum Optics for Engineers*, CRC Press, New York, 2014.
[a] prf values do not represent absolute limits.
[b] References relate to bandwidth exclusively.

TABLE 9.5

Transitions of CW He–Ne Laser

Transition[a]	λ (nm)[b]
$3s_2-2p_{10}$	543.30
$3s_2-2p_8$	593.93
$3s_2-2p_7$	604.61
$3s_2-2p_6$	611.80
$3s_2-2p_4$	632.82
$2s_4-2p_8$	1114.30
$2s_2-2p_4$	1152.35
$2s_4-2p_3$	1268.90
$2s_2-2p_1$	1523.10

Note: Partial list of available transitions.

[a] Transition assignment following Willett, C.S., *An Introduction to Gas Lasers: Population Inversion Mechanisms*, Pergamon Press, New York, 1974.

[b] Wavelength values from Beck, R., et al., *Table of Laser Lines in Gases and Vapors*, Springer-Verlag, Berlin, Germany, 1976.

from a few milliwatts to a few tens of milliwatts. Additional transitions available from the He–Ne laser are listed in Table 9.5. He–Ne lasers incorporating broadband mirrors and tuning optics can emit at several visible transitions.

Another CW gas laser that is widely used in the laboratory, due to its powerful blue-green transitions, is the Ar^+ laser. All the transitions listed in Table 9.6 are excited via electron impact (Willett 1974). The dominant CW transitions are those at 487.99 and 514.53 nm. It should be noted that relatively compact Ar^+ lasers can be engineered to deliver powers at the tens of milliwatts range, whereas large systems have been configured to yield as much as 175 W (Anliker et al. 1977). An additional feature of laboratory Ar^+ lasers is the option to tune from line to line using intracavity optics.

Additional CW metal vapor lasers are the He–Zn and He–Cd lasers. Ion transitions in these lasers are excited via Penning and Duffendack reactions (Piper and Gill 1975). In the first reaction, the helium metastable He^* ($3S_1$) interacts with the metal atoms, whereas in the second reaction, it is helium ion He^+ ($^2S_{1/2}$) that participates in the excitation. In the case of the He–Zn laser, the transitions at 491.16, 492.40, and 758.85 nm are excited via Duffendack reactions, which can only occur in hollow-cathode discharges that give origin to energetic electrons. Using a hollow-cathode $He–CdI_2$ discharge, Piper (1976) combined simultaneously 4 transitions from Cd^+ and 11 transitions from I^+ to produce a most striking TEM_{00} *white light laser beam*. All of the I^+ transitions listed in Table 9.5 participated in the emission plus the Cd^+ transitions at 441.56, 533.75, 537.80, and 806.70 nm. All of these transitions are excited via Duffendack reactions except the 441.56 nm transition from Cd^+ that results from Penning ionization.

TABLE 9.6
Ionized CW Gas Lasers

Laser	Transition[a]	λ (nm)[b]
Ar$^+$	$4p\,^2P_{3/2}^0 - 4s\,^2P_{3/2}$	454.50
	$4p\,^2S_{1/2}^0 - 4s\,^2P_{1/2}$	457.93
	$4p\,^2P_{1/2}^0 - 4s\,^2P_{3/2}$	465.79
	$4p\,^2D_{3/2}^0 - 4s\,^2P_{3/2}$	472.69
	$4p\,^2P_{3/2}^0 - 4s\,^2P_{1/2}$	476.49
	$4p\,^2D_{5/2}^0 - 4s\,^2P_{3/2}$	487.99
	$4p\,^2D_{3/2}^0 - 4s\,^2P_{1/2}$	496.51
	$4p\,^4D_{5/2}^0 - 4s\,^2P_{3/2}$	514.53
	$4p\,^4D_{3/2}^0 - 4s\,^2P_{1/2}$	528.69
Zn$^+$	$4f\,^2F_{5/2}^0 - 4d\,^2D_{3/2}$	491.16
	$4f\,^2F_{7/2}^0 - 4d\,^2D_{5/2}$	492.40
	$4s^2\,^2D_{3/2} - 4p\,^2P_{1/2}^0$	589.44
	$5d\,^2D_{5/2} - 5p\,^2P_{3/2}^0$	610.25
	$4s^2\,^2D_{5/2} - 4p\,^2P_{3/2}^0$	747.88
	$5p\,^2P_{3/2}^0 - 5s\,^2S_{1/2}$	758.85
Kr$^+$	$4p\,^4P_{3/2}^0 - 5s\,^4P_{1/2}$	520.83
	$5p\,^4P_{5/2}^0 - 5s\,^4P_{3/2}$	530.87
	$5p\,^4D_{5/2}^0 - 5s\,^2P_{3/2}$	568.19
	$5p\,^4P_{5/2}^0 - 5s\,^2P_{3/2}$	647.09
	$5p\,^4P_{3/2}^0 - 5s\,^2P_{1/2}$	752.55
	$5p\,^4P_{3/2}^0 - 4d\,^2D_{1/2}$	799.32
Cd$^+$	$5s^2\,^2D_{5/2} - 5p\,^2P_{3/2}^0$	441.56
	$4f\,^2F_{5/2}^0 - 5d\,^2D_{3/2}$	533.75
	$4f\,^2F_{7/2}^0 - 5d\,^2D_{5/2}$	537.80
	$6g\,^2G_{7/2} - 4f\,^2F_{5/2}^0$	635.48
	$6g\,^2G_{9/2} - 4f\,^2F_{7/2}^0$	636.00
	$6f\,^2F_{5/2}^0 - 6d\,^2D_{3/2}$	723.70
	$6p\,^2P_{3/2}^0 - 6s\,^2S_{1/2}$	806.70
	$6p\,^2P_{1/2}^0 - 6s\,^2S_{1/2}$	853.00
	$9s\,^2S_{1/2} - 7p\,^2P_{3/2}^0$	887.80
I$^+$	$6p'\,^3D_2 - 6s'\,^3D_2^0$	540.73
	$6p'\,^3F_2 - 6s'\,^3D_2^0$	567.81
	$6p'\,^3D_2 - 6s'\,^3D_1^0$	576.07
	$6p'\,^3D_1 - 6s'\,^3D_2^0$	612.75
	$6p'\,^3D_1 - 6s'\,^3D_1^0$	658.52
	$6p'\,^3F_2 - 5d'\,^3D_2^0$	703.30
	$6p'\,^3D_2 - 5d'\,^3D_3^0$	713.89
	$6p'\,^3F_2 - 5d'\,^3D_3^0$	761.85[c]

(Continued)

TABLE 9.6
(Continued) Ionized CW Gas Lasers

Laser	Transition[a]	λ (nm)[b]
	$6p'^3D_1 - 5d'^3D_2^0$	773.58[c]
	$6p'^3D_2 - 5d'^3F_3^0$	817.02[c]
	$6p'^3F_2 - 5d'^3F_3^0$	880.42
	$6p'^3F_3^3 - 5d'^3G_4^0$	887.78[c]

Note: Partial list of available transitions.

[a] Transition assignment following Willett, C.S., *An Introduction to Gas Lasers: Population Inversion Mechanisms*. Pergamon Press, New York, 1974.

[b] Wavelength values from Beck, R., et al., *Table of Laser Lines in Gases and Vapors*, Springer-Verlag, Berlin, Germany, 1976.

[c] Data from Piper, J.A., *J. Phys. D: Appl. Phys.*, 7, 323–328, 1974.

For a survey of CW CO_2 lasers, the reader should refer to Willett (1974) and Freed (1995). An excellent review on frequency selectivity in CO_2 lasers was provided by Tratt et al. (1985).

9.3 ORGANIC DYE LASERS

Dye lasers are inherently tunable and intrinsically apt for the generation of high average powers (Duarte 1991b). These lasers can be divided into two main categories: pulsed dye lasers and CW dye lasers. These lasers, using various dyes, can span the electromagnetic spectrum from the near ultraviolet to the near infrared (see Chapter 1). Some individual dyes, such as rhodamine 6G, can provide tuning ranges in excess of 50 nm, centered at $\lambda \approx 590$ nm, whereas coumarin 545T has been shown to be tunable in the $501 \leq \lambda \leq 574$ nm range (Duarte et al. 2006). For historical reviews, the reader should refer to Schäfer (1990) or Duarte and Hillman (1990).

9.3.1 PULSED ORGANIC DYE LASERS

Pulsed dye lasers are divided into two subclasses: laser-pumped pulsed dye lasers and flashlamp-pumped pulsed dye lasers.

As apparent from Table 9.7, laser-pumped dye lasers can provide either hundreds of joules per pulse at low repetition rates or very high average powers, well into the kilowatt regime at prfs in excess of 10 kHz. Excellent reviews on CVL-pumped dye lasers are given by Webb (1991) and on excimer laser-pumped dye lasers by Tallman and Tennant (1991).

The performance of laboratory-size laser-pumped dye lasers can be illustrated, considering the work of Bos (1981) who reported a linewidth of $\Delta\nu \approx 320$ MHz at $\lambda = 590$ nm with a telescopic oscillator incorporating an intracavity etalon. Using three stages of amplification, the output energy was 165 mJ at an overall conversion efficiency of 55% for excitation at 532 nm. Employing a HMPGI grating oscillator

TABLE 9.7
High-Power Laser-Pumped Organic Dye Lasers

Excitation Laser	Energy[a]	prf	Power[b]	η (%)[c]	Tuning Range (nm)	Reference
XeCl	800 J[d]	Low		27	475[e]	Tang et al. (1987)
XeCl	200 mJ	250 Hz	50 W	20	401.5[e]	Tallman and Tennant (1991)
CVL	190 mJ	13.2 kHz	2.5 kW	50	550–650	Bass et al. (1992)

Source: Duarte, F.J., *Quantum Optics for Engineers*, CRC Press, New York, 2014.

[a] Output energy per pulse.
[b] Average power.
[c] Efficiency.
[d] Pulse length quoted at 500 ns.
[e] Central wavelength.

and two stages of amplification, Dupre (1987) reported 3.5 mJ, and $\Delta v = 1.2$ GHz, at $\lambda = 440$ nm. The conversion energy efficiency was ~9% for excitation at 355 nm.

CVL excitation of HMPGI grating oscillators yielded $\Delta v \approx 600$ MHz, for $\lambda = 575$ nm, and a pulse length of 12 ns (FWHM) at a conversion efficiency of ~5% (Duarte and Piper 1984). The average output power was 80 mW at a prf of 8 kHz. Using the same class of multiple-prism oscillator, and one amplifier stage, Singh et al. (1994) reported $\Delta v \approx 1.5$ GHz and a conversion efficiency of 40% at a prf of 6.5 kHz. Emission characteristics of narrow-linewidth tunable dye laser oscillators are summarized in Table 9.8.

The performance of large flashlamp-pumped dye lasers is summarized in Table 9.9. For a comprehensive and authoritative review of flashlamp-pumped dye lasers, the reader should consult Everett (1991). In general, these lasers have been used to generate large energies in pulses up to the microsecond regime. The energy per pulse is so high that fairly modest prfs can generate average powers in the kilowatt regime. In this regard, organic dye lasers continue to provide the best alternative for high-energy pulse generation tunable directly in the visible spectrum (Duarte 2012), which is an attractive avenue for directed energy applications.

Besides their intrinsic ability to generate large pulsed energies, flashlamp-pumped dye lasers have also been configured in small-scale laboratory versions designed to yield narrow-linewidth, highly stable, tunable laser emission. For a review on this subject, see Duarte (1995b).

A ruggedized MPL grating coaxial flashlamp-pumped dye laser oscillator yielding a single-transverse-mode laser beam with $\Delta\theta \approx 0.5$ mrad, and a laser linewidth of $\Delta v \approx 300$ MHz, at $\lambda \approx 590$ nm, was reported by Duarte et al. (1991). The emission was extremely pure with ASE levels a few parts in 10^{-7}. This dispersive oscillator provided ~3 mJ per pulse, at a pulse length of $\Delta t \approx 50$ ns, using rhodamine 590 at a concentration of 0.01 mM. Its wavelength stability was measured to be

TABLE 9.8
Narrow Linewidth Liquid Organic Dye Lasers

Excitation Source	Cavity	Δν	Δt	Tuning Range (nm)	η (%)
CVL[a]	MPLG[b]	60 MHz[c]	30 ns	560–600	5
CVL[d]	HMPGIG	400 MHz[d]	12 ns	565–603	4–5
Nd:YAG[e,f]	HMPGIG	650 MHz	5 ns	425–457	~9[g]
Flashlamp[h]	MPLG[b]	300 MHz	70 ns	565–605	

Source: Duarte, F.J., *Quantum Optics for Engineers*, CRC Press, New York, 2014.
[a] Data from Bernhardt, A.F., and Rasmussen, P., *Appl. Phys. B*, 26, 141–146, 1981.
[b] Includes an intracavity etalon.
[c] SLM emission.
[d] Data from Duarte, F.J., and Piper, J.A., *Appl. Opt.*, 23, 1391–1394, 1984.
[e] Uses the third harmonics of the fundamental (3ν).
[f] Data from Dupre, P., *Appl. Opt.*, 26, 860–871, 1987.
[g] This efficiency includes amplification via two stages.
[h] Data from Duarte et al. (1991); the output laser energy is ~3 mJ per pulse.
HMPGIG, hybrid multiple-prism grazing-incidence grating configuration; MPLG, multiple-prism Littrow grating configuration.

TABLE 9.9
High-Energy Flashlamp-Pumped Organic Dye Lasers

Excitation	Δt (μs)	Energy (J)	η (%)	Reference
Linear[a,b]	7	40	0.4	Fort and Moulin (1987)
Transverse[b,c]	5	140[d]	1.8	Klimek et al. (1992)
Coaxial[b]	10	400	0.8	Baltakov et al. (1974)

Source: Duarte, F.J., *Quantum Optics for Engineers*, CRC Press, New York, 2014.
[a] Employs 12 flashlamps in a linear configuration.
[b] Uses rhodamine 6G dye.
[c] Employs 16 flashlamps in a transverse configuration.
[d] Yields an average power of 1.4 kW at a prf of 10 Hz.

$(\delta\lambda/\lambda) \approx 5 \times 10^{-7}$. This organic tunable laser was engineered in an all-Invar structure and its gain medium was flowed in a *hermetically sealed* stainless steel and polytetrafluoroethylene system. This ruggedized oscillator (also included in Table 9.8) was shown to maintain its wavelength and linewidth characteristics following displacement on a rugged terrain (Duarte et al. 1991).

The output from narrow-linewidth oscillators can be amplified using single-stage amplifiers to yield hundreds of millijoules per pulse. Flamant and Maillard (1984) used a two-etalon oscillator to excite a flat-mirror amplifier to attain $\Delta v = 346$ MHz and a pulse energy of 300 mJ at $\lambda = 590$ nm. Using a multiple-prism grating oscillator and a single-stage unstable-resonator amplifier, Duarte and Conrad (1987) achieved $\Delta v = 375$ MHz and a pulse energy of 600 mJ at $\lambda = 590$ nm.

9.3.1.1 Solid-State Tunable Organic Lasers

In addition to traditional liquid organic dye lasers, there has been considerable research and development activity in the area of solid-state dye lasers. Excellent reviews on dye-doped solid-state matrices was provided by Costela et al. (2009, 2013) and the photophysical properties of organic dye-doped polymer gain materials were characterized by Holzer et al. (2000). Dye-doped polymer-nanoparticle gain media was demonstrated to lase efficiently while exhibiting TEM_{00} and reduced beam divergence by Duarte and James (2003). The performance of broadband solid-state organic dye lasers is summarized in Table 9.10.

Although most of the recent activity has been centered around polymeric matrices, and hybrid silicate–polymer composite materials, there has also been work reported on crystalline dye lasers (Rifani et al. 1995; Braun et al. 2000). Reviews of organic lasers, mainly centered around optically pumped conjugate polymers, are given by Karnutsch (2007) and Samuel et al. (2007).

Optimized high-performance multiple-prism grating solid-state organic oscillators, as described in Chapter 7, have yielded tuning in the $550 \le \lambda \le 603$ nm range

TABLE 9.10
Broadband Solid-State Organic Dye Lasers

Excitation	Matrix	Energy[a] (mJ)	η (%)	Reference
Flashlamp	PMMA[b]	50		Pacheco et al. (1988)
Nd:YAG (2v)	MPMMA[c]		65	Maslyukov et al. (1995)
LPDL[d]	HEMA:MMA[b]	0.8	40	Duarte et al. (1998)
LPDL[d]	DDPN[b,e]		63	Duarte and James (2003)
FLPDL[f]	TEOS[b,g]	2.5		Duarte et al. (1993)
Nd:YAG (2v)	ORMOSIL	3.5	35	Larrue et al. (1994)

[a] Output energy per pulse.
[b] Doped with rhodamine 6G.
[c] Doped with rhodamine 11B.
[d] Using coumarin 152.
[e] The polymer is polymethyl methacrylate.
[f] Using coumarin 525.
[g] $Si(OC_2H_5)_4$.
DDPN, dye-doped polymer nanoparticle; FLPDL, flashlamp-pumped dye laser; HEMA:MMA, 2-hydroxyethyl methacrylate:methyl methacrylate; LPDL, laser-pumped dye laser; MPMMA, modified polymethyl methacrylate; ORMOSIL, organically modified silicate; PMMA, polymethyl methacrylate.

TABLE 9.11

Narrow-Linewidth Solid-State Organic Laser Oscillators

Cavity	Matrix	Δv	Δt	Tuning (nm)	Energy[a] (mJ)
MPLG[b,c]	MPMMA[d]	350 MHz[e]	3 ns	$550 \leq \lambda \leq 603$	~0.1[f]
MPLG[g,h]	HEMA:MMA[i]	650 MHz	105 ns	$564 \leq \lambda \leq 602$	~0.4

Source: Duarte, F.J., *Quantum Optics for Engineers*, CRC Press, New York, 2014.
[a] Output energy per pulse.
[b] Laser-pumped dye laser; optimized cavity.
[c] Data from Duarte, F.J., *Appl. Opt.*, 38, 6347–6349, 1999.
[d] Doped with rhodamine 6 G.
[e] Linewidth corresponds to SLM oscillation.
[f] Power per pulse ~33 kW.
[g] Flashlamp-pumped dye laser.
[h] Data from Duarte, F.J., et al., *Appl. Opt.*, 37, 3987–3989, 1998.
[i] HEMA:MMA doped with rhodamine 6 G.
HEMA:MMA, 2-hydroxyethyl methacrylate:methyl methacrylate; MPLG, multiple-prism Littrow grating configuration; MPMMA, modified polymethyl methacrylate.

with TEM_{00} laser beams with beam divergences ~1.5 times the diffraction limit. The emission is in a single longitudinal mode (SLM) at $\Delta v \approx 350$ MHz in pulses $\Delta t \approx 3$ ns (FWHM) with a near-Gaussian temporal profile. Conversion efficiency is reported at ~5% and the ASE levels are extremely low at ~10^{-6} (Duarte 1999).

Transverse excitation in the long-pulse regime of a four-prism grating solid-state dye laser oscillator has led to pulses a long as $\Delta t \approx 105$ ns (FWHM), and $\Delta v \approx 650$ MHz, in the $564 \leq \lambda \leq 602$ nm tuning range. The pulse energies are reported to be ~0.4 mJ per pulse (Duarte et al. 1998). The characteristics of these high-performance solid-state organic laser oscillators are summarized in Table 9.11.

9.3.2 CW Organic Dye Lasers

Dye lasers have had a significant impact in *high resolution spectroscopy*, and other iconic applications including *laser cooling*, given their tunability, excellent TEM_{00} beam quality, and intrinsic narrow linewidths that can readily reach the few megahertz level. CW dye lasers typically use Ar^+ and Kr^+ as excitation sources, although in principle they could use any compatible laser yielding TEM_{00} emission. It should be noted that CW dye lasers have been excited with a variety of lasers including diode lasers (see, e.g., Scheps 1993). Table 9.12 summarizes the performance of relatively high-power CW dye lasers, some of which yield SLM oscillation at linewidths in the megahertz regime. Stabilization techniques resulting in the demonstration of laser linewidths as low as 100 Hz (Drever et al. 1983) are discussed in an excellent review by Hollberg (1990).

A further towering contribution of the CW dye lasers to the field of optics and lasers was their use as principal tools in the development of ultrashort pulse lasers that gave origin to the femtosecond lasers. Among the important concepts developed

TABLE 9.12
Performance of High-Power CW Organic Dye Lasers

Cavity	Spectral Range (nm)	Δν	Power[a]	η (%)	Reference
Linear[b]			33 W[c,d]	30	Anliker et al. (1977)
Linear[b]	560–650	SLM[e]	33 W[f,g]	17	Baving et al. (1982)
Ring[b]	407–887[h]	SLM[e]	5.6 W[i]	23.3	Johnston and Duarte (2002)

[a] CW output power.
[b] Under Ar+ laser excitation.
[c] Maximum CW power quoted: 52 W for a pump power of 175 W.
[d] Using rhodamine 6G at 0.7 mM.
[e] Linewidth values can be in the few MHz range.
[f] Without intracavity tuning prism, the quoted output power is 43 W for a pump power of 200 W.
[g] Using rhodamine 6G at 0.94 mM.
[h] Using 11 dyes.
[i] Using rhodamine 6G.

TABLE 9.13
Femtosecond Pulse Dye Lasers

Pulse Compressor	Δt (fs)	Reference
Single prism	60	Dietel et al. (1983)
Double prism	50	Kafka and Baer (1987)
Double prism	18	Osvay et al. (2005)
Four prisms	29	Kubota et al. (1988)
Four prisms plus grating pair	6	Fork et al. (1987)

in this endeavor were the generation of bandwidth-limited ultrashort pulses (Ruddock and Bradley 1976), the colliding-pulse mode locking technique (Fork et al. 1981), and prismatic pulse compression (Dietel et al. 1983; Fork et al. 1984) (see Chapter 4). Using an extracavity pulse compressor, consisting of a four-grating array and a four-prism array, Fork et al. (1987) compressed a 50 fs pulse further to just 6 fs. Diels (1990) and Diels and Rudolph (2006) provided comprehensive reviews on this subject. Table 9.13 summarizes the performance of femtosecond pulse organic lasers.

9.4   SOLID-STATE LASERS

The solid-state laser field is a vast field that includes traditional crystalline materials and fiber gain media. These lasers can emit in both the pulse and CW regimes. In this section, some of the most well-known gain media are surveyed with emphasis on spectral characteristics.

9.4.1 Ionic Solid-State Lasers

Optically pumped ionic solid-state lasers include the well-known Nd laser that can exist either in a crystalline or in a glass host. These lasers are very well suited to be configured in various cavity arrangements, including unstable resonator arrangements, which yield single-transverse-mode emission. The linewidth of a TEM_{00} laser at 1064 nm is typically 15–30 GHz (Chesler and Geusic 1972). Frequency doubling using nonlinear crystals, intracavity or extracavity, yields efficient conversion into the visible. Originally, these lasers were excited using flashlamp pumping; however, diode laser pumping has become rather pervasive. Commercially available diode laser-pumped Nd:YAG lasers can yield tens of watts at prfs in the kilohertz regime. Individual laser pulse lengths can be in the 10–15 ns range. Ionic gain media, in crystalline hosts, has also lased in the CW regime.

Nd:glass lasers are mainly operated at very high peak powers, and low prfs, in the TW regime. Some of the most well-known ionic solid-state lasers, with their respective transitions, are listed in Table 9.14.

9.4.2 Transition Metal Solid-State Lasers

Transition metal solid-state lasers include the alexandrite and Ti:sapphire lasers that are widely tunable. This quality has made the Ti:sapphire lasers particularly applicable to the generation of ultrashort pulses in the femtosecond regime. Although the ruby laser is mainly operated in the pulsed regime, typically delivering a few joules of energy per pulse, the other two media are very versatile and are well suited to both pulsed and CW operations (Walling and Peterson 1980; Walling et al. 1980a; Moulton 1986). Tuning ranges are listed in Table 9.15. A comprehensive review of transition metal solid-state lasers was given by Barnes (1995a).

All-solid-state Ti:sapphire lasers are available commercially with TEM_{00} beam profiles, and emission in the SLM domain, delivering average powers in the watt regime at ~10 kHz. Using CVL pumping, narrow-linewidth emission has been demonstrated at average powers of 5 W at 6.2 kHz, at a conversion efficiency of ~26% (Coutts et al. 1998). Liquid nitrogen cooling of Ti:sapphire gain media has resulted in CW output powers of up to 43 W at an efficiency of ~42% for broadband lasing (Erbert et al. 1991). In the ultrashort pulse regime, Ti:sapphire lasers have been shown to deliver pulses as short as 5 fs (Ell et al. 2001).

TABLE 9.14
Ionic Solid-State Lasers

Ion	Transition	λ (nm)
Yb^{3+}	$^2F_{5/2}-^2F_{7/2}$	1015
Nd^{3+}	$^4F_{3/2}-^4I_{11/2}$	1064
Er^{3+}	$^4I_{13/2}-^4I_{15/2}$	1540
Tm^{3+}	$^3H_4-^3H_6$	2013
Ho^{3+}	$^5I_7-^5I_8$	2097

TABLE 9.15
Transition Metal Solid-State Lasers

Ion:Host Crystal	Transition	λ (nm)	Reference
$Cr^{3+}:Al_2O_3$	$^2E(\bar{E})-^4A_2$	694.3	Maiman (1960)
$Cr^{3+}:Be_3Al_2(SiO_3)^6$	$^4T_2-^4A_2$	$695 \leq \lambda \leq 835$	Shand and Walling (1982)
$Cr^{3+}:BeAl_2O_4$	$^4T_2-^4A_2$	$701 \leq \lambda \leq 818$	Walling et al. (1980b)
$Ti^{3+}:Al_2O_3$	$^2T_2-^2E$	$660 \leq \lambda \leq 986$	Moulton (1986)

TABLE 9.16
Ultrashort Pulse Solid-State Lasers

Laser	Post Laser	Δt (fs)	λ (nm)	Energy	Reference
$Ti^{3+}:Al_2O_3$	Noncollinear OPA[a]	6.9		20 µJ	Tavella et al. (2010)
$Ti^{3+}:Al_2O_3$	Chirped mirror compressor	3.64[b]	810[c]		Demmler et al. (2011)

[a] Optics includes a two-prism stretcher, a fiber stretcher, and two fiber preamplifiers.
[b] 1.3 cycle at 810 nm.
[c] Central wavelength.

Titanium sapphire lasers emitting in the ultrafast, or ultrashort pulse, regime are included in Table 9.16.

9.4.3 COLOR CENTER LASERS

Laser-excited color center lasers span the electromagnetic spectrum mainly in the near infrared at the $0.82 \leq \lambda \leq 3.3$ µm region. Lasing has been reported mainly in the CW regime at power levels in the 10 mW–2.7 W range. An excellent review article on this subject was written by Mollenauer (1985).

Using the center F_2^+ in a LiF host, lasing in the 0.82–1.05 µm region was reported with a maximum power of 1.8 W (Mollenauer and Bloom 1979). Using the same center in a KF host, the same authors reported laser emission in the $1.22 \leq \lambda \leq 1.50$ µm region at a maximum power of 2.7 W for an excitation power of ~5 W. Narrow-linewidth emission was reported by German (1981), using the FA(II) center in a RbCl:Li host, in a grazing-incidence cavity configuration. SLM oscillation was reported at power levels in excess of 10 mW in the $2.7 \leq \lambda \leq 3.1$ µm region.

9.4.4 DIODE LASER-PUMPED FIBER LASERS

An area of the solid-state laser field that has evolved rapidly over the past few years is the fiber laser subfield. Glass or silica fibers doped with elements such as Yb, Nd, Er, or Tm are the active media that are increasingly excited with high-power diode lasers. From the versatility of doping elements, it should be apparent that these lasers

TABLE 9.17
Diode Laser-Pumped Yb-Doped Fiber Lasers

Cavity	λ (nm)	Bandwidth	CW Power	η (%)	Reference
Linear	~1120	BB	110 W[a]	58	Dominic et al. (1999)
	~1100	BB	1.36 kW[b]	86	Jeong et al. (2004)

[a] Excitation wavelength at ~915 nm.
[b] Excitation wavelength at ~975 nm.
BB, broadband.

TABLE 9.18
Tunable Fiber Lasers

Tuning Technique	Tuning (nm)	Δν	CW Power	Reference
HTGIG[a]	$1032 \leq \lambda \leq 1124$	2.5 GHz	10 W[b]	Auerbach et al. (2002)
BG[c] (Er^{3+})	$1510 \leq \lambda \leq 1580$	100 MHz	0.5 mW	Chen et al. (2003)
GIG (Tm^{3+})	$2275 \leq \lambda \leq 2415$	210 MHz	6.0 mW	McAleavey et al. (1997)

[a] Hybrid telescope grazing-incidence grating configuration in a ring cavity.
[b] Excitation wavelength at 980 nm.
BG, Bragg grating; GIG, grazing-incidence grating.

cover the near infrared from beyond 1 μm to about 3 μm. Since the breadth and scope of this subfield is enormous, attention is only given to some of the important characteristics of fiber lasers. First, these lasers can be extremely efficient. Second, as indicated in Table 9.17, they can achieve high output CW powers under diode laser excitation. The third important quality of diode laser-pumped fiber lasers is broad continuous tunability. These qualities are well illustrated by fiber lasers included in Table 9.18.

In regard to the narrow-linewidth tunable laser reported by Auerbach et al. (2002), it should be mentioned that the authors reported extremely low levels of ASE. Also, the hybrid telescope grazing-incidence (HTGI) grating configuration (Yodh et al. 1984; Smith and DiMauro 1987) used to induce narrow-linewidth oscillation is a variant of the principle of the HMPGI grating configurations (Duarte and Piper 1981, 1984) that is described in Chapter 7. Tunable fiber lasers were reviewed in detail by Shay and Duarte (2009).

9.4.5 OPTICAL PARAMETRIC OSCILLATORS

Although the optical parametric oscillator (OPO) does not involve the process of population inversion in its excitation mechanism, it is included, nevertheless, since it is a source of spatially, and spectrally, coherent emission which is inherently tunable.

TABLE 9.19

Pulsed OPOs

Crystal	Optical Transmission[a]	λ_p (nm)	Tuning (µm)	Reference
KTP	0.35–4.5	532	$0.61 \leq \lambda \leq 4.0$	Orr et al. (1995)
BBO	0.20–2.2	355	$0.41 \leq \lambda \leq 3.0$	Orr et al. (1995)
LBO	0.16–2.3	355	$0.41 \leq \lambda \leq 2.47$	Schröder et al. (1994)
$LiNbO_3$	0.33–5.5	532	$0.61 \leq \lambda \leq 4.4$	Orr et al. (1995)
$AgGaS_2$	0.15–13	1064	$1.4 \leq \lambda \leq 4.0$	Fan et al. (1984)

[a] For the ordinary ray. (Data from Barnes, N.P., *Tunable Lasers Handbook*, Academic Press, New York, 1995a.)

For detailed review articles on this subject, the reader is referred to Barnes (1995b) and Orr et al. (1995, 2009). The nonlinear optics aspects of OPOs are examined in Chapter 8. Spectral characteristics of several well-known OPOs are given in Table 9.19.

OPOs are optically pumped devices that make wide use of Nd:YAG lasers and the different harmonics that can be provided with these excitation sources. A fairly instructive case study on the performance of pulsed OPOs was provided by Schröder et al. (1994) who reported an output energy of 77 mJ for an excitation energy of 170 mJ, at 355 nm, for a 15 mm-long LiB_3O_5 crystal. The emission bandwidth was ~0.5 nm (~366 GHz) at $\lambda \approx 640$ nm. This performance was obtained using excitation pulses 5–6 ns in duration at a prf of 10 Hz. Pulsed OPOs for analytical applications tend to operate at prfs of ~10 Hz (He and Orr 2001). Barnes and Williams-Byrd (1995) discuss thermally induced phase mismatch and thermally induced lensing that tend to limit the average power in OPOs and optical parametric amplifiers.

Certainly, all the linewidth narrowing techniques described in Chapter 7 are applicable to OPOs and further details are given in Chapter 8. For example, a multiple-prism grating cavity, incorporating an intracavity etalon, in a $LiNbO_3$ OPO yielded a linewidth of $\Delta v \approx 30$ MHz at $\lambda \approx 3.4$ µm. The energy conversion efficiency was reported at ~0.74%.

Initially, OPOs in the CW regime were longitudinally excited by CW frequency-doubled Nd:YAG lasers (Smith et al. 1968). An improved approach consists in using diode laser excitation in intracavity configurations. Using fiber-coupled excitation from a diode laser, at 940 nm, Jensen et al. (2002) excited a Yb:YAG crystal whose emission was focused on a periodically poled $LiNbO_3$ crystal. For a diode laser power of 13.5 W, tunable emission in the $3.82 \leq \lambda \leq 4.57$ µm range was reported at output powers of 200 mW (Jensen et al. 2002).

Femtosecond OPOs in linear and ring cavities, incorporating multiple-prism pulse compressors, were discussed by Powers et al. (1993). An authoritative and detailed review on the performance and output characteristics of OPOs for spectroscopic applications was given by Orr et al. (2009).

9.5 SEMICONDUCTOR LASERS

The area of semiconductor lasers is a rapidly evolving field that can be classified into high-power diode lasers, external cavity lasers, and miniature lasers. Semiconductor lasers work via direct electrical excitation, are compact, offer wavelength tunability, and can be very stable. The approximate tuning ranges available from various semiconductor laser materials are listed in Table 9.20. Their compactness, with aperture dimensions in the micrometer regime, results in large beam divergences that require external, or intracavity, beam shaping optics to reduce the divergence (see Chapter 7).

High-power diode lasers, as those used in the excitation of crystalline solid-state lasers and fiber lasers, integrate many individual diode lasers to increase the overall output power. Arrays of numerous diode lasers in parallel form what is known as a *bar*. The divergent emission from these bars is collected and guided using external beam shaping optics. As indicated in Table 9.21, these bars can yield powers in the 10–40 W range. For uses requiring even higher output powers, such as industrial applications, the bars are further integrated into *stacks*. These stacks can yield powers in the kilowatt regime.

TABLE 9.20
Wavelength Coverage of Semiconductor Laser Materials

Semiconductor	Wavelength Coverage (nm)[a]
II–VI materials	$450 \leq \lambda \leq 530$
AlGaInP/GaAs	$610 \leq \lambda \leq 690$
AlGaAs/GaAs	$780 \leq \lambda \leq 880$
InGaAs/GaAs	$880 \leq \lambda \leq 1100$
InGaAsP/InP	$1100 \leq \lambda \leq 1600$
InGaAs/InP	$1600 \leq \lambda \leq 2100$
Quantum cascade[b]	Deep infrared[c]

[a] Approximate range; some ranges are not continuous.
[b] Semiconductors used in QCL configurations include GaInAs and GaAs/ AlGaAs quantum-well heterostructures.
[c] The emission range of these lasers is quoted in the 3–24 μm range. (Data from Silfvast, W.T., *Laser Fundamentals*, 2nd edn., Cambridge University Press, Cambridge, 2008.)

TABLE 9.21
CW Diode Laser Arrays

λ (nm)	CW Output Power	Reference
791	10 W	Srinivasan et al. (1999)
~915	45 W	Dominic et al. (1999)
975	1.2 kW	Jeong et al. (2004)

One area of research that has received considerable attention is the line narrowing and tunability of semiconductor devices. External cavity tunable semiconductor lasers utilize grating and multiple-prism grating cavity architectures similar to those described in Chapter 7. The performance of these lasers, which have been widely applied in areas such as *laser cooling*, and Bose–Einstein condensation, is summarized in Table 9.22. Note that albeit many authors use, for convenience, open cavity configurations, only *closed cavity* configurations are considered here, given their superior signal-to-noise characteristics and inherent protection against undesired external optical coupling (Zorabedian 1992; Duarte 1993). For reviews on this subject, the reader is referred to Zorabedian (1995), Duarte (1995c, 2009), and Fox et al. (1997).

One further variant, in the field of tunable semiconductor lasers, consists in building miniature lasers for applications in optics communications. One such approach utilizes a silicon microelectromechanical systems (MEMS) grazing-incidence grating external cavity (Berger et al. 2001). This miniature dispersive laser has been shown to yield narrow-linewidth emission tunable for over a 26 nm range in the near infrared. A second approach tunes an integrated vertical-cavity surface-emitting laser that varies its wavelength by changing the length of the cavity (Kner et al. 2002). The length of the cavity is changed by displacing the voltage-driven output mirror, as explained in Chapter 7. The performance of these lasers is summarized in Table 9.23.

Ultrashort pulse semiconductor lasers are presented in Table 9.24. These femtosecond lasers utilize a variety of saturable absorber techniques including multiple quantum wells. Pulse compression techniques include the use of grating pairs and multiple-prism arrays, as described in Chapter 4.

9.5.1 TUNABLE QUANTUM CASCADE LASERS

A relatively new class of tunable lasers are the quantum cascade lasers often simply designated by the acronym QCLs. The principle of operation of the QCLs is outlined in Chapter 1. These lasers have had a significant impact in a broad spectral region as

TABLE 9.22
External Cavity Tunable Semiconductor Lasers

Semiconductor	Cavity	Tuning (nm)	Δv	CW Power (mW)	Reference
InGaAsP/InP	LG[a]	$1285 \leq \lambda \leq 1320$	31 kHz	1	Shan et al. (1991)
InGaAsP/InP	MPL	$1255 \leq \lambda \leq 1335$	100 kHz		Zorabedian (1992)
GaAlAs	LG[a]	$815 \leq \lambda \leq 825$	1.5 MHz	5	Fleming and Mooradian (1981)

Note: All entries in this table use closed cavity configurations.
[a] Tuning is performed by an LG.
LG, Littrow grating.

TABLE 9.23
MEMS Tunable Semiconductor Lasers

Semiconductor	Cavity	Tuning (nm)	Δν	CW Power (mW)	Reference
InGaAsP/InP	GIG[a]	$1532 \leq \lambda \leq 1557$	2 MHz	20	Berger et al. (2001)
InGaAsP/InP	GIG[a,b]	$1529 \leq \lambda \leq 1571$	2 MHz	70	Berger and Anthon (2003)
GaAs/AlGaAs[c]	Mirror[d]	$1533 \leq \lambda \leq 1555$	SM	0.9	Kner et al. (2002)
	GIG[e]	$1530 \leq \lambda \leq 1570$	50 kHz		Zhang et al. (2012)

[a] Uses a silicon MEMS-driven GIG cavity.
[b] Open cavity configuration.
[c] VCEL.
[d] Tuning is achieved by displacing a voltage-driven micromirror that is supported by a cantilever.
[e] Open cavity GIG and etalon.
GIG, grazing-incidence grating; SM, single mode.

TABLE 9.24
Ultrashort Pulse External Cavity Semiconductor Lasers

Laser	Cavity	Mode-Locking Technique	Δt (fs)	λ (nm)	Reference
InGaAsP[a]	Etalon[b]	Active	580	1300	Corzine et al. (1988)
AlGaAs	Four prisms[c]	Hybrid MQW SA	200	~838	Delfyett et al. (1992)
AlGaAs[a]	Six prisms[b]	Active, SA	650	805	Pang et al. (1992)
MQW[a]	Grating pair[c,d]	Passive	260	~846	Salvatore et al. (1993)

Source: Duarte, F.J., *Tunable Laser Applications*, CRC Press, New York, 2009.
[a] AR coating of the internal facet next to the frequency-selective optics.
[b] Closed cavity configuration.
[c] Open cavity configuration.
[d] Uses a Littrow grating for tuning.
SA, saturable absorber.

they have introduced a compact alternative, with all the advantages of the semiconductor technology, to produce broadly tunable coherent radiation in the mid and deep infrared. As indicated in Table 9.25, external cavity tunable QCLs have been shown to operate in the $7 \leq \lambda \leq 12$ μm region.

9.5.2 TUNABLE QUANTUM DOT LASERS

Quantum dot lasers engineered with external cavities have demonstrated tuning ranges within the $1000 \leq \lambda \leq 1300$ nm region as indicated in Table 9.26. A tunable

TABLE 9.25

Tunable External Cavity Quantum Cascade Lasers

Stages in Cascade	Cavity	Tuning (µm)	Δv	CW Power	Reference
20	Littrow	$8.2 \leq \lambda \leq 10.4$	SM		Maulini et al. (2006)
74	Littrow	$7.6 \leq \lambda \leq 11.4$	~3.59 GHz	15 mW	Hugi et al. (2009)
	DFB[a]		5 GHz[b]	65 µW	Lu et al. (2013)

[a] Primary emissions around 9 and 10.22 µm.
[b] Single mode.
DFB, distributed feedback; SM, single mode.

TABLE 9.26

Tunable External Cavity Quantum Dot Lasers

QD	Cavity	Tuning (nm)	Δv (kHz)	CW Power (mW)	Reference
InAs	Littrow	$1033 \leq \lambda \leq 1234$			Varangis et al. (2000)
InAs	Littrow	$1125 \leq \lambda \leq 1288$	200 kHz	200	Nevsky et al. (2008)
InAs/GaAs	Littrow	$1122 \leq \lambda \leq 1324$		480	Fedorova et al. (2010)

QD, quantum dot semiconductor.

quantum dot laser, using InAs as gain medium and employing a diffraction grating deployed in Littrow configuration, is reported to yield a laser linewidth of $\Delta v \approx 200$ kHz and a tuning range of $1125 \leq \lambda \leq 1288$ nm. Frequency stabilization reduced the linewidth to $\Delta v \approx 30$ kHz (Nevsky et al. 2008).

9.6 ADDITIONAL LASERS

An important source of widely tunable coherent radiation, not included in this survey, is the free-electron laser (FEL). The reason for this choice is that FELs are fairly large high-power devices that require accelerator technology. In his review of FELs, Benson (1995) reported that various devices around the world cover, with their primary emission wavelengths, the electromagnetic spectrum from 2 to 2500 µm. However, more recent advances in this field have demonstrated FEL radiation well into the extreme ultraviolet (see, e.g., Allaria et al. 2012). A book on FELs was written by Brau (1990). Sources of discretely tunable soft X-ray radiation, using Ni-like ionic transitions, in the 16.5–10.9 nm range were demonstrated by Rocca and colleagues in table-top laser configurations (Wang et al. 2005).

Additional lasers not included in this survey are chemical lasers, the far-infrared lasers, and nuclear-pumped lasers. Chemical lasers are large and powerful sources of coherent radiation suitable for military applications. The hydrogen flouride laser, which results from a reaction between H_2 and F, emits in the 2.640–2.954 µm region

(Miller 1988). The iodine laser, which originates from a reaction of oxygen and iodine, emits at 1.3 μm (Yoshida et al. 1988).

Far-infrared lasers are optically pumped molecular lasers that cover the spectrum deep in the infrared from 0.1 to 1 mm (see, e.g., James et al. 1988). The active media in these lasers are molecules such as HCOOH and CH_3OH (Farhoomand and Pickett 1988) and the excitation is provided by tunable CO_2 lasers.

Nuclear-pumped lasers use reactors as excitation sources and various gaseous mixtures as the lasing media (Schneider and Cox 1988).

PROBLEMS

9.1 A fixed wavelength metrology application in the red portion of the visible spectrum $(600 \le \lambda \le 700 \text{ nm})$ requires homogeneous single-transverse-mode illumination, in a circular beam, with a Gaussian intensity profile polarized parallel to the plane of propagation, at a linewidth in the $1 \le \Delta\nu \le 3 \text{ GHz}$ range. The required CW power is in the few milliwatts range and the space constraints *are not* important. Using the information given in this chapter, and other relevant chapters, select an appropriate laser that would meet the requirements using a minimum of optical components.

9.2 A fixed wavelength metrology application in the red portion of the visible spectrum $(600 \le \lambda \le 700 \text{ nm})$ requires homogeneous single-transverse-mode illumination, in a circular beam polarized parallel to the plane of propagation, at a linewidth in the $1 \le \Delta\nu \le 3 \text{ MHz}$ range. The required CW power is in the few milliwatts range and the space constraints *are* important. Using the information given in this chapter, and other relevant chapters, select an appropriate laser that would meet the requirements using a minimum of optical components.

9.3 A sequential selective excitation application in the orange-red portion of the visible spectrum $(560 \le \lambda \le 610 \text{ nm})$ requires homogeneous single-transverse-mode illumination, in a TEM_{00} beam, polarized parallel to the plane of propagation, at a linewidth of $\Delta\nu \approx 350 \text{ MHz}$. The required average power is 100 W and the space constraints *are not* important. Using the information given in this chapter, and other relevant chapters, select the most direct and efficient laser that would meet the requirements using a minimum of optical complexity.

9.4 A directed energy application in the orange-red portion of the visible spectrum $(560 \le \lambda \le 610 \text{ nm})$ requires homogeneous single-transverse-mode illumination, in a circular beam polarized parallel to the plane of propagation, at a linewidth in the $350 \le \Delta\nu \le 700 \text{ MHz}$ range. The required average power is 10 kW, at a prf of 100 Hz. The space constraints *are not* important. Using the information given in this chapter, and other relevant chapters, select the most direct and efficient laser system that would meet the requirements using a minimum of optical complexity. An oscillator-amplifier system is allowed.

9.5 A fixed wavelength industrial application in the near-infrared portion of the spectrum $(1000 \le \lambda \le 2000 \text{ nm})$ requires homogeneous single-transverse-mode illumination, in a circular beam polarized parallel to the plane of propagation, at a linewidth of $\Delta v \approx 1$ GHz range. The required CW average power is 1 kW and the space constraints *are* important. Using the information given in this chapter, and other relevant chapters, select the most direct and efficient laser system that would meet the requirements using a minimum of optical complexity. An oscillator-amplifier system is allowed.

9.6 Identify the most suitable type of laser to be incorporated in a miniaturized instrumentation system for optical coherence tomography applications centered at $\lambda \approx 1300$ nm.

9.7 Consider at least two different fixed wavelength laser options for photolithography applications requiring radiation in the $150 \leq \lambda \leq 200$ nm. The specified average power is 100 W at a prf of 1 kHz. A master oscillator power amplifier system is allowed. Compare the cost, optical complexity, and energetic differences.

10 The *N*-Slit Laser Interferometer
Optical Architecture and Applications

10.1 INTRODUCTION

In this chapter, attention is given to one particular optical architecture that integrates several concepts introduced in Chapters 2, 4 and 6 and has various applications in imaging and metrology. That particular optical architecture is that of the *N*-slit laser interferometer (*N*SLI). Depending on the application, the *N*SLI can be configured with a narrow-linewidth tunable laser or a narrow-linewidth fixed-frequency laser.

10.2 OPTICAL ARCHITECTURE OF THE *N*SLI

The *N*SLI is illustrated in Figure 10.1. In its basic configuration, this interferometer requires the illumination from a laser emitting in a single transverse mode (TEM$_{00}$) with narrow-linewidth characteristics, which ideally should be confined to a single longitudinal mode (SLM). The beam is magnified by a two-dimensional transmission telescope, such as a Galilean telescope, and then propagates through an optional convex lens, which is necessary for microscopic and microdensitometry applications. Transmission through a telescope–convex lens combination yields a very tightly focused beam with an excellent depth of focus. Propagation of this beam, through a multiple-prism beam expander, maintains the focusing in the plane transverse to the plane of propagation and expands the beam in the plane of propagation. The result is an extremely elongated near-Gaussian beam typically 25 μm in height and 25,000 μm in width (see Figure 10.2). Note that this telescope–lens–multiple-prism configuration can easily yield additional beam expansion; for instance, Duarte (1987) reported on extremely elongated Gaussian beams 20 μm in height and 60,000 μm in width, that is, a spatial height-to-width ratio of 1:3000.

Albeit we refer here to these beam profiles as extremely elongated Gaussian beams, it should also be mentioned that particularly in the fields of microscopy, and nanoscopy, this type of illumination is also known as thin light-sheet illumination and selective plane illumination.

An alternative configuration includes the deployment of the focusing convex lens post multiple-prism expander. Also, straightforward interferometric configurations

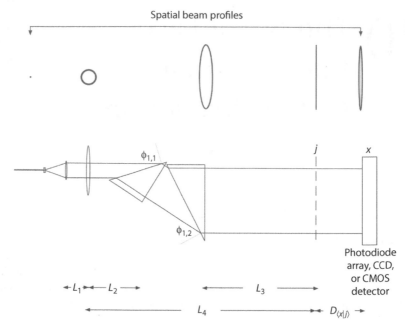

FIGURE 10.1 NSLI depicting the Galilean telescope, the focusing lens, the multiple-prism beam expander, the position of the N-slit array (j), or transmission surface of interest, and the interferometric plane (x). The intra-interferometric distance from j to x is $D_{\langle x|j\rangle}$. A depiction of approximate beam profiles, at various propagation stages, is included on top. This drawing is not to scale.

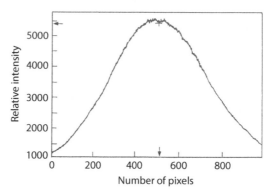

FIGURE 10.2 Intensity profile of an extremely elongated, approximately 1:1000 (height: width), near-Gaussian laser beam.

do not require the convex lens as part of the architecture. However, for the sake of completeness in the following analysis, the convex lens is included.

At this stage, it should be noted that the one-dimensional beam expansion parallel to the plane of propagation (or plane of incidence) is provided by a multiple-prism beam expander (Duarte 1987, 1991, 1993b) since this form of beam expansion does not introduce further focusing variables. Further, the multiple-prism beam expander can be designed to yield zero dispersion at a chosen wavelength of design (Duarte and Piper 1982; Duarte 1985) as described in Chapter 4.

As illustrated in Figure 10.1, the expanded laser beam illuminates the N-slit array at j where N sub-beams are created and proceed to propagate toward the detection screen at x, which is configured by a digital detector. Following some spatial displacement after j, the N sub-beams, due to divergence mandated by the uncertainty principle, begin to undergo interference since all these beams are composed of undistinguishable photons. The pattern of interference is recorded by the detector at x at the distance $D_{\langle x|j \rangle}$ from the N-slit array. Note that although the detector of choice is a digital detector, such as a photodiode array, *charge-coupled device* (CCD), or *complementary metal–oxide–semiconductor* (CMOS) array, photographic detection can also be used. If detection is provided by a linear array of microscopic detectors, then, as explained in Chapter 2, the spatial distribution of the interference signal can be described by

$$\left| \langle x|s \rangle \right|^2 = \sum_{j=1}^{N} \Psi(r_j) \sum_{m=1}^{N} \Psi(r_m) e^{i(\Omega_m - \Omega_j)} \tag{10.1}$$

which can be expressed as (Duarte and Paine 1989; Duarte 1991)

$$\left| \langle x|s \rangle \right|^2 = \sum_{j=1}^{N} \Psi(r_j)^2 + 2 \sum_{j=1}^{N} \Psi(r_j) \left[\sum_{m=j+1}^{N} \Psi(r_m) \cos(\Omega_m - \Omega_j) \right] \tag{10.2}$$

Interference in two dimensions is described by (Duarte 1995)

$$\left| \langle x|s \rangle \right|^2 = \sum_{z=1}^{N} \sum_{y=1}^{N} \Psi(r_{zy}) \sum_{q=1}^{N} \sum_{p=1}^{N} \Psi(r_{qp}) e^{i(\Omega_{qp} - \Omega_{zy})} \tag{10.3}$$

Equation 10.2 has been successfully applied to characterize the interference resulting from the interaction of expanded narrow-linewidth laser beams slit arrays of various dimensions and N in the range of $2 \leq N \leq 2000$ (see, e.g., Duarte 1993b, 1995).

10.2.1 Beam Propagation in the *N*SLI

The first stage of design is to optimize the transmission of the laser beam to the N-slit array. Although the telescope–convex lens system is polarization neutral, the multiple-prism beam expander has a strong transmission preference for radiation polarized parallel to the plane of propagation (see Chapter 5). The first step consists in matching the polarization of the laser to the polarization preference of the multiple-prism beam expander. For the mth prism, this transmission can be characterized using the expression

$$T_{1,m} = 1 - L_{1,m} \tag{10.4}$$

for the incidence surfaces and

$$T_{2,m} = 1 - L_{2,m} \qquad (10.5)$$

for the exit surfaces. The equations for the respective losses are given in Chapter 5 and are

$$L_{1,m} = L_{2,(m-1)} + \left[1 - L_{2,(m-1)}\right]\mathfrak{R}_{1,m} \qquad (10.6)$$

and

$$L_{2,m} = L_{1,m} + (1 - L_{1,m})\mathfrak{R}_{2,m} \qquad (10.7)$$

where:

$\mathfrak{R}_{1,m}$ and $\mathfrak{R}_{2,m}$ are given by

$$\mathfrak{R}_{\parallel} = \left[\frac{\tan^2(\phi - \psi)}{\tan^2(\phi + \psi)}\right] \qquad (10.8)$$

Note that given the inherent high intensity of laser sources, for most applications the use of antireflection coatings at the optics is not necessary.

The ray transfer matrix at the plane of propagation is given by (Duarte 1993a)

$$\begin{pmatrix} M_t\left(M - (\zeta/f)\right) & B_t\left(M - (\zeta/f)\right) + L_1(M/M_t) + (\zeta/M_t)\left(1 - (L_1/f)\right) \\ \\ -(M_t/Mf) & (MM_t)^{-1}\left(1 - (L_1/f)\right) - (B_t/Mf) \end{pmatrix} \qquad (10.9)$$

where the following quantities correspond to the multiple-prism beam expander:

$$M = M_1 M_2 \qquad (10.10)$$

$$M_1 = \prod_{m=1}^{r} k_{1,m} \qquad (10.11)$$

$$M_2 = \prod_{m=1}^{r} k_{2,m} \qquad (10.12)$$

and

$$B = M_1 M_2 \sum_{m=1}^{r-1} L_m \left(\prod_{j=1}^{m} k_{1,j} \prod_{j=1}^{m} k_{2,j}\right)^{-2} + \frac{M_1}{M_2}\sum_{m=1}^{r}\frac{l_m}{n_m}\left(\prod_{j=1}^{m} k_{1,j}\right)^{-2}\left(\prod_{j=m}^{r} k_{2,j}\right)^{2} \qquad (10.13)$$

where:

L_m is the distance separating the prisms

l_m is the path length at the mth prism

Also, M_t and B_t correspond to the A and B terms of the transfer matrix for the Galilean telescope given in Chapter 6 and

$$\zeta = ML_2 + B + \frac{L_3}{M} \tag{10.14}$$

In the above equations, L_1 is the distance between the telescope and the lens, L_2 is the distance between the lens and the multiple-prism beam expander, and L_3 is the distance between the multiple-prism beam expander and the *N*-slit array (see Figure 10.1).

For the vertical component, the ray transfer matrix is given by

$$\begin{pmatrix} M_t\left(1-(L_4/f)\right) & \left(1-(L_4/f)\right)\left(B_t+(L_1/M_t)\right)+(L_4/M_t) \\ \\ -(M_t/f) & (M_t)^{-1}\left(1-(L_1/f)\right)-(B_t/f) \end{pmatrix} \tag{10.15}$$

where:
L_4 is the distance between the lens and the *N*-slit array

In the absence of the convex lens following the two-dimensional telescope, Equations 10.9 and 10.15 reduce to

$$\begin{pmatrix} M_t M & B_t M + L_1(M/M_t) + (\zeta/M_t) \\ \\ 0 & (MM_t)^{-1} \end{pmatrix} \tag{10.16}$$

and

$$\begin{pmatrix} M_t & B_t + (L_1/M_t) + (L_4/M_t) \\ \\ 0 & (M_t)^{-1} \end{pmatrix} \tag{10.17}$$

The width of the Gaussian beam can be calculated using the expression given by Turunen (1986)

$$w(B) = w_0 \left[A^2 + \left(\frac{B}{L_{\mathcal{R}}} \right)^2 \right]^{1/2} \tag{10.18}$$

where:
the A and B terms are given by Equations 10.9 and 10.15, if using a convex lens, or by Equations 10.16 and 10.17 in the absence of a convex lens
$L_{\mathcal{R}} = (\pi w_0^2 / \lambda)$ is the Rayleigh length

The focused extremely elongated near-Gaussian coherent illumination is used in nanoscopic, microscopic, and microdensitometry applications (Duarte 1993a, 1993b) requiring illumination of N-slit arrays with transverse nanoscopic or microscopic dimensions. The unfocused elongated near-Gaussian illumination is used in conventional interferometric measurements (Duarte 2002, 2005; Duarte et al. 2010, 2011, 2013).

10.2.1.1 Example

In a propagation example discussed by Duarte (1995), $\lambda = 632.82$ nm, $w_0 = 250$ μm, $M = 5.75$, $M_t = 20$, $f = 30$ cm, and the elongated near-Gaussian beam becomes 53.4 mm wide by 32.26 μm high at the focal plane. Notice that the parameters in Equation 10.14 are chosen so that condition $\zeta \approx f$ is met and the width of the beam is dominated by the product MM_t. Appropriate selection of the distance from the lens to the focal plane also makes $L_4 \approx f$ so that the vertical dimension of the beam, determined by the A and B terms of Equation 10.15, becomes very small. The intensity profile of an expanded near-Gaussian beam with a spatial height-to-width ratio of approximately 1:1000 is shown in Figure 10.2. Removal of the convex lens yields a near-Gaussian beam approximately 53.4 mm wide by 10 mm high.

10.3   AN INTERFEROMETRIC COMPUTER

Interferograms recorded with the NSLI have been compared for numerous geometrical and wavelength parameters with interferograms calculated via Equation 10.1 or 10.2. One such case is reproduced in Figure 10.3. In this regard, it should be mentioned that good agreement, between theory and experiment, exists from the near to the far field. Slight differences, especially at the baseline, are due to thermal noise in the digital detector that is used at room temperature. The interferometric calculations, using Equations 10.1 through 10.3, require the following input information:

1. Slit dimensions, w
2. Standard deviation of the slit dimensions, Δw
3. Interslit dimensions
4. Standard deviation of interslit dimensions
5. Wavelength, λ
6. N-slit array, or grating, screen distance, $D_{\langle x|j\rangle}$
7. Number of slits, N

The program also gives options for the illumination profile and allows for multiple-stage calculations. That is, it allows for the propagation through several sequential N-slit arrays prior to arrival at x as considered in Chapter 2.

An interesting aspect of comparisons, between theory and experiment, is that for a given wavelength, a set of slit dimensions, and a distance from j to x, calculations in a conventional universal computer take longer as the number of slits N increases. In fact, the computational time $t(N)$ behaves in a nonlinear fashion as N increases. This is clearly illustrated in Figure 10.4, where $t = 0.96$ s for $N = 2$ and $t = 3111.2$s for $N = 1500$ (Duarte 1996). By contrast, all of these calculations can be performed in the NSLI at a constant time of ~30 ms, which is a time mainly imposed by the integration time of the digital detector.

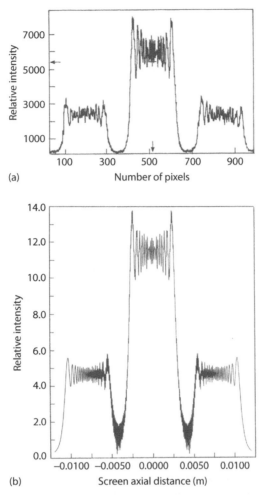

FIGURE 10.3 Measured interferogram (a) and calculated interferogram (b). Slits are 30 μm wide, separated by 30 μm, and $N = 100$. The intra-interferometric distance is $D_{\langle x|j\rangle} = 75$ cm and $\lambda = 632.82$ nm. This calculation assumes uniform illumination (see Chapter 2). (Reprinted from *Opt. Commun.*, 103, Duarte, F.J., On a generalized interference equation and interferometric measurements, 8–14, Copyright 1993b, with permission from Elsevier.)

In this regard, following the criteria outlined by Deutsch (1992), the *N*SLI can be classified as a physical, or interferometric, computer that can perform certain specific computations at times orders of magnitude below the computational time required by a universal computer. Among the computations that the *interferometric computer* can perform are as follows:

1. *N*-slit array interference calculations
2. Near- or far-field diffraction calculations
3. Beam divergence calculations
4. Wavelength calculations

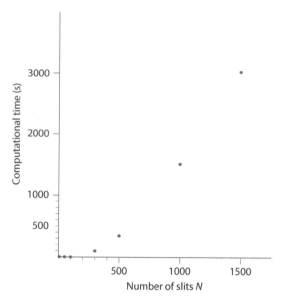

FIGURE 10.4   Computational time, in a universal computer, as a function of the number of slits. For these calculations, the slit width is 30 μm, the interslit width 30 μm, the *j*-to-*x* distance is $D_{\langle x|j\rangle} = 75$ cm, and $\lambda = 632.28$ nm. (Data from Duarte, F.J., *Proceedings of the International Conference on Lasers '95*, STS Press, McLean, VA, 1996.)

For these tasks, the interferometric computer based on the *N*SLI outperforms, by orders of magnitude, universal computers. Hence, it can be classified as a very fast, albeit limited in scope, optical computer. The advantage of the universal computer remains its versatility and better signal-to-noise ratio. Also, in the universal computer, there is access to intermediate results at all stages of the computation. This is not allowed in the *N*SLI where access is strictly limited to the input stage and the final stage of the computation. Attempts to acquire information about the intermediate stages of the computation destroy the final answer.

See Duarte (2014) for a discussion of quantum computing using *qubits*.

10.4   SECURE INTERFEROMETRIC COMMUNICATIONS IN FREE SPACE

Optical signals in free space have been used in the field of communications since ancient times. An example to optical communications in a modern context is the use of Morse code. More recent interest in this field has produced a variety of laser-based optical architectures and approaches (Yu and Gregory 1996; Boffi et al. 2000; Willebrand and Ghuman 2001). Prevalent among the secure approaches offering secure communications is quantum cryptography (Bennett and Brassard 1984; Duarte 2014). Here, an alternative approach to secure optical communications in free space, based on interferometric communications, is introduced and described.

For a given set of geometrical parameters and wavelength, the *N*SLI yields a unique interferogram that can be accurately matched to its theoretical counterpart given by

$$\left| \langle x|s \rangle \right|^2 = \sum_{j=1}^{N} \Psi(r_j)^2 + 2 \sum_{j=1}^{N} \Psi(r_j) \left[\sum_{m=j+1}^{N} \Psi(r_m) \cos(\Omega_m - \Omega_j) \right]$$

which was introduced in detail in Chapter 2. This feature can be utilized in the field of optical communications to perform secure communications in *free space*. The optical architecture of the *N*SLI used for this application is as described earlier with one modification that the distance from the *N*-slit array (*j*) to the digital detector (*x*) can be very large and allowances are made for a beam splitter, representing a possible intruder, to be inserted in the optical path between *j* and *x*, that is, $D_{\langle x|j \rangle}$. This modified, long-path length, *N*SLI is depicted in Figure 10.5. Again, the critical parameters determining the interferometric distribution are the narrow-linewidth laser wavelength λ, the slit width, the number of slits, and the intra-interferometric distance from the *N*-slit array (positioned at *j*) to the interference plane (positioned at *x*), which is denoted by the quantity $D_{\langle x|j \rangle}$ in Figure 10.5.

The principle of operation is extraordinarily simple: any optical distortion introduced in the intra-interferometric optical path $D_{\langle x|j \rangle}$ alters the predetermined interferogram recorded at *x*. Thus, the receiver at *x* immediately detects the presence of an intruder or eavesdropper in the optical path of communications. This means that interferometric communications provide a simple alternative to secure communications in free space. The method is particularly suited to provide secure communications in outer space.

As described by Duarte (2002), secure interferometric communications using the *N*SLI relies on an interferometric alphabet where an alphabetic character such as an *a* is related to a specific interferogram. Four possible *interferometric characters* corresponding to *a*, *b*, *c*, and *z* are shown in Figure 10.6. Here, the letter *a* is represented by two slits (*N* = 2), the letter *b* by three slits (*N* = 3), the letter *c* by four slits (*N* = 4), and so on. In these calculations, the slits are 50 μm wide and are separated by 50 μm at λ = 632.82 nm. Certainly, there is a limitless choice of alphabetic characters.

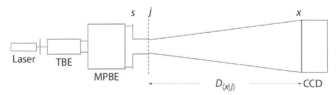

FIGURE 10.5 Very large *N*SLI configured without the focusing lens. In this class of interferometer, the intra-interferometric path is in the $7 \leq D_{\langle x|j \rangle} \leq 527$ m range, although larger configurations are possible. In these *N*SLIs, the TEM_{00} beam from the He–Ne laser is transmitted via a spatial filter. (Reproduced from Duarte, F.J., et al., *J. Opt.*, 12, 015705, 2010. With permission from the Institute of Physics.)

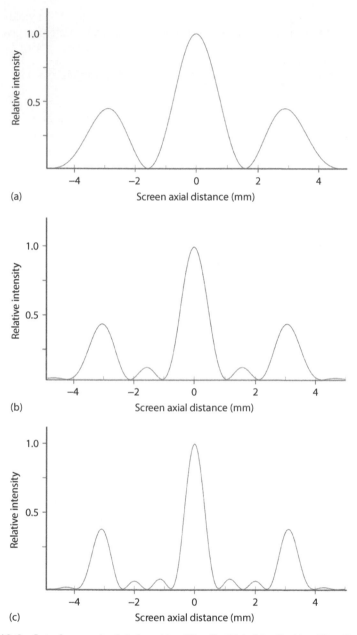

(a)

(b)

(c)

FIGURE 10.6 Interferometric alphabet: (a) *a* (*N* = 2); (b) *b* (*N* = 3); (c) *c* (*N* = 4).

Transmission integrity is demonstrated in Figure 10.7 for the case of the interferometric character *a* (*N* = 2). To this effect, an optically smooth surface with an average thickness of ~150 μm is introduced, at an angle, in the optical path to cause a reflection of the character *a*. It should be noted that insertion of the beam splitter normal to the optical axis produces no measurable spatial optical distortions except a decrease in the intensity of ~8%.

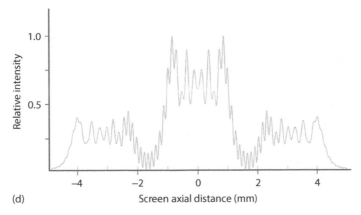

FIGURE 10.6 (Continued) (d) z ($N = 26$). For these calculations, the slits are 50 µm, separated by 50 µm, $D_{\langle x|j \rangle} = 50$ cm, and $\lambda = 632.82$ nm.

The angle of incidence of the interferometric character on the beam splitter was selected to be close to the Brewster angle of incidence to reduce transmission losses while still being able to reflect a measurable fraction of the signal. In the sequence of measurements, Figure 10.7a shows the original undistorted character a. The severe distortions depicted in Figure 10.7b–d show the effect of introducing the thin beam splitter in the optical path. Figure 10.7e depicts the intercepted interferometric character a. Although the severe distortions are no longer present, close scrutiny of the

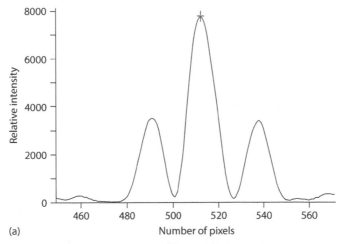

FIGURE 10.7 Interception sequence for the interferometric character a ($N = 2$) the slits are 50 µm, separated by 50 µm, $D_{\langle x|j \rangle} = 10$ cm, and $\lambda = 632.82$ nm. (a) Original interferometric character. (b–d) Show the distortion sequence of the interferometric character due to the insertion of a thin beam splitter into the optical path. (e) Interferometric character a showing slight distortion and displacement due to the stationary beam splitter. (Reprinted from *Opt. Commun.*, 205, Duarte, F.J., Secure interferometric equations in free space, 313–319, Copyright 2002, with permission from Elsevier.)

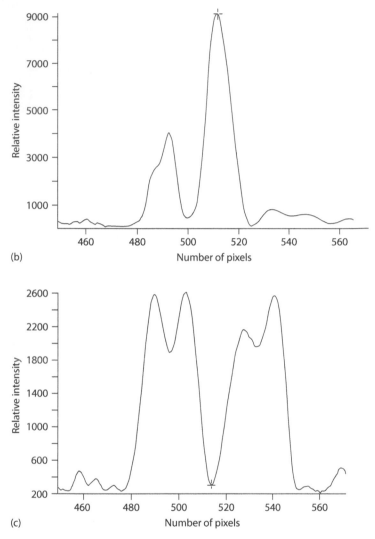

(b)

(c)

FIGURE 10.7   (Continued)

interferogram reveals a decrease by ~3.7% in the intensity of the signal relative to the original character shown in Figure 10.7a. The intercepted signal is also displaced by approximately 50 μm, in the frame of reference of the detector, due to the refraction induced at the beam splitter. In addition, there is a slight obliqueness in the intensity distribution as determined from the secondary maxima. Hence, in comparison with the original interferometric character or a theoretically generated character, it can be concluded that the integrity of the intercepted character *a* has been distinctly compromised.

Although the measurements considered above were performed over short propagation path lengths in the laboratory, of 0.1 and 1 m, Duarte (2002) also discussed propagation over larger distances. Using interferometric calculations, via

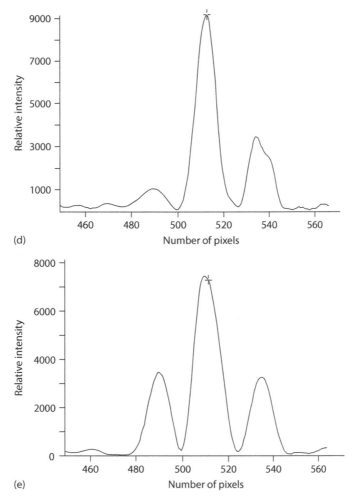

(d)

(e)

FIGURE 10.7   (Continued)

Equation 10.2 or 10.3, it can be shown that communications in free space can proceed over long path lengths using visible wavelengths and a detector comprising a few tiled photodiode arrays. One specific example involves the generation of the interferometric character *a* using two 1 mm slits separated by 1 mm. For $\lambda = 632.82$ nm, this arrangement produces an interferometric distribution bound within 10 cm for a propagation path length of 100 m. The interferometric character *z* is produced by an array of $N = 26$ slits of 1 mm, separated by 1 mm. For $\lambda = 632.82$ nm, this arrangement produces an interferometric distribution bound within 14 cm for a propagation path length of 100 m. This can be accomplished using two linear photodiode arrays (each 72 mm long) tiled together. If the dimensions of the slits are increased to 3 mm, at $\lambda = 441.16$ nm, interferometric characters could be propagated over distances of 1000 m using four such tiled photodiode arrays (Duarte 2002).

10.4.1 VERY LARGE NSLIs FOR SECURE INTERFEROMETRIC COMMUNICATIONS IN FREE SPACE

The examples considered up to now assume a propagation intra-interferometric path $D_{\langle x|j\rangle}$ characterized by a single homogeneous propagation medium, such homogeneous air or vacuum, between j and x.

Free-space communications in terrestrial environments of interferometric characters would need to account for the inherent atmospheric turbulence present in such surroundings. This would certainly detract from the simplicity of the method. This could still be accomplished, noting that atmospheric distortions are stochastic in nature compared to systematic distortions introduced by optical interception.

$D_{\langle x|j\rangle}$ has been extended experimentally to the meter range (Duarte 2005), to 35 m (Duarte et al. 2010), and to 527 m (Duarte et al. 2011).

The c interferometric character ($N = 4$), generated with 570 μm-wide slits separated by 570 μm, using $\lambda = 632.82$ nm, at $D_{\langle x|j\rangle} = 7.235$ m is shown in Figure 10.8. Interception of this interferometric character by a high-optical-quality ultrathin beam splitter at Brewster's angle produces the collapse of the interferogram as illustrated in Figure 10.9. Interception of the interferometric character by mild air turbulence, generated by a heat source, yields mild distortions of the c interferometric character as illustrated in Figure 10.10.

The c interferometric character ($N = 4$), generated with 1000 μm-wide slits separated by 1000 μm, using $\lambda = 632.82$ nm, at $D_{\langle x|j\rangle} = 35$ m in open air at $T \approx 30°$C is shown in Figure 10.11. Visible here is a very slight distortion of the interferometric character due to very mild atmospheric turbulence. This led to the observation that the NSLI is applicable as an effective detector of *clear air turbulence* and could be deployed, using infrared lasers, at the thresholds of aviation runways to alert pilots of any possible detrimental turbulence (Duarte et al. 2010).

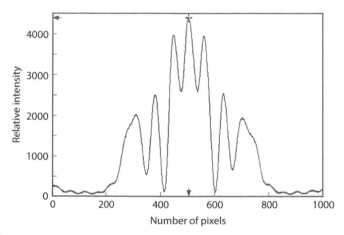

FIGURE 10.8 The interferometric character c ($N = 4$). Here, the slits are 570 μm, separated by 570 μm, $D_{\langle x|j\rangle} = 7.235$ m, and $\lambda = 632.82$ nm. (Reproduced from Duarte, F.J., *J. Opt. A: Pure Appl. Opt.*, 7, 73–75, 2005. With permission from the Institute of Physics.)

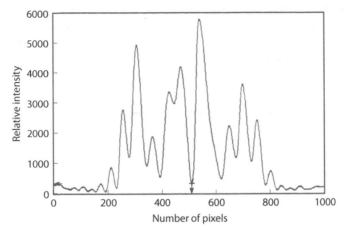

FIGURE 10.9 The interferometric character *c* (*N* = 4), as described in Figure 10.8, destroyed by optical interception. See text for further details. (Reproduced from Duarte, F.J., *J. Opt. A: Pure Appl. Opt.*, 7, 73–75, 2005. With permission from the Institute of Physics.)

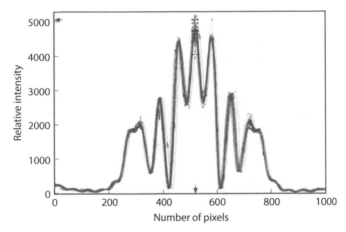

FIGURE 10.10 The interferometric character *c* (*N* = 4), as described in Figure 10.8, distorted due to turbulence generated by a thermal source. See text for further details. (Reproduced from Duarte, F.J., *Tunable Laser Applications*, CRC Press, New York, 2009. With permission from Taylor & Francis.)

Experiments at $D_{\langle x|j \rangle} = 527$ m also led to the discovery that the interferometric characters could be intercepted and modified in a controlled nondestructive, or nondemolition, manner using transparent spider silk web fibers (Duarte et al. 2011). Indeed, accurate knowledge of the position of the fiber relative to the propagating interferometric character leads to an accurate prediction of the interferometric character showing a superimposed diffraction signature created by the interaction of the spider web fiber (with a diameter in the 25–30 μm range) with the *N*-slit interferogram.

For the *b* interferometric character (*N* = 3) generated with 570 μm-wide slits separated by 570 μm, using λ = 632.82 nm, at $D_{\langle x|j \rangle} = 7.235$ the control interferogram is shown in Figure 10.12. Inserting the spider web fiber 15 cm from the x plane (or CCD detector) at $D_{\langle x|j \rangle} = 7.235 - 0.150$ yields a measured interferogram as illustrated in Figure 10.13. Representation of the spider web fiber as two wide slits separated by 25 µm while using

$$\left| \langle x|s \rangle \right|^2 = \sum_{j=1}^{N} \Psi(r_j)^2 + 2 \sum_{j=1}^{N} \Psi(r_j) \left[\sum_{m=j+1}^{N} \Psi(r_m) \cos(\Omega_m - \Omega_j) \right]$$

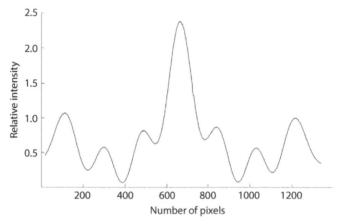

FIGURE 10.11 The interferometric character c ($N = 4$) showing a slight unevenness due to incipient atmospheric turbulence. Here, the slits are 1000 µm, separated by 1000 µm, $D_{\langle x|j \rangle} = 35$ m, and $\lambda = 632.82$ nm. (Reproduced from Duarte, F.J., et al., *J. Opt.*, 12, 015705, 2010. With permission from the Institute of Physics.)

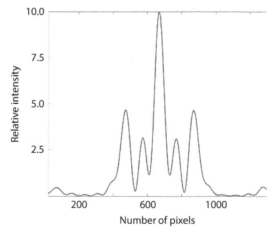

FIGURE 10.12 The interferometric character b ($N = 3$) used as a control. Here, the slits are 570 µm, separated by 570 µm, $D_{\langle x|j \rangle} = 7.235$ m, and $\lambda = 632.82$ nm. (Reproduced from Duarte, F.J., et al., *J. Mod. Opt.*, 60, 136–140, 2013. With permission from Taylor & Francis.)

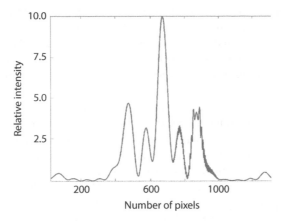

FIGURE 10.13   The interferometric character b ($N = 3$) intercepted by a spider web silk fiber positioned at $D_{\langle x|j \rangle} = 7.235 - 0.150$ m. The slits are 570 μm, separated by 570 μm; the interferogram is recorded at $D_{\langle x|j \rangle} = 7.235$ m and $\lambda = 632.82$ nm. (Reproduced from Duarte, F.J., et al., *J. Mod. Opt.*, 60, 136–140, 2013. With permission from Taylor & Francis.)

in a cascade approach (Duarte 1993b) where the predicted interferogram at $D_{\langle x|j \rangle} = 7.235 - 0.150$ m becomes the input distribution at the new array yields the calculated interferogram illustrated in Figure 10.14, which agrees closely with the measured interferogram of Figure 10.13 (Duarte et al. 2013; Duarte 2014). These results illustrate that albeit it is possible to interact nondestructively with the propagating interferogram, the presence of the interception is still detected. More subtle, harder to detect nondemolition interactions are discussed by Duarte et al. (2013) and Duarte (2014).

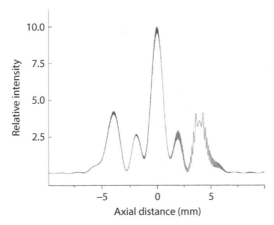

FIGURE 10.14   Theoretical interferometric character b ($N = 3$) assuming interception by a spider web silk fiber positioned at $D_{\langle x|j \rangle} = 7.235 - 0.150$ m (see text). The slits are 570 μm, separated by 570 μm; the interferometric plane is positioned at $D_{\langle x|j \rangle} = 7.235$ m and $\lambda = 632.82$ nm. (Reproduced from Duarte, F.J., et al., *J. Mod. Opt.*, 60, 136–140, 2013. With permission from Taylor & Francis.)

One modification applicable to these large and very large NSLIs would be the introduction of a distortionless high-fidelity beam expander, such as an optimized multiple-prism beam expander. In Chapter 4, it was shown that these expanders can easily provide beam magnification factors of $M \approx 100$. Deployment of such multiple-prism beam expander next to the slit array would reduce the beam divergence significantly, thus reducing the requirements on the dimensions of the digital detectors. From a technological viewpoint, it is important to emphasize the use of TEM_{00} lasers with narrow-linewidth, preferably SLM, emission since that characteristic is essential in providing well-defined sharp high-visibility interferometric characters close to their theoretical counterparts. The characters could be changed in real time either by using a tunable laser or by incorporating precision variable slit arrays. The use of narrow band-pass filters could allow transmission during daylight.

Quantum cryptography provides secure optical communications guaranteed by the quantum entanglement of polarizations (Pryce and Ward 1947; Ward 1949) and has been shown to be applicable over distances of tens of kilometers [see Duarte (2014) for a recent review]. Interferometric communications using the NSLI provides security using the principle of interference. As outlined in Chapter 3, the uncertainty principle itself can be formulated from interferometric arguments. Advantages of free-space communications using interferometric characters include a very simple optical architecture and the use of relatively high-power narrow-linewidth SLM lasers, although the method also applies to single-photon emission.

As a final point of interest, it can be shown that the quantum entanglement of polarizations also has an interferometric origin (Duarte 2013, 2014) as outlined in Chapter 3.

10.5   APPLICATIONS OF THE NSLI

In this section, various applications of the NSLI are described. First, it is discussed as a digital laser microscope (DLM) and as a tool to perform light modulation measurements in imaging. Next, its application in secure optical communications is considered. The section concludes with a discussion on wavelength and temporal measurements.

10.5.1   Digital Laser Micromeasurements

Micromeasurements are widely applied in the field of imaging. One such class of measurements, called microdentitometry, was described by Dainty and Shaw (1974). In a traditional microdensitometer, a beam of light, with a diameter typically in the 10–50 µm range, is used to illuminate a transmission surface. The ratio of the transmitted intensity (I_t), through the surface, over the incident intensity (I_i) is a measure of the transmission and the density is defined as (Dainty and Shaw 1974)

$$D = \log_{10}\left(\frac{I_i}{I_t}\right) \qquad (10.19)$$

Thousands of measurements over the surface yield an average density and a standard deviation. The standard deviation is a measure of the so-called *granularity*, or σ and is a parameter widely used to evaluate the microdensity characteristics of transmission imaging materials such as photographic films. Typically, the optical density of photographic films varies in the $0.1 \leq D \leq 3.0$ range. A low granularity value, indicating a fine film, would be in the $0.001 \leq \sigma \leq 0.005$ range. Traditional high-speed microdensitometers using incoherent illumination sources face various challenges, including adverse signal-to-noise ratios, to determine microdensity variations in very fine imaging surfaces and very short depths of focus.

The use of the NSLI as a DLM was introduced by Duarte (1993b, 1995). In a DLM, an extremely elongated near-Gaussian beam illuminates, at its focal point, the width of the imaging surface of interest. The interaction of the expanded illumination beam with the imaging surface at j yields an interference pattern at x. In this regard, the imaging surface at j can be considered as a regular, or an irregular, transmission grating, and the interference pattern at x is unique to that imaging surface. If the smooth expanded illumination distribution, illustrated in Figure 10.15a,

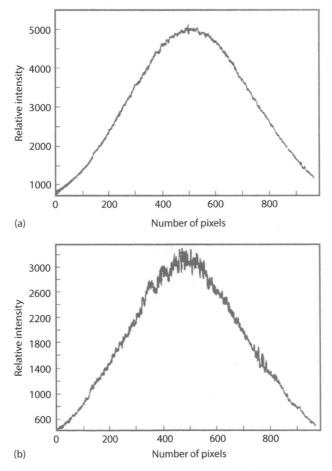

FIGURE 10.15 Transmission signal showing no interference from an optical homogeneous imaging surface (a) and interferogram from an imaging surface including relatively fine particles (b).

is defined as $I_i(x,\lambda)$, and the transmitted interferometric signal as $I_t(x,\lambda)$, then the optical density can be defined as

$$D(x,\lambda) = \log_{10}\left[\frac{\int I_i(x,\lambda)\mathrm{d}x}{\int I_t(x,\lambda)\mathrm{d}x}\right] \qquad (10.20)$$

This equation provides the macrodensity of the surface under illumination. The microdensity is obtained by exploiting the spatial dependence of the transmitted interferogram in conjunction with the spatial discrimination available from the digital detector. In order to illustrate this mode of operation, consider a unilayer film of very fine grains. The slits of very small dimensions cause, according to the uncertainty principle, a high divergence, which implies that the interferometric pattern at x is characterized by fine features of moderate modulation. By contrast, a unilayer film of coarser grains, or slits of larger dimensions, causes less divergence so that the interference pattern at x is characterized by coarser features and larger modulation. This comparison is illustrated in Figure 10.15.

The average microdensity, at a given wavelength, is obtained from

$$D(x,\lambda) = \left[\sum_{x=1}^{N}\frac{I_i(x,\lambda)}{I_t(x,\lambda)}\right]N^{-1} \qquad (10.21)$$

The standard deviation of this quantity is a measure of the granularity, or σ, of the imaging surface. The number N is determined by the number of pixels in the digital detector, which is typically 1024 or 2048. The size of the sampling depends on the dimensions of the pixels that vary from a few micrometers to 25 μm.

Detailed cross-over measurements between traditional microdensitometers, with incoherent illumination, and the DLM have yielded very good agreement of macrodensities and similar behavior in σ as a function of D. Absolute values of σ in the DLM tend to be higher than the values determined with traditional means. One essential advantage of the DLM is that from interferograms such as those displayed in Figure 10.15 it is possible to determine the average size of the slits causing the interference. Further characteristics that make DLMs very attractive are as follows:

1. A dynamic range approaching 10^9
2. A signal-to-noise ratio $\sim\!10^7$
3. A depth of focus greater than 1 mm
4. Simultaneous collection of a large number of data points

From an imaging perspective, it should be mentioned that the mathematical form of Equation 10.3 is similar to the equation of power spectrum, which is widely applied in traditional studies of microdensitometry.

A simple modification of the optical architecture transforms the DLM from a transmission mode to a reflection mode as illustrated in Figure 10.16. The same physics applies. This configuration is useful to determine surface characteristics of imaging surfaces.

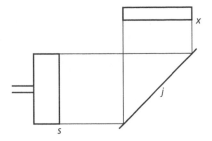

FIGURE 10.16   *N*SLI configured in the reflection mode.

10.5.2   Light Modulation Measurements

Modulation transfer measurements are extensively used in imaging to determine the spatial resolution of transmission gratings configured by coatings of various materials. In principle, the technique is quite simple and consists in coating regular *N*-slit gratings, with a given material, for a series of spatial frequencies. Then the near-field modulation of the light transmitted via these gratings is recorded as a function of spatial frequency. In transmission gratings comprising imaging materials of a crystalline nature, the spatial resolution decreases as the spatial frequency increases. This is manifested in a deterioration of the light modulation as the spatial frequency increases.

The *N*SLI can be applied in a straightforward manner to quantify the modulation of light, by a given transmission grating, by configuring the interferometer with a fairly short *j*-to-*x* distance. This is demonstrated in Figure 10.17. Here, a grating made from a metallic coating with slits 100 μm wide, separated by 100 μm, and $N = 23$ is illuminated with an expanded near-Gaussian beam at $\lambda = 632.82$ nm. The *j*-to-*x* distance, that is, $D_{\langle x|j \rangle}$, is 1.5 cm. Note that, at this grating-to-detector distance, for the slit dimensions given, interference is rather weak and does not dominate the modulation of the signal. A theoretical version of the signal is given in Figure 10.17b.

Comparison between theory and experiment shows that the depth of modulation even for a metallic coating, ~90% in this case, is less than the theoretical modulation. Transmission gratings made from photographic coatings can show a significant deterioration in modulation for spatial frequencies beyond ~40 lines/mm.

10.5.3   Wavelength Meter and Broadband Interferograms

Generalized *N*-slit interference equations, such as Equations 10.2 and 10.3, are inherently wavelength dependent, since the interference term is a function of wavelength as explained in Chapter 2. Thus, it is straightforward to predict that, for a fixed set of geometrical parameters, the measured interferogram depends uniquely on the wavelength of the laser. This feature can be applied to use the *N*SLI as a wavelength meter as explained in detail in Chapter 11.

Albeit emphasis has been made up to now on the desirability of using narrow-linewidth lasers, in conjunction with the *N*SLI, the scope of the measurements can also be extended to include broadband emission. For broadband emission sources,

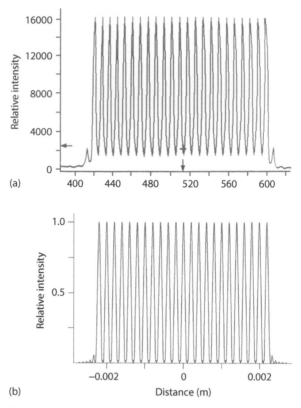

(a)

(b)

Distance (m)

FIGURE 10.17 Near-field modulation signal in a weak interferometric domain arising from the interaction of laser illumination, at $\lambda = 632.82$ nm, and a grating composed of $N = 23$ slits 100 μm wide separated by 100 μm. The intra-interferometric distance is $D_{\langle x|j \rangle} = 1.5$ cm. Measured modulation signal (a) and calculated signal (b). Each pixel is 25 μm wide. (Reprinted from *Opt. Commun.*, 103, Duarte, F.J., On a generalized interference equation and interferometric measurements, 8–14, Copyright 1993b, with permission from Elsevier.)

the measured interferogram represents a cumulative interferogram, resulting from a series of individual wavelengths, as illustrated in Figure 10.18. This concept is central to this measurement approach and it is based on Dirac's dictum on interference (Dirac 1978). That is, interference occurs between undistinguishable photons only. In other words, blue photons do not interfere with green or red photons. Hence, an interferogram with broad features, as illustrated in Figure 10.18, is a cumulative signal integrated by a series of individual interferograms arising from a series of different wavelengths (see Chapter 11). Once the central wavelength of emission of the broadband interferometer is determined, using a standard spectrometer or suitable wavelength meter, a theoretical cumulative interferogram can be constructed to match the measured signal and determine its bandwidth.

In principle, for broadband ultrashort pulsed lasers, once the bandwidth of the emission is determined, an approximate estimate of the temporal pulse duration is possible using the time–frequency uncertainty relation $\Delta v \Delta t \approx 1$. This simple

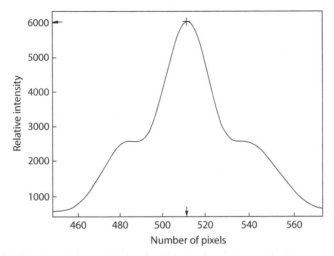

FIGURE 10.18 Measured double-slit interferogram generated using a broadband visible light source. The slits are 50 μm wide separated by 50 μm and the intra-interferometric distance is $D_{\langle x|j \rangle} = 10$ cm.

concept is only applicable to ultrashort pulse lasers emitting pulses and spectral distributions obeying the time–frequency uncertainty limit.

10.5.4 IMAGING LASER PRINTERS

Traditional sensitometry and sensitometers are described by Altman (1977). In essence, a sensitometer is an instrument that illuminates an unexposed imaging material to produce a series of exposures at various light intensity levels. A laser sensitometer uses stable lasers yielding TEM_{00} beams and various optical techniques to *print* a scale of exposures that can then be optically characterized to determine the sensitivity of the imaging material. In this section, the optical architecture of a multiple-laser sensitometer, or multiple-laser printer, is described.

Laser sensitometers work on the principle of exposing a line by displacing a focused near-Gaussian beam with a beam waist in the $60 \le w \le 100$ μm range. This line is exposed on the imaging material that is deployed at a plane perpendicular to the optical axis and to the plane of propagation. The imaging medium is displaced, orthogonal to the plane of propagation, at a velocity allowing for an overlap (usually 50%) of the near-Gaussian beam. The movement of the laser beam provides the temporal component of the exposure. Once an exposure of certain dimensions is produced, usually 10 mm in width, the intensity of the laser beam is adjusted, using electro-optical means, and a new series of line exposures is produced. Eventually a scale of rectangular exposures, at different laser intensity levels, is rendered. Three optical channels, corresponding to blue, green, and red lasers, converging to a single exposure plane, are often employed.

An industrial laser printer, for sensitometry applications, is depicted in Figure 10.19. This is a single-channel, multiple-laser, multiple-prism printer using polarization to

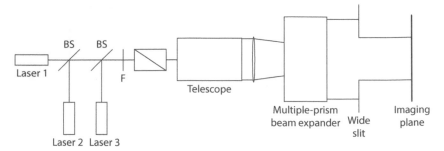

FIGURE 10.19 Three-color industrial PMPML printer used to expose a scale of images at various laser intensities for sensitometric measurements. The telescope expands the beam in two dimensions, whereas the multiple-prism beam expander magnifies in only one dimension parallel to the plane of propagation.

vary the intensity of the laser exposure (Duarte 2001). In this description, this laser printer is referred to as a polarizer multiple-prism multiple-laser (PMPML) printer.

The PMPML printer described in Figure 10.19 uses a single optical channel with a principal laser as the first element defining the optical axis and secondary lasers adding their radiation via beam splitters. All lasers are polarized parallel to the plane of incidence. A variable broadband neutral density filter is inserted as a coarse intensity control prior to the polarizer. The polarizer is a Glan–Thompson prism pair, with an extinction coefficient of 1×10^{-6} or better, mounted on a high-precision annular rotational stage capable of a 0.001 arc sec angular resolution. As described in Chapter 5, rotation of this polarizer causes the transmission of the lasers to decrease from nearly full transmission to total extinction (Duarte 2001). For the lasers polarized parallel to the plane of propagation, optimum transmission is accomplished with the Glan–Thompson polarizer deployed as depicted in Figure 10.19. Rotation of the polarizer by $\pi/2$ radians causes complete extinction of the combined laser beam, and thus no exposure. The use of Glan–Thompson polarizers to vary and fine-tune the intensity of lasers used in laser cooling experiments was reported by Olivares et al. (2009).

Following the polarizer, the beams enter a telescope–lens system and a multiple-prism beam expander as described in Figure 10.1. The elongated near-Gaussian beams are then propagated through a wide aperture so as to produce a diffractive profile as depicted in Figure 10.20. Note that the diffractive profile is wider than the width of the exposures needed so that the intensity variation caused by the *ears* of the profile does not affect the wanted area. The slight variations toward the center of the distribution have a negligible effect.

Using the method just described, a line exposure is instantaneously printed, thus eliminating the need to displace the laser beams and the associated electromechanical means necessary to accomplish this task. Since the line exposure is horizontal, the imaging material is displaced in a plane orthogonal to the plane of incidence of the instrument. In PMPML printers, the temporal exposures are provided by using the lasers in a pulsed mode. Thus, depending on the lasers, it is possible to vary the exposure time from less than 1 to 1000 ns, thus greatly increasing the range of exposures available for sensitometry and imaging applications (Duarte et al. 2005).

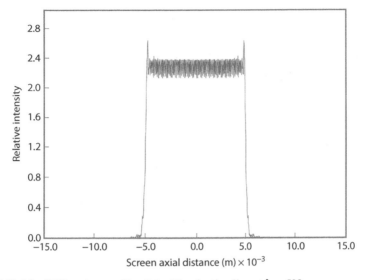

FIGURE 10.20 Diffraction profile of the illumination line at $\lambda = 532$ nm.

It should also be mentioned that careful selection of the beam profiles of the lasers and their respective distances to the main optical axis enables spatial overlapping of the laser beams to within 1 μm at the focal plane with a minimal use of extra beam shaping optics. As indicated by Duarte (2001), lasers suitable for illumination include mode-locked diode-pumped frequency-doubled Nd:YAG lasers and pulsed semiconductor lasers.

PROBLEMS

10.1 For the telescope, lens, multiple-prism, architecture depicted in Figure 10.1, derive its propagation ray transfer matrix given in Equation 10.9.

10.2 Show that in the absence of a convex lens, Equation 10.9 reduces to Equation 10.16.

10.3 Show that in the absence of a convex lens, Equation 10.15 reduces to Equation 10.17.

10.4 Design a double-prism beam expander yielding zero dispersion at the wavelength of design and $M = 5$ for an optical system as depicted in Figure 10.1. Using a Galilean telescope with $M_t = 20$, and a convex lens with $f = 30$ cm, calculate the width and the height of the resulting extremely elongated near-Gaussian beam at the focal plane.

Assume a TEM_{00} He–Ne laser at $\lambda = 632.82$ nm and $w_0 = 250$ μm. For the material of the multiple-prism beam expander, use fused silica.

10.5 For a beam splitter made of fused silica and a thickness of 0.4 mm, determine the lateral displacement, from its original path, of a TEM_{00} He–Ne laser beam at $\lambda = 632.82$ nm, immediately following the beam splitter if the angle of incidence is at the Brewster angle.

11 Interferometry

11.1 INTRODUCTION

In this chapter, some widely applied interferometric configurations in the measurement of wavelength and linewidth are described. Attention is focussed on two-beam, and multiple-beam, interferometers. See Steel (1967) for a detailed treatment on the subject of interferometry. The presentation given here follows that of the first edition of *Tunable Laser Optics* (Duarte 2003) while adopting modifications and improvements introduced mainly in the work of Duarte (2014).

11.2 TWO-BEAM INTERFEROMETERS

Two-beam interferometers are optical devices that divide and then recombine a light beam. It is on recombination of the beams that interference occurs. The most well-known two-beam interferometers are the Sagnac interferometer, the Mach–Zehnder interferometer, and the Michelson interferometer. For a highly coherent light beam, such as the beam from a narrow-linewidth laser, the coherence length

$$\Delta x \approx \frac{c}{\Delta \nu} \tag{11.1}$$

can be rather large, thus allowing a relatively large optical path length in the two-beam interferometer of choice. Alternatively, this relation provides an avenue to accurately determine the linewidth of a laser by increasing the optical path length until interference ceases to be observed.

11.2.1 THE SAGNAC INTERFEROMETER

The Sagnac, or cyclic, interferometer is illustrated in Figure 11.1. In this interferometer, the incident light beam is divided into two sub-beams by a beam splitter (BS). The reflected beam, on the incidence BS, is then sent into a path defined by the reflections on M_1, M_2, and M_3 mirrors. The transmitted beam, on the incidence BS, is sent into a path defined by the reflections on M_3, M_2, and M_1 mirrors. Both counter-propagating beams are recombined at the BS. The interference mechanics of the counter-propagating round trips can be described using Dirac's notation via the probability amplitude:

$$\langle x|s \rangle = \langle x|j \rangle \langle j|M_3 \rangle \langle M_3|M_2 \rangle \langle M_2|M_1 \rangle \langle M_1|j \rangle \langle j|s \rangle$$
$$+ \langle x|j' \rangle \langle j'|M_1 \rangle \langle M_1|M_2 \rangle \langle M_2|M_3 \rangle \langle M_3|j' \rangle \langle j'|s \rangle \tag{11.2}$$

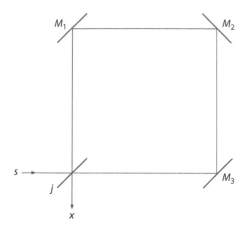

FIGURE 11.1 Sagnac interferometer. All three mirrors M_1, M_2, and M_3 are assumed to be identical.

where:
  j refers to the reflection mode of the BS
  j' refers to the transmission mode of the BS

Assuming that

$$\langle j|M_3\rangle\langle M_3|M_2\rangle\langle M_2|M_1\rangle\langle M_1|j\rangle = 1 \qquad (11.3)$$

and

$$\langle j'|M_1\rangle\langle M_1|M_2\rangle\langle M_2|M_3\rangle\langle M_3|j'\rangle = 1 \qquad (11.4)$$

Then, Equation 11.2 reduces to

$$\langle x|s\rangle = \langle x|j\rangle\langle j|s\rangle + \langle x|j'\rangle\langle j'|s\rangle \qquad (11.5)$$

If $j' = 1$ represents the BS in the transmission mode and $j = 2$ in the reflection mode, then Equation 11.5 can be written as (Duarte 2003)

$$\langle x|s\rangle = \langle x|2\rangle\langle 2|s\rangle + \langle x|1\rangle\langle 1|s\rangle \qquad (11.6)$$

which, for $N = 2$, can be expressed as

$$\langle x|s\rangle = \sum_{j=1}^{N=2}\langle x|j\rangle\langle j|s\rangle \qquad (11.7)$$

An alternative triangular Sagnac interferometer, with only two mirrors (M_1 and M_2), is shown in Figure 11.2.

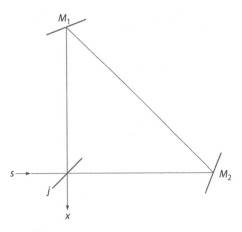

FIGURE 11.2 Triangular Sagnac interferometer.

11.2.2 THE MACH–ZEHNDER INTERFEROMETER

The Mach–Zehnder interferometer is illustrated in Figure 11.3. In this interferometer, the incident light beam is divided into two sub-beams by a BS. The reflected
beam, on the incidence BS, is then sent into a path defined by the reflection on M_1
toward the exit BS. The transmitted beam, on the incidence BS, is sent into a path
defined by the reflection on M_2 toward the exit BS. Both counter-propagating beams
are recombined at the exit BS. The interference mechanics of the counter-propagating
beams can be described using Dirac's notation via the probability amplitude

$$\langle x|s\rangle = \langle x|k'\rangle\langle k'|M_1\rangle\langle M_1|j\rangle\langle j|s\rangle + \langle x|k\rangle\langle k|M_2\rangle\langle M_2|j'\rangle\langle j'|s\rangle \qquad (11.8)$$

which can be abstracted to

$$\langle x|s\rangle = \langle x|k'\rangle\langle k'|j\rangle\langle j|s\rangle + \langle x|k\rangle\langle k|j'\rangle\langle j'|s\rangle \qquad (11.9)$$

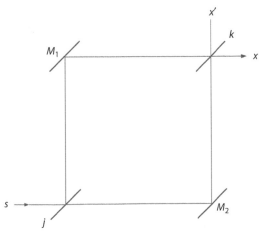

FIGURE 11.3 Mach–Zehnder interferometer.

If $j' = k' = 1$ represents the BS in the transmission mode and $j = k = 2$ in the reflection mode, then Equation 11.9 can be written as (Duarte 2003)

$$\langle x|s\rangle = \langle x|1\rangle\langle 1|2\rangle\langle 2|s\rangle + \langle x|2\rangle\langle 2|1\rangle\langle 1|s\rangle \tag{11.10}$$

The same result can be obtained from

$$\langle x|s\rangle = \sum_{k=1}^{N}\sum_{j=1}^{N}\langle x|k\rangle\langle k|j\rangle\langle j|s\rangle \tag{11.11}$$

which leads to

$$\begin{aligned}\langle x|s\rangle &= \langle x|1\rangle\langle 1|2\rangle\langle 2|s\rangle + \langle x|1\rangle\langle 1|1\rangle\langle 1|s\rangle \\ &\quad + \langle x|2\rangle\langle 2|2\rangle\langle 2|s\rangle + \langle x|2\rangle\langle 2|1\rangle\langle 1|s\rangle\end{aligned} \tag{11.12}$$

However, since $\langle 1|1\rangle$ and $\langle 2|2\rangle$ illuminate x' rather than x, the probability amplitude, for this geometry, reduces to that given in Equation 11.10.

A prismatic Mach–Zehnder interferometer is illustrated in Figure 11.4. In this prismatic version of the Mach–Zehnder, there is asymmetry in regard to the intra-interferometric beam dimensions. The $P_1 - M_2 - P_2$ beam is expanded relatively to the $P_1 - M_1 - P_2$ beam. Also, in this particular example (based on a prism with a magnification of $k_{1,1} \approx 5$), there is also a power asymmetry since the unexpanded beam propagating in the $P_1 - M_1 - P_2$ arm has about 30% of the incident power, whereas the expanded beam $P_1 - M_2 - P_2$ carries the remaining 70% of the incident power, for light polarized parallel to the plane of incidence. In this regard, it should be possible to design a prismatic Mach–Zehnder where the power density (W m^{-2}) in each arm is balanced. Applications for this type of interferometer include imaging and microscopy. Additional Mach–Zehnder interferometric configurations include transmission gratings as BS (Steel 1967).

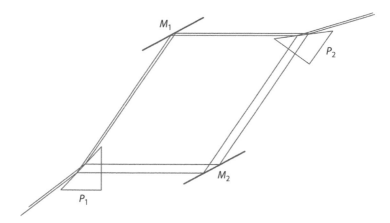

FIGURE 11.4 Prismatic Mach–Zehnder interferometer. (Reproduced from Duarte, F.J., *Quantum Optics for Engineers*, CRC Press, New York, 2014. With permission.)

11.2.3 The Michelson Interferometer

The Michelson interferometer (Michelson 1927) is illustrated in Figure 11.5. In this interferometer, the incident light beam is divided into two sub-beams by a BS that serves as both input and output elements. The reflected beam, on the incidence BS, is then sent into a path defined by the reflection on M_1 and back toward the exit BS. The transmitted beam, on the incidence BS, is sent into a path defined by the reflection on M_2 and back toward the exit BS. Both beams are recombined interferometrically at the BS. For the Michelson interferometer, the interference can be characterized using a probability amplitude of the form:

$$\langle x|s\rangle = \langle x|j\rangle\langle j|M_2\rangle\langle M_2|j'\rangle\langle j'|s\rangle + \langle x|j'\rangle\langle j'|M_1\rangle\langle M_1|j\rangle\langle j|s\rangle \qquad (11.13)$$

which can be abstracted to

$$\langle x|s\rangle = \langle x|j\rangle\langle j|j'\rangle\langle j'|s\rangle + \langle x|j'\rangle\langle j'|j\rangle\langle j|s\rangle \qquad (11.14)$$

If $j' = 1$ represents the function of the BS in the transmission mode and $j = 2$ in the reflection mode:

$$\langle x|s\rangle = \langle x|2\rangle\langle 2|1\rangle\langle 1|s\rangle + \langle x|1\rangle\langle 1|2\rangle\langle 2|s\rangle \qquad (11.15)$$

It is clear that substitution of the appropriate wave functions for the various terms in Equations 11.5, 11.9, and 11.14, and multiplication of these equations with their respective complex conjugates yield probability equations of an interferometric character. A variant of the Michelson interferometer uses retroreflectors (Steel 1967).

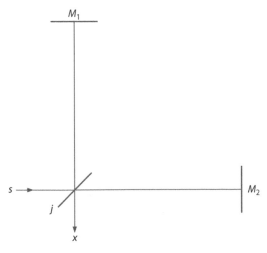

FIGURE 11.5   Michelson interferometer.

11.3   MULTIPLE-BEAM INTERFEROMETERS

An N-slit interferometer, which can be considered as a multiple-beam interferometer, was introduced in Chapter 2 and is depicted in Figure 11.6. In this configuration, an expanded beam of light illuminates simultaneously N slits. Following propagation, the N sub-beams interfere at a plane perpendicular to the plane of propagation. The probability amplitude is given by the Dirac principle

$$\langle x|s \rangle = \sum_{j=1}^{N} \langle x|j \rangle \langle j|s \rangle \tag{11.16}$$

and the probability is

$$\left| \langle x|s \rangle \right|^2 = \sum_{j=1}^{N} \Psi(r_j) \sum_{m=1}^{N} \Psi(r_m) e^{i(\Omega_m - \Omega_j)} \tag{11.17}$$

which can also be expressed as (Duarte 1991, 1993)

$$\left| \langle x|s \rangle \right|^2 = \sum_{j=1}^{N} \Psi(r_j)^2 + 2\sum_{j=1}^{N} \Psi(r_j) \left[\sum_{m=j+1}^{N} \Psi(r_m)\cos(\Omega_m - \Omega_j) \right] \tag{11.18}$$

The explicit expansion of the above equation for higher values of N is illustrated in Chapter 2. Two-dimensional and three-dimensional versions of Equation 11.18 are also given in Chapter 2.

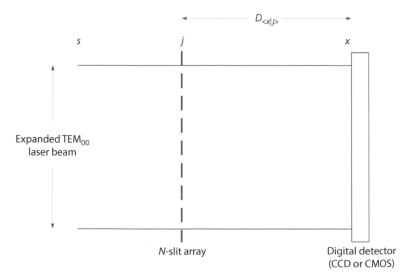

FIGURE 11.6 N-slit interferometer. CMOS, complementary metal–oxide–semiconductor; TEM_{00}, single transverse mode.

11.3.1   THE HANBURY BROWN–TWISS INTERFEROMETER

The Hanbury Brown–Twiss effect originates in interferometric measurements performed by an *intensity interferometer* used for astronomical observations (Hanbury Brown and Twiss, 1956). A diagram of the stellar intensity interferometer used to determine the diameter of stars is depicted in Figure 11.7. Feynman in one of his exercises to *The Feynman Lectures on Physics* (Feynman et al. 1965) explains that the electrical currents from the two detectors are mixed in a coincidence circuit where the currents become indistinguishable. Feynman then asks to show that the coincidence counting rate, in the Hanbury Brown–Twiss configuration, is proportional to an expression of the form:

$$2 + 2\cos k(R_2 - R_1) \tag{11.19}$$

where:

R_1 and R_2 are the distances from detector 1 and detector 2 to the source

Using the *N*-slit interferometric equation (Duarte 1991, 1993), that is, Equation 11.18, with $N = 2$, one immediately arrives at

$$\left|\langle x|s\rangle\right|^2 = \Psi(r_1)^2 + \Psi(r_2)^2 + 2\Psi(r_1)\Psi(r_2)\cos(\Omega_2 - \Omega_1) \tag{11.20}$$

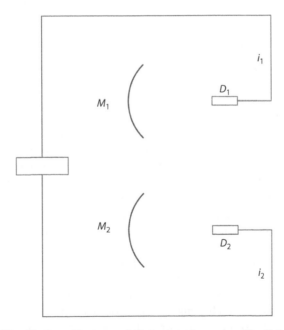

FIGURE 11.7   The Hanbury Brown and Twiss interferometer. The light, from an astronomical source, is collected at mirrors M_1 and M_2 and focused onto detectors D_1 and D_2. The current generated at these detectors, i_1 and i_2, interfere at the electronics to produce an interference signal characterized by an equation of the form of Equation 11.18 with $N = 2$.

and setting $\Psi(r_1) = \Psi(r_2) = 1$

$$\left|\langle x|s\rangle\right|^2 = 2 + 2\cos(\Omega_2 - \Omega_1) \qquad (11.21)$$

Now, using (as suggested by Feynman) $\Omega_1 = kR_1$ and $\Omega_2 = kR_2$

$$\left|\langle x|s\rangle\right|^2 = 2 + 2\cos k(R_2 - R_1) \qquad (11.22)$$

which is the result given by Feynman in his problem book (Feynman 1965). From the measured signal distribution, and these equations, the angular spread of the emission can be determined, and knowing the distance from the source to the detector, it becomes possible to estimate the diameter of the aperture at the emission, in other words, the diameter of the star under observation.

Finally, since all the physics of the Hanbury Brown–Twiss interferometer follows from the generalized interferometric equation:

$$\left|\langle x|s\rangle\right|^2 = \sum_{j=1}^{N} \Psi(r_j)^2 + 2\sum_{j=1}^{N} \Psi(r_j) \left[\sum_{m=j+1}^{N} \Psi(r_m)\cos(\Omega_m - \Omega_j)\right]$$

it can be easily seen that the experimental arrangement of the astronomical telescope should not be limited to just two parabolic mirrors and two detectors ($N = 2$), but can be extended to an array of N parabolic mirrors with their corresponding detectors.

11.3.2 THE FABRY–PÉROT INTERFEROMETER

The second multiple-beam interferometer is the Fabry–Pérot interferometer depicted in Figure 11.8. This interferometer has already been introduced in Chapter 7 as an *intracavity etalon*. Generally, intracavity etalons are a solid slab of optical glass, or fused silica, with highly parallel surfaces coated to increase reflectivity (Figure 11.8b). These are also known as Fabry–Pérot etalons. Fabry–Pérot interferometers, however, are constituted by two separate slabs of optical flats with their inner surfaces coated as shown in Figure 11.8a. The space between the two coated surfaces is filled with air or other inert gas. The optical flats in a Fabry–Pérot interferometer are mounted on rigid metal bars, with a low thermal expansion coefficient, such as Invar. The plates can be moved, with micrometer precision or better, to vary the free spectral range (FSR). These interferometers are widely used to characterize and quantify the laser linewidth.

The mechanics of multiple-beam interferometry can be described in some detail, considering the multiple reflection, and refraction, of a beam incident on two parallel surfaces separated by a region of refractive index n as illustrated in Figure 11.9. In this configuration, at each point of reflection and refraction, a fraction of the beam, or a sub-beam, is transmitted toward the boundary region. Following propagation, these sub-beams interfere. In this regard, the physics is similar to that of the N-slit interferometer with the exception that each parallel beam has less intensity due to the increasing number of reflections. Here, for transmission, interference can be

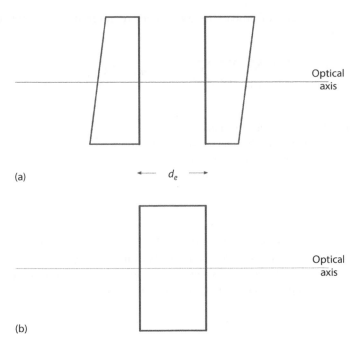

(a)

← d_e →

(b)

FIGURE 11.8 Fabry–Pérot interferometer (a) and Fabry–Pérot etalon (b). Dark lines represent coated surfaces. Focusing optics is often used with these interferometers when used in linewidth measurements.

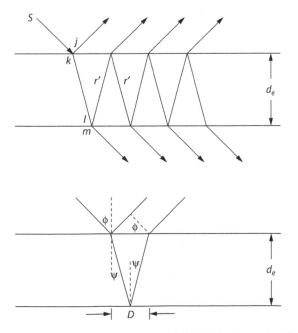

FIGURE 11.9 Multiple-beam interferometer: (a) Multiple internal reflection diagram; (b) detailed view depicting the angles of incidence and refraction.

described using a series of probability amplitudes representing the events depicted in Figure 11.9

$$\langle x|s \rangle = \sum_{m=1}^{N} \sum_{l=1}^{N} \sum_{k=1}^{N} \sum_{j=1}^{N} \langle x|m \rangle \langle m|l \rangle \langle l|k \rangle \langle k|j \rangle \langle j|s \rangle \qquad (11.23)$$

where:
  j is at the reflection surface of incidence
  k is immediately next to the surface of reflection
  l is at the second surface of reflection
  m is immediately next to the second surface of reflection as illustrated in Figure 11.9

The problem can be simplified considerably if the incident beam is considered as a narrow beam incident at a single point j. Propagation of the single beam then proceeds to l and is represented by the incidence amplitude A_i, which is a complex number, attenuated by the transmission factor t, so that the first three probability amplitudes can be represented by an expression of the form:

$$\langle l|k \rangle \langle k|j \rangle \langle j|s \rangle = A_i t \qquad (11.24)$$

and Equation 11.23 reduces to

$$\langle x|s \rangle = A_i t \sum_{m=1}^{N} \langle x|m \rangle \langle m|l \rangle \qquad (11.25)$$

which, using the notation of Born and Wolf (1999), can be expressed as

$$A_t(p) = A_i t \left[t' + t'r'^2 e^{i\delta} + t'r'^4 e^{i2\delta} + \cdots + t'r'^{2(p-1)} e^{i(p-1)\delta} \right] \qquad (11.26)$$

defining

$$\mathfrak{I} = tt' \qquad (11.27)$$

and

$$\mathfrak{R} = r'^2 \qquad (11.28)$$

and taking the limit as $p \to \infty$, Equation 11.25 reduces to (Born and Wolf 1999)

$$A_t = \mathfrak{I} \left(1 - \mathfrak{R} \, e^{i\delta} \right)^{-1} A_i \qquad (11.29)$$

and multiplication with its complex conjugate yields an expression for the intensity

$$I_t = \mathfrak{I}^2 \left(1 + \mathfrak{R}^2 - 2\mathfrak{R} \cos\delta \right)^{-1} I_i \qquad (11.30)$$

which is known as the *Airy formula* or *Airy function*. Plotting the ratio of the two intensities, as a function of

$$\delta = 2\pi m \qquad (11.31)$$

shows that the contrast of the fringes increases as the reflectivity increases (Born and Wolf 1999). From the geometry of Figure 11.9b, the path difference between the first reflected beam and the first beam that undergoes internal reflection, followed by refraction, is (see Problem 11.4)

$$\Delta L = 2nd_e \left(\frac{1}{\cos \psi_e} - \tan \psi_e \sin \psi_e \right) \qquad (11.32)$$

which reduces to

$$\Delta L = 2nd_e \cos \psi_e \qquad (11.33)$$

Hence, using $\Delta L = m\lambda$, the path difference can be expressed as

$$m\lambda = 2nd_e \cos \psi_e \qquad (11.34)$$

where:
   n is the refractive index of the medium
   d_e is the distance between the reflection surfaces

It also follows that the phase difference, using Equations 11.31 and 11.33, can be found to be

$$\delta = \frac{4\pi nd_e}{\lambda} \cos \psi_e \qquad (11.35)$$

The device just described is an uncoated interferometer. If the two surfaces of the plate are coated with metal films of equal reflectivity, the phase term is modified so that (Born and Wolf 1999)

$$\delta = \frac{4\pi nd_e}{\lambda} \cos \psi_e + 2\varphi \qquad (11.36)$$

where:
   φ represents a phase change

Under these circumstances, the multiple-beam interferometer is classified as a Fabry–Pérot etalon. An interferogram produced by the interaction of narrow-linewidth laser emission and a Fabry–Pérot etalon followed by a convex lens, of focal length f, is shown in Figure 11.10.
   Using $\delta = 2\pi m$ in Equation 11.36, we get

$$m = \frac{2nd_e}{\lambda} \cos \psi_e + \frac{\varphi}{\pi}$$

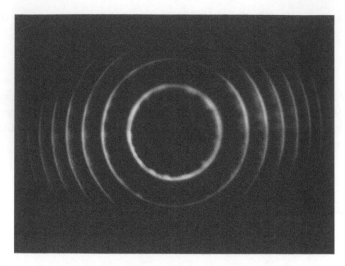

FIGURE 11.10 Fabry–Pérot interferogram depicting single-longitudinal-mode oscillation, at $\Delta v \approx 700$ MHz, from a tunable multiple-prism grating solid-state oscillator. (Reprinted from *Opt. Commun.*, 117, Duarte, F.J., Solid-sate dispersive dye laser oscillator: Very compact cavity, 480–484, Copyright 1995, with permission from Elsevier.)

Here, the brightest central ring corresponds to the maximum value of m_0, which is

$$m_0 = \frac{2nd_e}{\lambda} + \frac{\varphi}{\pi} \qquad (11.37)$$

which can be expressed as

$$m_0 = \frac{2nd_e}{\lambda} + \varepsilon \qquad (11.38)$$

where:
   ε is the fractional fringe order

Using these equations, the diameter of the rings is given by (Born and Wolf 1999)

$$D_p^2 = \left(\frac{4n\lambda f^2}{n_0^2 d_e} \right)(p - 1 + \varepsilon) \qquad (11.39)$$

where:
   $p = 1,\ 2,\ 3...$ enumerates the successive rings from the center

11.3.3 Design of Fabry–Pérot Etalons

The FSR of a Fabry–Pérot interferometer, or a Fabry–Pérot etalon, corresponds to the difference in wavelength of two adjacent orders. From Equation 11.34, for a small angle of incidence,

$$m \approx \frac{2nd_e}{\lambda} \qquad (11.40)$$

and for an infinitesimal wavelength difference, corresponding to two adjacent orders, Δm becomes

$$\Delta m \approx 2nd_e\left(\frac{\lambda_2 - \lambda_1}{\lambda_1\lambda_2}\right) \tag{11.41}$$

$$\Delta m \approx 2nd_e\left(\frac{\delta\lambda}{\lambda^2}\right) \tag{11.42}$$

Since $\Delta m = 1$, the wavelength difference corresponds to

$$\delta\lambda \approx \frac{\lambda^2}{2nd_e} \tag{11.43}$$

which has the same form of $\Delta\lambda \approx \lambda^2/\Delta x$ derived in Chapter 3 from interferometric principles while discussing Heisenberg's uncertainty principle.

Renaming the wavelength difference as the FSR, we can write

$$\mathrm{FSR}_e \approx \frac{\lambda^2}{2nd_e} \tag{11.44}$$

which in the frequency domain becomes

$$\mathrm{FSR}_e \approx \frac{c}{2nd_e} \tag{11.45}$$

Here, the approximations $\delta\lambda \approx (\lambda_1 - \lambda_2)$ and $\lambda^2 \approx \lambda_1\lambda_2$ are justified since $\lambda \gg \Delta\lambda$.

The FSR corresponds to the separation of the rings in Figure 11.10 and a measure of the width of the rings determines the linewidth of the emission being observed. The minimum resolvable linewidth is given by

$$\Delta v_e = \frac{\mathrm{FSR}_e}{\mathcal{F}} \tag{11.46}$$

where:
$\mathcal{F}$ is the *effective finesse*.

Thus, a Fabry–Pérot etalon with an $\mathrm{FSR}_e = 7.49$ GHz and $F = 50$ provides discrimination down to $\Delta v_{\mathrm{FRS}} \approx 150$ MHz. The finesse is a function of the flatness of the surfaces (often in the $\lambda/100-\lambda/50$ range), the dimensions of the aperture, and the reflectivity of the surfaces. The effective finesse is given by (Meaburn 1976)

$$\mathcal{F}^{-2} = \mathcal{F}_R^{-2} + \mathcal{F}_F^{-2} + \mathcal{F}_A^{-2} \tag{11.47}$$

where:
$\mathcal{F}_R, \mathcal{F}_F$, and $\mathcal{F}_A$ are the reflective, flatness, and aperture finesses, respectively

The reflective finesse is given by (Born and Wolf 1999)

$$\mathcal{F}_R = \frac{\pi\sqrt{\mathcal{R}}}{(1-\mathcal{R})} \tag{11.48}$$

11.3.3.1   Example

A Fabry–Pérot etalon, with $d_e = 10.27$ mm and capable of resolving laser linewidths down to $\Delta v_e \approx 100$ MHz, needs to be coated with only the necessary reflectivity. The flatness of the surfaces is $\lambda/100$, the material is fused silica ($n = 1.4583$ at $\lambda = 590$ nm), and the dominating finesse parameter is $\mathscr{F}_R$. With this information, it is found that

$$\mathrm{FSR}_e \approx \frac{c}{2nd_e} \approx 10.00 \text{ GHz}$$

$$\mathscr{F} = \frac{\mathrm{FSR}_e}{\Delta v_e} \approx 100$$

which, from Equation 11.48, requires $R \approx 0.97$.

11.4   COHERENT AND SEMICOHERENT INTERFEROGRAMS

The topic of coherence and semicoherence of N-slit interferograms is discussed in more detail by Duarte (2014). Here, a summary of the main concepts is given.

The generalized interferometric equation in one dimension

$$\left| \langle x | s \rangle \right|^2 = \sum_{j=1}^{N} \Psi(r_j)^2 + 2 \sum_{j=1}^{N} \Psi(r_j) \left[\sum_{m=j+1}^{N} \Psi(r_m) \cos(\Omega_m - \Omega_j) \right]$$

was derived for single-photon propagation (Duarte 1991, 1993, 2004), which means that, strictly speaking, it applies only to monochromatic sources. To emphasize this point, this equation should really be written as (Duarte 2014)

$$\left| \langle x | s \rangle \right|^2_\lambda = \sum_{j=1}^{N} \Psi(r_j)^2_\lambda + 2 \sum_{j=1}^{N} \Psi(r_j)_\lambda \left[\sum_{m=j+1}^{N} \Psi(r_m)_\lambda \cos(\Omega_m - \Omega_j) \right] \qquad (11.49)$$

In practice, however, it has been found that this equation can be applied to either reproduce or predict N-slit interferograms generated using narrow-linewidth lasers that emit ensembles of indistinguishable photons (Duarte 1993). These interferograms are characterized by sharp well-defined features with a high degree of visibility as defined by (Michelson 1927)

$$\mathcal{V} = \frac{I_1 - I_2}{I_1 + I_2} \qquad (11.50)$$

Also, from experiments, we know that broadband light sources produce N-slit interferograms with broad features that are characterized by lack of sharpness (Duarte 2007, 2008, 2010). As explained in Duarte (2014), interferograms produced with broadband light are generated by a multitude of wavelengths, and the digital

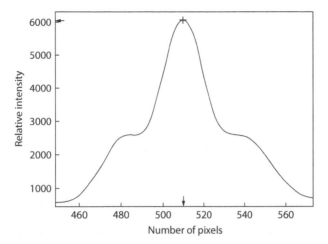

FIGURE 11.11 Measured double-slit ($N = 2$) interferogram generated with a broadband light source. The slits are 50 μm wide separated by 50 μm and $D_{\langle x|j\rangle} = 10$ cm. Comparison with other two-slit interferograms generated with coherent sources (see Figures 11.12 and 11.13) reveals lack of spatial definition and low visibility as defined by Equation 11.50.

detector, or photographic plate, used to record the interference yields an integrated version of a multitude of interferograms, all of which have different interferometric features. Therefore, interferograms generated using non-narrow-linewidth emission, or broadband radiation, have broad features and low visibility as defined by Equation 11.50 (see Figure 11.11). Thus, the correct interferometric equation for semicoherent radiation, and broadband radiation, is modified to include a sum over the wavelength range involved so that (Duarte 2014)

$$\sum_{\lambda=\lambda_1}^{\lambda_n}\left|\langle x|s\rangle\right|_\lambda^2 = \sum_{\lambda=\lambda_1}^{\lambda_n}\left\{\sum_{j=1}^{N}\Psi(r_j)_\lambda^2 + 2\sum_{j=1}^{N}\Psi(r_j)_\lambda\left[\sum_{m=j+1}^{N}\Psi(r_m)_\lambda\cos(\Omega_m-\Omega_j)\right]\right\} \quad (11.51)$$

From a computational perspective, the broadly featured interferogram can be approximated via the generation of a large number of single-wavelength interferograms at fine intervals of wavelength. Each of these interferograms will have slightly different intensity-spatial features and the overall interferometric profile will be an integrated, or cumulative, interferogram.

11.4.1 Example

In Figure 11.12, the double-slit interferogram produced with narrow-linewidth emission from the $3s_2 - 2p_{10}$ transition of a He–Ne laser, at $\lambda \approx 543.3$ nm, is displayed. The visibility of this interferogram is calculated, using Equation 11.50, to be $\mathcal{V} \approx 0.95$. In Figure 11.13, the double-slit interferogram produced, under identical geometrical conditions, but with the emission from an electrically excited coherent organic semiconductor interferometric emitter (Duarte et al. 2005), at $\lambda \approx 540$ nm,

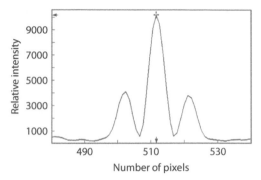

FIGURE 11.12 Double-slit interferogram from the narrow-linewidth emission from the $3s_2 - 2p_{10}$ transition of a He–Ne laser, at $\lambda \approx 543.3$ nm. The visibility of this interferogram is $\mathcal{V} \approx 0.95$. Here, $N = 2$, the slits are 50 μm wide separated by 50 μm, and $D_{\langle x | j \rangle} = 5$ cm. (Reproduced from Duarte, F.J., et al., *Opt. Lett.*, 30, 3072–3074, 2005. With permission from the Optical Society.)

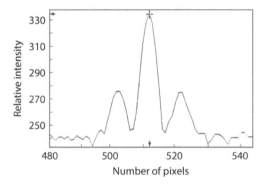

FIGURE 11.13 Double-slit interferogram from an electrically excited organic semiconductor interferometric emitter at $\lambda \approx 540$ nm. The visibility of this interferogram is less than $\mathcal{V} \approx 0.90$. Here, $N = 2$, the slits are 50 μm wide separated by 50 μm, and $D_{\langle x | j \rangle} = 5$ cm. (Reproduced from Duarte, F.J., et al., *Opt. Lett.*, 30, 3072–3074, 2005. With permission from the Optical Society.)

is displayed. Here, the visibility is lower ($\mathcal{V} \approx 0.90$). A comparison between the two interferograms reveals that the second interferogram has slightly broader spatial features relative to the interferogram produced with illumination from the $3s_2 - 2p_{10}$ transition of the He–Ne laser. The differences in spatial distributions between these two interferograms have been used to estimate the linewidth of the emission from the interferometric emitter (Duarte 2008).

11.5 INTERFEROMETRIC WAVELENGTH METERS

Interferometric signals and profiles are a function of the wavelength of the radiation that produces them. Thus, interferometers are well suited to be applied as wavelength meters, especially when a digital detector array is used to record the resulting interferogram. As such, a variety of interferometric configurations have been used in the measurement of tunable laser wavelengths. For a review in this subject, the reader should refer to the work of Demtröder (2008).

The wavelength sensitivity of multiple-beam interferometry has its origin in the phase information of the equations describing the behavior of the interferometric signal. In the case of the N-slit interferometer, the interferometric profile is characterized by Equation 11.18, which includes a phase difference term that, as explained in Chapter 2, can be expressed as

$$\cos\left[(\theta_m - \theta_j) \pm (\phi_m - \phi_j)\right] = \cos\left(|\,l_m - l_{m-1}\,|\,k_1 \pm |\,L_m - L_{m-1}\,|\,k_2\right) \qquad (11.52)$$

where:

$$k_1 = \frac{2\pi n_1}{\lambda_v} \qquad (11.53)$$

and

$$k_2 = \frac{2\pi n_2}{\lambda_v} \qquad (11.54)$$

where:

λ_v is the vacuum wavelength

n_1 and n_2 are the corresponding indexes of refraction

Here, $\lambda_1 = \lambda_v/n_1$ and $\lambda_2 = \lambda_v/n_2$ (Wallenstein and Hänsch 1974; Born and Wolf 1999). See Chapter 2 for further details. Hence, it is easy to see that different wavelengths will produce different interferograms. To illustrate this point in Figure 11.14, four calculated interferograms, using Equation 11.18, for the N-slit interferometer, with $N = 50$, are shown. For a given set of geometrical parameters, measured

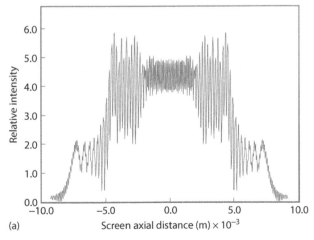

(a)

FIGURE 11.14 Interferograms at $\lambda_1 = 580\,\mathrm{nm}$ (a), $\lambda_2 = 585\,\mathrm{nm}$ (b), $\lambda_3 = 590\,\mathrm{nm}$ (c), and $\lambda_4 = 591\,\mathrm{nm}$ (d). These calculations are for slits 100 μm wide, separated by 100 μm, and $N = 50$. The j-to-x distance is $D_{\langle x|j\rangle} = 100\,\mathrm{cm}$.

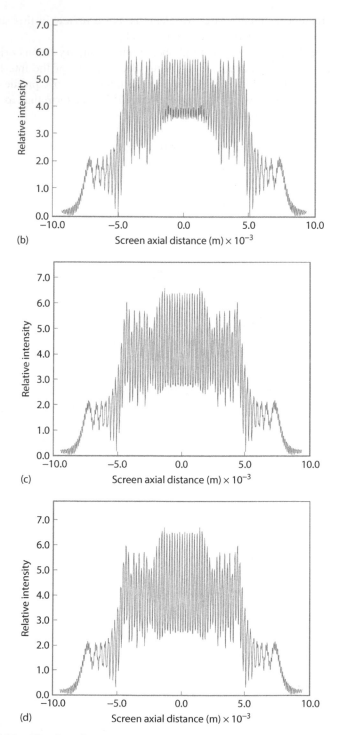

(b)

(c)

(d)

FIGURE 11.14 (Continued)

interferograms can be matched, in an iterative process, with theoretical interference patterns to determine the wavelength of the radiation. Again, resolution depends on the optical path length between the slit array and the digital detector, and on the size of the pixels and the linearity of the detector.

11.5.1 FABRY–PÉROT WAVELENGTH METERS

For a multiple-beam interferometer, the transmission intensity is given by Equation 11.30, where, in reference to Figure 11.9, the phase term δ can be expressed by Equation 11.36, which depends on the reciprocal of the wavelength. Thus, recording of the transmission interferometric signal by a photodiode array, or charge-coupled device (CCD), yields information on the wavelength of the radiation being measured.

Wavelength meters based on Fabry–Pérot interferometers generally involve configurations with multiple etalons in parallel (Byer et al. 1977; Fischer et al. 1981; Konishi et al. 1981). In the multiple-etalon configuration depicted in Figure 11.15, each etalon has a different FSR that is compatible with the FSR and the finesse of the next etalon. For instance, Fischer et al. (1981) used three etalons with FSRs of 1000, 67, and 3.3 GHz. Briefly, the methodology of this measurement consists in the application of Equations 11.38 and 11.39 to determine λ with reduced uncertainty at each etalon. Using this approach, Fischer et al. (1981) reported the measurements of laser frequencies with an accuracy of 60 MHz.

A simple interferometric configuration that is widely used in the measurement of laser wavelengths is that of the Fizeau, or optical wedge, *two-beam interferometer* first introduced by Snyder (1977). In this configuration, the incident laser beam propagates on an axis that is at an angle to the optical axis of the digital detector (see Figure 11.16). Also, the incident beam illuminates a wide area of the interferometer using some beam expansion method in conjunction with a spatial filter. The interference of the *two beams* is recorded by a digital detector with a 25 μm resolution or better. The method employs two predetermined parameters: the angle of the wedge and the wedge separation at a reference position as shown in Figure 11.16b.

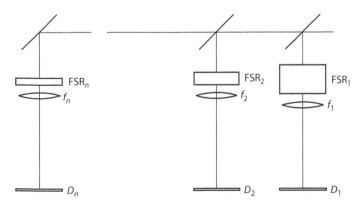

FIGURE 11.15 Multiple-etalon wavelength meter.

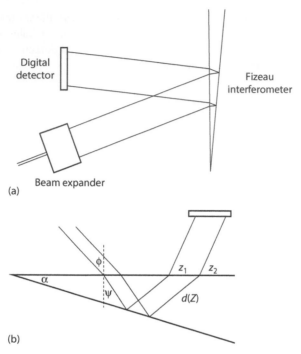

FIGURE 11.16 (a) Fizeau wavelength meter. (b) Geometrical details of a Fizeau interferometer.

The spacing of the Fizeau fringes Δz provides an approximate value for the wavelength according to

$$\Delta z = \left[\frac{d(z_2) - d(z_1)}{\tan \alpha} \right] \tag{11.55}$$

Assuming a very small wedge angle,

$$m\lambda \approx 2nd(z_1)\cos\psi \tag{11.56}$$

and

$$(m+1)\lambda \approx 2nd(z_2)\cos\psi \tag{11.57}$$

so that (Demtröder 2008)

$$\lambda \approx \Delta z \ (2n\cos\psi\tan\alpha) \tag{11.58}$$

thus allowing the determination of the order of interference at a minimum. The method then compares, using a computer program, the periodicity of the interference pattern with that predetermined from two-beam interference intensity functions. As indicated in Equation 11.18, these interferometric functions depend on the cosine of a phase term that, in this case, depends on Δz, which in turn is a function of the wavelength and the geometry of the wedge. Thus, accurate calibration of the spacing of the wedge allows a determination of the wavelength compatible with the accuracy to which the angle α is known. This static two-beam interferometric approach offers a wide wavelength range and an accuracy better than two parts in 10^6 (Gardner 1985) in very compact configurations.

PROBLEMS

11.1 A laser beam fails to provide interference fringes when the distance from the BS to the mirrors, in a Michelson interferometer, is 1 m. Estimate the linewidth of the laser.

11.2 Use the usual complex-wave representation for probability amplitudes and Equation 11.10 to arrive at an equation for the probability of transmission in a Mach–Zehnder interferometer.

11.3 List the simplifying assumptions that lead from Equation 11.23 through 11.26.

11.4 Use the geometry of Figure 11.9 to derive Equation 11.32.

11.5 Use Equation 11.50 to estimate the visibility of the two-slit interferogram displayed in Figure 11.11.

12 Spectrometry

12.1 INTRODUCTION

Spectrometry and spectrophotometry are two important subfields that are part of the larger field of optical metrology. Spectrometry mainly utilizes dispersive-based instrumentation that can be based on prismatic configurations, diffraction grating configurations, or a combination of these. The principal exercise here is to obtain a record of intensity versus wavelength of the material of interest, which provides a unique signature for the atomic or molecular structure of the medium under examination. If a tunable, or broadband, light source illuminates a material sample in the transmission mode, as illustrated in Figure 12.1, then a wavelength scan of the spectrometer yields a record of the absorption spectrum. Alternatively, as illustrated in Figure 12.2, a laser-excited atomic or molecular medium can be used to yield a fluorescence spectrum and an emission spectrum.

In Figure 12.3, a fluorescence spectrum from $B^3\Pi_{ou}^+ - X^1\Sigma_g^+$ electronic system of the I_2 molecule is displayed. This spectrum was generated using a narrow-linewidth tunable laser emitting at $\lambda \approx 589$ nm (Duarte 1981). Here the basics of dispersive spectrometry, or spectrophotometry, are presented, while indicating that most of the recent advancements in this area are related to compactness, miniaturized electronics, digital detector resolution improvements, digital detection wavelength range extensions, digital detector sensitivity improvements, and software augmentations, whereas the optics basics remain the same.

12.2 SPECTROMETRY

Spectrometry, in principle, depends on the interaction of a light beam with a dispersive element, or dispersive elements, and on spatial discrimination following post-dispersive propagation. Hence, the higher the dispersive power, the longer the optical path of the postdispersive propagation, the higher the spatial resolution of the digital detector, the higher the wavelength resolution. As described in earlier chapters, dispersion can be provided by prisms, gratings, or prism–grating combinations. For a detailed treatment on the subject of spectrometry, the reader should refer to the work of Meaburn (1976).

12.2.1 PRISM SPECTROMETERS

Prism spectrometry has its origin in the experiments reported by Newton (1704) in his book *Opticks*. The generalized single-pass multiple-prism dispersion equation (Duarte and Piper 1982, 1983)

$$\nabla_\lambda\phi_{2,m} = \mathcal{H}_{2,m}\nabla_\lambda n_m + (k_{1,m}k_{2,m})^{-1}\left[\mathcal{H}_{1,m}\nabla_\lambda n_m \pm \nabla_\lambda\phi_{2,(m-1)}\right] \tag{12.1}$$

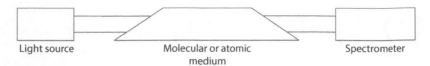

FIGURE 12.1 Simple optical configuration for absorption measurements using a spectrometer.

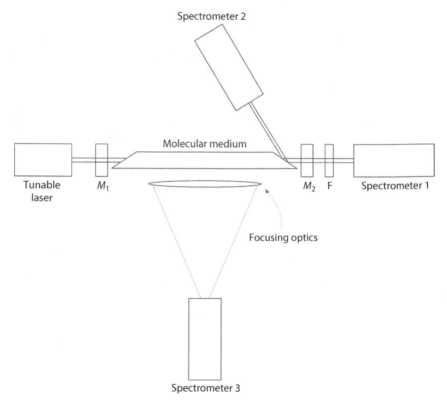

FIGURE 12.2 Simplified optical configuration for fluorescence and emission measurements, in an optically pumped laser system, using spectrophotometers. A narrow-linewidth tunable laser excites longitudinally a molecular medium, such as I_2, and the emission is detected by spectrometer 1 via a filter F used to stop residual emission from the pump tunable laser. In addition, the emission can also be detected via the reflection of the Brewster window of the optical cell in spectrometer 2. The fluorescence and emission are detected orthogonally to the optical axis in spectrometer 3 as described by Duarte. (Adapted from Duarte, F.J., *Investigation of the Optically-Pumped Iodine Dimer Laser in the Visible*, Macquarie University, Sydney, Australia, 1981.)

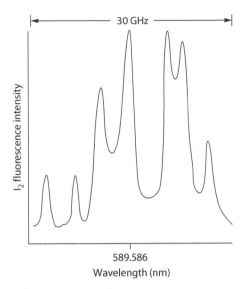

|← ——————— 30 GHz ——————— →|

I₂ fluorescence intensity

589.586
Wavelength (nm)

FIGURE 12.3 Fluorescence spectrum of the I_2 molecule generated using pulsed narrow-linewidth tunable laser excitation in the vicinity of $\lambda = 589.586$ nm. The tuning range is 30 GHz or ~1 cm^{-1} as described by Duarte. (Data from Duarte, F.J., *Investigation of the Optically-Pumped Iodine Dimer Laser in the Visible*, Macquarie University, Sydney, Australia, 1981.)

and its explicit version (Duarte 1985, 1990)

$$\nabla_\lambda \phi_{2,r} = \sum_{m=1}^{r} (\pm 1) \mathcal{H}_{1,m} \left(\prod_{j=m}^{r} k_{1,j} \prod_{j=m}^{r} k_{2,j} \right)^{-1} \nabla_\lambda n_m$$

$$+ (M_1 M_2)^{-1} \sum_{m=1}^{r} (\pm 1) \mathcal{H}_{2,m} \left(\prod_{j=1}^{m} k_{1,j} \prod_{j=1}^{m} k_{2,j} \right) \nabla_\lambda n_m \qquad (12.2)$$

can be used to quantify the dispersion of any multiple-prism dispersive arrangement applicable to various instruments of practical interest. This is important since the dispersion determines the linewidth resolution of the instrument via

$$\Delta \lambda \approx \Delta \theta (\nabla_\lambda \phi_{2,r})^{-1} \qquad (12.3)$$

Prism spectrometers usually deploy equilateral, or isosceles, prisms, in series to augment the dispersion of the instrument. As described in Chapter 4, for an array of r identical isosceles, or equilateral, prisms deployed symmetrically, in an additive configuration, so that $\phi_{1,1} = \phi_{1,2} = \cdots = \phi_{1,m}$ and $\phi_{1,m} = \phi_{2,m}$, the cumulative dispersion (Equation 12.1) reduces to (Duarte 1990)

$$\nabla_\lambda \phi_{2,r} = r \nabla_\lambda \phi_{2,1} \qquad (12.4)$$

and

$$\nabla_\lambda \phi_{2,1} = \left[\frac{\sin \psi_{2,1}}{\cos \phi_{2,1}} + \left(\frac{\cos \psi_{2,2}}{\cos \phi_{2,1}} \right) \tan \psi_{1,1} \right] \nabla_\lambda n \qquad (12.5)$$

where all angular parameters are described in Chapter 4. These equations indicate that dispersion can be augmented by a combination of two factors: increasing the number of prisms and using prisms with a high material dispersion. For some materials, $\nabla_\lambda n$ is given in Table 4.1.

Prism spectrometers assume various configurations. Two of these are considered here. The first configuration, depicted in Figure 12.4, uses two prisms in series with a considerable distance between the two prisms to increase the overall angular spread of the emerging beam. Long postdispersive optical paths, of up to 3 m, in conjunction with narrow apertures, provide wavelength resolutions in the nanometer range. The second architecture, described by Meaburn (1976), includes a sequence of several prisms in series as depicted in Figure 12.5. In this configuration, dispersion is simply augmented by a larger factor r in Equation 12.4. The same observations about the postdispersive optical path, and spatial discrimination, are relevant.

12.2.2 DIFFRACTION GRATING SPECTROMETERS

As described in Chapter 2, the diffraction grating equation is given by

$$d\,(\sin\Theta \pm \sin\Phi) = m\lambda \qquad (12.6)$$

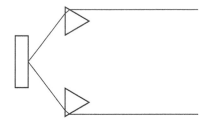

FIGURE 12.4 Long-optical path double-prism spectrometer.

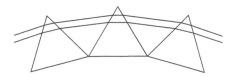

FIGURE 12.5 Dispersive assembly of multiple-prism spectrometer. The angle of incidence $\phi_{1,m}$ and the angle of emergence $\phi_{2,m}$ are identical for all prisms. In addition to the cumulative dispersion, resolution is determined by the path length toward the exit slit and the dimensions of the slit.

and, the angular dispersion is obtained by differentiating this equation so that

$$\frac{\partial \Theta}{\partial \lambda} = \frac{m}{d \cos \Theta} \tag{12.7}$$

or alternatively

$$\frac{\partial \Theta}{\partial \lambda} = \frac{\sin \Theta \pm \sin \Phi}{\lambda \cos \Theta} \tag{12.8}$$

For a grating deployed in Littrow configuration, $\Theta = \Phi$ and the grating equation becomes

$$2d \, \sin \Theta = m\lambda \tag{12.9}$$

so that the dispersion can be expressed as

$$\frac{\partial \Theta}{\partial \lambda} = \frac{2 \tan \Theta}{\lambda} \tag{12.10}$$

As with prismatic spectrometers, the three factors that increase wavelength resolution in a diffraction grating spectrometer are the dispersion, the optical path length, and the dimensions of the aperture at the detection plane. From Equation 12.7, it is clear that one avenue to increase dispersion is to use gratings deployed at a higher order or to employ gratings with a high groove density. Since deployment of gratings at high orders might lead to a decrease of diffraction efficiency, the alternative of using high-density gratings with 3000 lines/mm, or more, is a practical alternative to increase dispersion. Selection of a particular grating should consider the spectral region of desired operation since high-density gratings might cease to diffract toward the red end of the spectrum. The electromagnetic theory of diffraction gratings is considered in detail by Maystre (1980). A fundamental form of grating spectrometer is shown in Figure 12.6 in order to illustrate the basic concept of spectrometry. The use of high-dispersion gratings in conjunction with high-resolution digital detectors, as outlined in Figure 12.6b, can lead to the configuration of miniature dispersive spectrometers or wavelength meters. Among the most widely used diffraction grating spectrometer configurations is the Czerny–Turner spectrometer, as shown in Figure 12.7, where two mirrors are used to increase the optical path length, and thus the resolution. A modified Czerny–Turner spectrometer (Meaburn 1976) includes two gratings to increase the dispersion (Figure 12.8). A simple modification to increase resolution is to increase the optical path directly or by adding further stages of reflection. Various spectrometer design alternatives are discussed by Born and Wolf (1999) and Meaburn (1976). It should be noted that the use of curved gratings and curved mirrors to compensate for losses is prevalent in many designs. Notable among the curved grating approaches are the *Rowland* and the *Paschen* configurations (Born and Wolf 1999).

Modern Czerny–Turner spectrometer designs, with a ~4 m folded optical path, provide resolutions in the 0.01 nm region, in the visible spectrum, when using slits a few micrometers wide. These spectrometers are very useful in providing the first approach to determine the value of a laser wavelength.

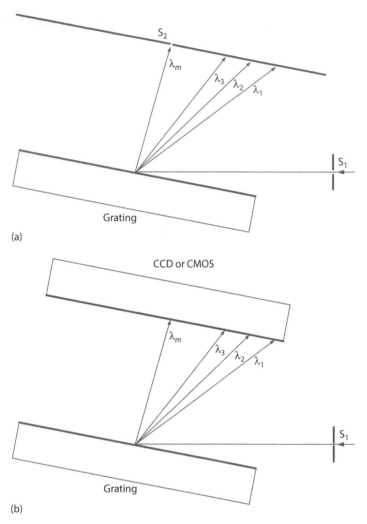

(a)

(b)

FIGURE 12.6 (a) Basic grating spectrometer. The resolution is mainly determined by the dispersion of the grating, deployed in a non-Littrow configuration, and by the optical path length toward the exit slit (S_2). (b) A variation on the basic grating spectrometer consists in deploying a high-dispersion diffraction grating in conjunction with a high spatial resolution digital detector (CCD or CMOS) to register the diffracted light.

12.2.2.1 Example

Using a 3000 lines/mm diffraction grating deployed at $\Theta = 60.0000°$, m = 1, at $\lambda = 590.00$ nm, from Equation 12.6

$$d\,(\sin\Theta \pm \sin\Phi) = m\lambda$$

it can be established that the angle of diffraction is $\Phi = 64.685530°$. Next, if the angle of incidence is maintained at $\Theta = 60.000000°$ and the wavelength is varied

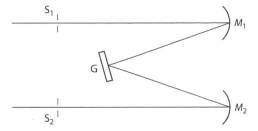

FIGURE 12.7   Czerny–Turner spectrometer. Mirrors M_1 and M_2 provide the necessary curvature to focus the diffracted beam at the exit slit.

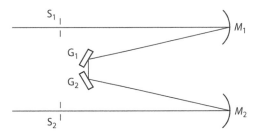

FIGURE 12.8   Double-grating Czerny–Turner spectrometer. Mirrors M_1 and M_2 provide the necessary curvature to focus the diffracted beam at the exit slit.

to $\lambda = 590.10$ nm, then the angle of diffraction becomes $\Phi = 64.725759°$, so that $\delta\Phi = 0.042293$ for $\delta\lambda = 0.10$ nm (see Problem 12.3).

12.3   DISPERSIVE WAVELENGTH METERS

Dispersive wavelength meters are in essence dispersive spectrometers with the conventional slit-detector arrangement replaced by a linear photodiode array, charge-coupled device (CCD), or complementary metal–oxide–semiconductor (CMOS) detector. Their main function is to determine the wavelength from the emission of tunable lasers. The resolution is determined by the available dispersion, the propagation distance between the dispersive element and the detector array, and the dimensions of the individual pixels at the digital array. Depending on the type of detector, individual pixels vary in size from a few micrometers to ~25 μm in width. The typical widths of these digital arrays vary from ~25 to ~50 mm. This type of wavelength meter is used in conjunction with well-known, and preferably stabilized, laser sources for calibration. Morris and McIlrath (1979) used a 40 cm spectrometer, incorporating an echelle grating deployed at a high order, in conjunction with a photodiode array integrated by 1024 elements (each ~25 mm wide) to determine wavelengths within 0.01 nm in the visible region.

   A wavelength meter based on a high-dispersion prism grating configuration is illustrated in Figure 12.9. As discussed in Chapter 4, the return-pass dispersion of a multiple-prism assembly and a diffraction grating is given by (Duarte and Piper 1982, 1984).

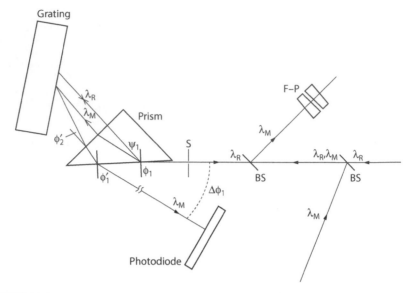

FIGURE 12.9 Architecture of the prism grating wavelength meter. A reference wavelength (λ_R) beam and the measurement wavelength (λ_M) beam are combined at the entrance beam splitter (BS). The reference beam is used to provide the initial reference wavelength (λ_R) and for alignment purposes ($\lambda_M > \lambda_R$). A second BS sends light from (λ_M) beam to a Fabry–Pérot (F–P). The (λ_M) and (λ_R) beams are separated at the prism. The (λ_M) beam undergoes deflection at the prism grating assembly, whereas the (λ_R) beam returns back on its original path. The measurement beam (λ_M) undergoes augmented angular deflection at the prism grating system and its position is registered by a photodiode array or CCD. Resolution is a function of the overall prism grating dispersion, the distance between the prism and the photodiode array detector, and the spatial resolution (pixel dimensions) of the digital detector. (Reproduced from Duarte, F.J., *J. Phys. E: Sci. Instrum.*, 16, 599–601, 1983. With permission from the Institute of Physics.)

$$\nabla_\lambda \phi'_{1,m} = \mathcal{H}'_{1,m} \nabla_\lambda n_m + \left(k'_{1,m} k'_{2,m} \right) \left[\mathcal{H}'_{2,m} \nabla_\lambda n_m \pm \nabla_\lambda \phi'_{1,(m+1)} \right] \tag{12.11}$$

where:

$$\nabla_\lambda \phi'_{1,(m+1)} = \left(\nabla_\lambda \Theta_G \pm \nabla_\lambda \phi_{2,r} \right) \tag{12.12}$$

so that

$$\nabla_\lambda \phi'_{1,m} = \mathcal{H}'_{1,m} \nabla_\lambda n_m + \left(k'_{1,m} k'_{2,m} \right) \left[\mathcal{H}'_{2,m} \nabla_\lambda n_m \pm \left(\nabla_\lambda \Theta_G \pm \nabla_\lambda \phi_{2,r} \right) \right] \tag{12.13}$$

For a single prism, $r = 1$, and Equation 12.13 becomes

$$\nabla_\lambda \phi'_{1,1} = \mathcal{H}'_{1,1} \nabla_\lambda n_1 + \left(k'_{1,1} k'_{2,1} \right) \left[\mathcal{H}'_{2,1} \nabla_\lambda n_1 + \left(\nabla_\lambda \Theta_G \pm \nabla_\lambda \phi_{2,1} \right) \right] \tag{12.14}$$

which, for orthogonal beam exit, that is, $\phi_{2,1} = \psi_{2,1} \approx 0$, reduces to

$$\nabla_\lambda \phi'_{1,1} \approx \mathcal{H}'_{1,1} \nabla_\lambda n_1 + k'_{1,1} \left(\nabla_\lambda \Theta_G \pm \nabla_\lambda \phi_{2,1} \right) \qquad (12.15)$$

Since for that condition

$$\nabla_\lambda \phi_{2,1} \approx \tan \psi_{1,1} \, \nabla_\lambda n \qquad (12.16)$$

then we have

$$\nabla_\lambda \phi'_{1,1} = 2k_{11} \tan \psi_{11} \nabla_\lambda n_1 + k_{1,1} \nabla_\lambda \Theta_G \qquad (12.17)$$

where:
$\nabla_\lambda \Theta_G$ is the grating dispersion
$\nabla_\lambda \phi_{2,1}$ is the single-pass prism dispersion

Equation 12.17 neatly illustrates that the original diffraction grating dispersion $\nabla_\lambda \Theta_G$ is now multiplied by the beam expansion $k_{1,1}$ experienced at the prism and the overall dispersion also adds the contribution by the prism.

Using a 600 lines/mm echelle grating deployed in the third order, and a BK-7 optical glass prism ($n = 1.512$ at $\lambda = 568.8$ nm) with an apex angle of 41.5°, the sensitivity of the system becomes 0.3 nm^{-1} (Duarte 1983). For a 1000 mm distance, from the prism to the detector, this dispersive combination can measure wavelengths with a 0.01 nm resolution. Absolute wavelength calibration was provided by the spectrum of molecular iodine and the I_2 spectral atlas of Gerstenkorn and Luc (1978). Once calibrated, this simple prism grating wavelength meter configuration offers static operation and can be used to characterize either pulsed or continuous-wave laser radiation. Resolution is a function of geometry and the intrinsic resolution of the digital detector employed.

PROBLEMS

12.1 Use Equation 12.5 to calculate the overall dispersion of an equilateral 60° prism. The prism is fused silica with $n = 1.4583$ at $\lambda = 590$ nm and assume incidence at the Brewster angle.

12.2 Assume you have three identical fused silica prisms as described in Problem 12.1.
Assume further that these prisms are deployed in series in an additive configuration (see Figure 12.5) for incidence at the Brewster angle. Calculate the overall dispersion of this triple-prism configuration

12.3 For a diffraction grating with 3000 lines/mm, as described in Section 12.2.2.1, calculate the minimum pixel resolution necessary for a linear CCD or CMOS detector to discriminate a wavelength change of $\delta\lambda = 0.10$ nm at $\Theta = 60.000000°$ and $\lambda = 590.00$ nm. Assume that the digital detector is positioned parallel to the surface of the diffraction grating (see Figure 12.6) at a distance of 20 mm, and that the divergence of the incident beam, with $2w = 10\,\mu m$, is negligible.

12.4 For a 600 lines/mm echelle grating deployed in Littrow configuration, at $\lambda_1 = 441.563$ nm, calculate the angular deviation experienced by a collinear, secondary laser beam, at $\lambda_2 = 568.8$ nm. Assume $m = 3$.

12.5 Using the prism grating configuration, depicted in Figure 12.9, estimate the new angular deviation $\Delta\phi_1$ between two collinear laser beams at $\lambda_1 = 441.563$ nm and $\lambda_2 = 568.8$ nm. Assume that the prism has a 41.5° apex angle and that $n = 1.512$ at $\lambda_2 = 568.8$. Also assume that the 600 lines/mm echelle grating is deployed in Littrow configuration, for the shorter wavelength, at $m = 3$.

13 Physical Constants and Optical Quantities

13.1 FUNDAMENTAL PHYSICAL CONSTANTS

Physical constants useful in optics are listed in Table 13.1. The values of these constants are those listed by the National Institute of Science and Technology (NIST) available at the time of publication.

13.2 CONVERSION QUANTITIES

Conversion quantities often used in optics are listed in Table 13.2. The conversion values for the electron volt and the atomic mass unit are the values listed by the NIST available at the time of publication.

In narrow-linewidth tunable laser design and spectroscopy, the units on linewidth and linewidth conversion units are important. From the approach to the uncertainty principle, given in Chapter 3, the following linewidth expressions are obtained:

$$\Delta\lambda = \frac{\lambda^2}{\Delta x} \tag{13.1}$$

given in meters (m), and its equivalent in the frequency domain

$$\Delta\nu = \frac{c}{\Delta x} \tag{13.2}$$

given in hertz (Hz). In spectroscopy, a widely used unit for linewidth is the reciprocal centimeter (cm^{-1}) (Herzberg 1950), as indicated in Table 13.2. This spectroscopist's linewidth follows from Equation 13.2 since

$$\frac{\Delta\nu}{c} = \frac{1}{\Delta x} \tag{13.3}$$

where:
  the units of Δx are in meters (m)

Conversion of the linewidth to units of cm^{-1} can be done via the identity

$$\frac{1}{\Delta x'} = \frac{1}{100}\left(\frac{1}{\Delta x}\right) \tag{13.4}$$

so that the equivalent of $\Delta\nu \approx 30$ GHz, which is $(1/\Delta x) \approx 100$ m^{-1}, becomes $(1/\Delta x') \approx 1$ cm^{-1}. A more specific example for the conversion of $\Delta\nu \approx 350$ MHz, at $\lambda \approx 590$ nm, is given in Table 13.3.

TABLE 13.1
Fundamental Physical Constants

Name	Symbol	Value	Unit
Boltzmann's constant	k_B	$1.3806488 \times 10^{-23}$	J K^{-1}
Elementary charge	e	$1.602176565 \times 10^{-19}$	C
Electron mass	m_e	$9.10938291 \times 10^{-31}$	kg
Newtonian constant of gravitation	G	6.67384×10^{-11}	m^3 kg^{-1} s^{-2}
Permeability of vacuum[a,b]	μ_0	$4\pi \times 10^{-7}$	N A^{-2}
Permittivity of vacuum[c]	ε_0	$8.854187817 \times 10^{-12}$	F m^{-1}
Planck's constant	h	$6.62606957 \times 10^{-34}$	J s
Speed of light in vacuum	c	2.99792458×10^8	m s^{-1}

Source: Duarte, F.J., *Tunable Laser Optics*, Elsevier Academic Press, New York, 2003.
[a] Also known as magnetic constant.
[b] $\pi = 3.141592654...$
[c] Also known as electronic constant.

TABLE 13.2
Laser Optics Conversion Quantities

Name	Symbol	Value	Unit
Electron volt	eV	$1.602176565 \times 10^{-19}$	J
Atomic mass unit	u	$1.660538921 \times 10^{-27}$	kg
Wavelength	$\lambda = c/\nu$		m
Wavenumber	$k = 2\pi/\lambda$		m^{-1}
Linewidth	$\Delta\nu = c/\Delta x$		Hz
Linewidth	$\Delta\lambda = \lambda^2/\Delta x$		m
1 cm^{-1}	$(1/\Delta x') = 1$ cm^{-1}	2.99792458×10^1	GHz
1 GHz	$\Delta\nu = 1$ GHz	$(2.99792458)^{-1} \times 10^{-1}$	cm^{-1}
1 GHz (at 547.533066 nm)	$\Delta\nu_{547.53\,nm} = 1$ GHz	1×10^{-12}	m
1 nm (at 547.533066 nm)	$\Delta\lambda_{547.53\,nm} = 1$ nm	1×10^3	GHz
1 nm (at 547.533066 nm)	$\Delta\lambda_{547.53\,nm} = 1$ nm	$(2.99792458)^{-1} \times 10^2$	cm^{-1}

Source: Duarte, F.J., *Tunable Laser Optics*, Elsevier Academic Press, New York, 2003.

TABLE 13.3
Linewidth Equivalence for $\Delta\lambda \approx 0.0004064$ nm at $\lambda \approx 590$ nm

Linewidth Domain	Value
Wavelength	$\Delta\lambda \approx 0.0004064$ nm at $\lambda \approx 590$ nm
Frequency	$\Delta\nu \approx 350$ MHz
Spatial	$(1/\Delta x') \approx 0.0116747$ cm^{-1}

Source: Duarte, F.J., *Tunable Laser Applications*, CRC Press, New York, 2009.

TABLE 13.4
Photon-Energy Wavelength Equivalence

Photon Energy	Wavelength (nm)[a]
1 eV	~1239.842
10 eV	~123.9842
100 eV	~12.39842
1 keV	~1.239842
10 keV	~0.123984

Source: Duarte, F.J., *Tunable Laser Applications*, CRC Press, New York, 2009.
[a] Using $h = 6.62606957 \times 10^{-34}$ Js and 1 eV = $1.602176565 \times 10^{-19}$ J.

The conversion between photon energy in eV units and wavelength is carried out using the identity

$$\lambda = \frac{hc}{E} \tag{13.5}$$

This equivalence is given in Table 13.4 for the energy range $1 \leq E \leq 10{,}000$ eV.

13.3   UNITS OF OPTICAL QUANTITIES

The units of optical quantities used throughout this book are listed in Table 13.5.

13.4   DISPERSION CONSTANTS OF OPTICAL MATERIALS

The Sellmeier dispersion equation applicable to various optical materials is given by

$$n(\lambda)^2 = A_0 + \sum_{i=1}^{N} A_i \lambda^2 \left(\lambda^2 - B_i^2 \right)^{-1} \tag{13.6}$$

where:
λ is the wavelength at which the refractive index n is to be calculated

The constants for fused silica, SF10, calcium fluoride, and zinc selenide are given in Table 13.6. For the constants given in this table, λ is in μm units.

An important parameter is the dispersion of the prism material, or $\partial n/\partial \lambda$ (see Chapter 4), that can be obtained by differentiating Equation 13.6 so that

$$\frac{\partial n}{\partial \lambda} = \sum_{i=1}^{N} \frac{A_i}{n} \left[\frac{\lambda}{(\lambda^2 - B_i^2)} - \frac{\lambda^3}{(\lambda^2 - B_i^2)^2} \right] \tag{13.7}$$

TABLE 13.5
Units of Optical Quantities

Name	Symbol	Unit[a]
Angular dispersion	$\nabla_\lambda \phi$	rad m^{-1}
n dispersion	$\nabla_\lambda n$	m^{-1}
Angular frequency	$\omega = 2\pi\nu$	Hz
Beam divergence	$\Delta\theta$	rad
Beam magnification	M	Dimensionless
Beam waist	W	m
Cross section	σ	m^2
Diffraction limited $\Delta\theta$	$\Delta\theta = \lambda/\pi w$	rad
Energy	E	J
Frequency	ν	Hz
Intensity	I	J s^{-1} m^{-2}
Laser linewidth	$\Delta\nu$	Hz
Laser linewidth	$\Delta\lambda$	m
Power	P	W = J s^{-1}
Rayleigh length	$L_{\mathscr{R}} = \pi w^2/\lambda$	m
Refractive index	n	Dimensionless
Wavelength	λ	m
Wavenumber	$k = 2\pi/\lambda$	m^{-1}
Wavenumber	$k = \omega/c$	m^{-1}

Source: Duarte, F.J., *Tunable Laser Optics*, Elsevier Academic Press, New York, 2003.

[a] Quantities such as I and σ are also used in cgs units.

TABLE 13.6
Dispersion Constants for Optical Materials

Material	A_0	A_i	B_i	Reference
Fused silica	1.00	$A_1 = 0.6961663$	$B_1 = 0.0684043$	Wolfe (1978)
		$A_2 = 0.4079426$	$B_2 = 0.1162414$	
		$A_3 = 0.8974794$	$B_3 = 9.896161$	
SF10	1.00	$A_1 = 1.62153902$	$B_1 = 0.11056286$	Schott (2001)
		$A_2 = 0.256287842$	$B_2 = 0.24407719$	
		$A_3 = 1.64447552$	$B_3 = 12.1436729$	
CaF	1.00	$A_1 = 0.5675888$	$B_1 = 0.050263605$	Wolfe (1978)
		$A_2 = 0.4710914$	$B_2 = 0.1003909$	
		$A_3 = 3.8484723$	$B_3 = 34.649040$	
ZnSe	3.71	$A_1 = 2.19$	$B_1 = 0.324$	Marple (1964)

Source: Duarte, F.J., *Tunable Laser Optics*, Elsevier Academic Press, New York, 2003.

TABLE 13.7
Refractive Index and $\partial n/\partial T$ of Laser and Nonlinear Optical Materials

Material	λ (nm)	n (λ)	$\partial n/\partial T$ (K⁻¹)	Reference
Al_2O_3 (a)		1.7654	1.31×10^{-5}	Barnes (1995a)
$BeAl_2O_4$ (a)		1.7422	0.94×10^{-5}	Barnes (1995a)
YAG		1.8289	1.04×10^{-5}	Barnes (1995a)
$AgGaS_2$ (o)	1064	2.4508	1.72×10^{-5}	Barnes (1995b)
AgGaSe (o)	1064	2.7005	7.7×10^{-5}	Barnes (1995b)
BBO (o)	1064	1.6551	-1.66×10^{-5}	Barnes (1995b)
KDP (o)	1064	1.4938	-3.40×10^{-5}	Barnes (1995b)
KTP (x)	1064	1.7386	2.20×10^{-5}	Barnes (1995b)
Quartz (o)	589.3	1.54424	-0.530×10^{-5}	Wolfe (1978)
Quartz (e)	589.3	1.54794	-0.642×10^{-5}	Wolfe (1978)
Fused silica	594.00	1.45824	1.19×10^{-5}	
CaF	580.262	1.43381	-1.04×10^{-5}	Wolfe (1978)
Rhodamine 6G in MPMMA	594.48	1.4953	$-1.4 \pm 0.2 \times 10^{-4}$	Duarte et al. (2000)
Rhodamine 6G in P(HEMA:MMA)	594.48	1.5039	$-1.3 \pm 0.2 \times 10^{-4}$	Duarte et al. (2000)
SiO_2-PMMA (30–70)	632.82	1.4755	-0.88×10^{-4}	Duarte and James (2003)
SiO_2-PMMA (50–50)	632.82	1.4659	-0.65×10^{-4}	Duarte and James (2003)

Source: Duarte, F.J., *Tunable Laser Optics*, Elsevier Academic Press, New York, 2003.
MPMMA, modified polymethyl methacrylate; P(HEMA:MMA), poly(2-hydroxyethyl methacrylate: methyl methacrylate); PMMA, polymethyl methacrylate.

which can be expressed more succinctly as

$$\nabla_\lambda n = \sum_{i=1}^{N} \frac{\lambda A_i}{n \Lambda_i} \left(1 - \lambda^2 \Lambda_i^{-1} \right) \tag{13.8}$$

where:

$$\nabla_\lambda n = \frac{\partial n}{\partial \lambda} \tag{13.9}$$

$$\Lambda_i = (\lambda^2 - B_i^2) \tag{13.10}$$

Thus, the second derivative can be written as

$$\nabla_\lambda^2 n = \sum_{i=1}^{N} \frac{A_i}{n \Lambda_i} \left[\left(1 - 5\lambda^2 \Lambda_i^{-1} + 4\lambda^4 \Lambda_i^{-2} \right) - \left(\frac{\lambda \nabla_\lambda n}{n} \right) \left(1 - \lambda^2 \Lambda_i^{-1} \right) \right] \tag{13.11}$$

The values for $n(\lambda)$, $\nabla_\lambda n$, and $\nabla_\lambda^2 n$, for various materials of interest, are given in Table 4.1.

13.5 ∂_N/∂_T OF LASER AND OPTICAL MATERIALS

An important parameter in the design of solid-state lasers, tunable laser oscillators, and optical systems is the $\partial n/\partial T$ factor. This is given in Table 13.7 for a collection of optical materials and gain media.

PROBLEMS

13.1 For a tunable laser emitting at the wavelength of $\lambda = 510.00$, with a linewidth of $\Delta v = 300$ MHz, express this linewidth in the wavelength domain $(\Delta\lambda)$ and also in the spatial domain $(1/\Delta x')$.

13.2 Discuss the advantages and disadvantages of expressing the laser linewidth in the wavelength domain $(\Delta\lambda)$, the frequency domain (Δv), and the spatial domain $(1/\Delta x')$.

13.3 Find the wavelength equivalence, in nanometers, of 300 keV.

13.4 Starting from Equation 13.8 show that $\nabla_\lambda^2 n$ is given by Equation 13.11.

13.5 Calculate for fused silica: $n(\lambda)$, $\nabla_\lambda n$, and $\nabla_\lambda^2 n$, at $\lambda = 510.554$ nm.

13.6 Calculate for fused silica: $n(\lambda)$, $\nabla_\lambda n$, and $\nabla_\lambda^2 n$, at $\lambda = 308$ nm.

References

CHAPTER 1

Byer, R. L., Herbst, R. L., Kildal, H., and Levenson, M. D. (1972). Optically pumped molecular iodine vapor-phase laser. *Appl. Phys. Lett.* **20**, 463–466.

Chutjian, A., and James, T. C. (1969). Intensity measurements in the $B^3\Pi_u^+ - X^1\Sigma_g^+$ system of I_2. *J. Chem. Phys.* **51**, 1242–1249.

de Broglie, L. (1923). Waves and quanta. *Nature* **112**, 540.

Diels, J.-C. (1990). Femtosecond dye lasers. In *Dye Laser Principles* (Duarte, F. J. and Hillman, L. W., eds.). Academic Press, New York, Chapter 3.

Diels, J.-C., and Rudolph, W. (2006). *Ultrashort Laser Pulse Phenomena*, 2nd edn. Academic Press, New York.

Dienes, A., and Yankelevich, D. R. (1998). Tunable dye lasers. In *Encyclopedia of Applied Physics*, Vol. 22 (Trigg, G. L., ed.). Wiley-VCH, New York, pp. 299–334.

Dirac, P. A. M. (1978). *The Principles of Quantum Mechanics*, 4th edn. Oxford University Press, London.

Dong, L., Sugunan, A., Hu, J., Zhou, S., Li, S., Popov, S., Toprak, M. S., Friberg, A. T., and Muhammed, M. (2013). Photoluminescence from quasi-type-II spherical CdSe-CdS core-shell quantum dots. *Appl. Opt.* **52**, 105–109.

Duarte, F. J. (1990). Technology of pulsed dye lasers. In *Dye Laser Principles* (Duarte, F. J. and Hillman, L. W., eds.). Academic Press, New York, Chapter 6.

Duarte, F. J. (ed.). (1995a). Dye lasers. In *Tunable Lasers Handbook*. Academic Press, New York, Chapter 5.

Duarte, F. J. (1995b). Solid-state dispersive dye laser oscillator: Very compact cavity. *Opt. Commun.* **117**, 480–484.

Duarte, F. J. (1998). Interference of two independent sources. *Am. J. Phys.* **66**, 662–663.

Duarte, F. J. (2003). *Tunable Laser Optics*. Elsevier Academic Press, New York.

Duarte, F. J. (2014). *Quantum Optics for Engineers*. CRC Press, New York.

Duarte, F. J., Liao, L. S., Vaeth, K. M., and Miller, A. M. (2006). Widely tunable green laser emission using the coumarin 545 tetramethyl dye as gain medium. *J. Opt. A: Pure Appl. Opt.* **8**, 172–174.

Dujardin, G., and Flamant, P. (1978). Amplified spontaneous emission and spatial dependence of gain in dye amplifiers. *Opt. Commun.* **24**, 243–247.

Everett, P. N. (1991). Flashlamp-excited dye lasers. In *High Power Dye Lasers* (Duarte, F. J., ed.). Springer-Verlag, Berlin, Germany, Chapter 6.

Faist, J., Capasso, F., Sivco, D. L., Sirtori, C., Hutchinson, A. L., and Cho A. Y. (1994). Quantum cascade laser. *Science* **264**, 553–556.

Feynman, R. P., Leighton, R. B., and Sands, M. (1965). *The Feynman Lectures on Physics*, Vol. III. Addison-Wesley, Reading, MA.

Fork, R. L., Greene, B. I., and Shank, C. V. (1981). Generation of optical pulses shorter than 0.1 psec by colliding pulse mode locking. *Appl. Phys. Lett.* **38**, 671–672.

Ganiel, U., Hardy, A., Neumann, G., and Treves, D. (1975). Amplified spontaneous emission and signal amplification in dye-laser systems. *IEEE J. Quant. Electron.* **QE-11**, 881–892.

Haag, G., Munz, M., and Maroowski, G. (1983). Amplified spontaneous emission (ASE) in laser oscillators and amplifiers. *IEEE J. Quant. Electron.* **QE-19**, 1149.

Haken, H. (1970). *Light and Matter*. Springer-Verlag, Berlin, Germany.

Haken, H. (1981). *Light*. North Holland, Amsterdam, The Netherlands.

Hammond, P. (1979). Spectra of the lowest excited singlet states of rhodamine 6G and rhodamine B. *IEEE J. Quant. Electron.* **QE-15**, 624–632.

Hargrove, R. S., and Kan, T. K. (1980). High power efficient dye amplifier pumped by copper vapor lasers. *IEEE J. Quant. Electron.* **QE-16**, 1108–1113.

Hillman, L. W. (1990). Laser dynamics. In *Dye Laser Principles* (Duarte, F. J. and Hillman, L. W., eds.). Academic Press, New York, Chapter 2.

Hollberg, L. (1990). CW dye lasers. In *Dye Laser Principles* (Duarte, F. J. and Hillman, L. W., eds.). Academic Press, New York, Chapter 5.

Javan, A., Benett, W. R., and Herriott, D. R. (1961). Population inversion and continuous optical maser oscillation in a gas discharge containing a He-Ne mixture. *Phys. Rev. Lett.* **6**, 106–110.

Jensen, C. (1991). Pulsed dye laser gain analysis and amplifier design. In *High Power Dye Lasers* (Duarte, F. J., ed.). Springer-Verlag, Berlin, Germany, Chapter 3.

Kittel, C. (1971). *Introduction to Solid State Physics*. John Wiley, New York.

Maiman, T. H. (1960). Stimulated optical radiation in ruby. *Nature* **187**, 493–494.

Munz, M., and Haag, G. (1980). Optimization of dye-laser output coupling by consideration of the spatial gain distribution. *Appl. Phys.* **22**, 175–184.

Nair, L. G., and Dasgupta, K. (1985). Amplified spontaneous emission in narrow-band pulsed dye laser oscillators—Theory and experiment. *IEEE J. Quant. Electron.* **21**, 1782–1794.

Penzkofer, A., and Falkenstein, W. (1978). Theoretical investigation of amplified spontaneous emission with picosecond light pulses in dye solutions. *Opt. Quant. Electron.* **10**, 399–423.

Planck, M. (1901). Ueber das gesetz der energieverteilung im normalspectrum. *Ann. Phys.* **309** (3), 553–563.

Sargent, M., Scully, M. O., and Lamb, W. E. (1974). *Laser Physics*. Addison-Wesley, Reading, MA.

Schäfer, F. P. (ed.). (1990). *Dye Lasers*, 2nd edn. Springer-Verlag, Berlin, Germany.

Schäfer, F. P., Schmidt, W., and Volze, J. (1966). Organic dye solution laser. *Appl. Phys. Lett.* **9**, 306–309.

Schrödinger, E. (1926). An undulatory theory of the mechanics of atoms and molecules. *Phys. Rev.* **28**, 1049–1070.

Siegman, A. E. (1986). *Lasers*. University Science Books, Mill Valley, CA.

Silfvast, W. T. (2008). *Laser Fundamentals*, 2nd edn. Cambridge University Press, Cambridge.

Sorokin, P. P., and Lankard, J. R. (1966). Stimulated emission observed from an organic dye, chloro-aluminum phthalocyanine. *IBM J. Res. Dev.* **10**, 162–163.

Teschke, O., Dienes, A., and Whinnery, J. R. (1976). Theory and operation of high-power CW and long-pulse dye lasers. *IEEE J. Quant. Electron.* **QE-12**, 383–395.

Townes, C. H. (1999). *How the Laser Happened*. Oxford University Press, Oxford.

Willett, C. S. (1974). *An Introduction to Gas Lasers: Population Inversion Mechanisms*. Pergamon Press, New York.

CHAPTER 2

Born, M., and Wolf, E. (1999). *Principles of Optics*, 7th edn. Cambridge University Press, New York.

Dirac, P. A. M. (1939). A new notation for quantum mechanics. *Math. Proc. Cam. Phil. Soc.* **35**, 416–418.

Dirac, P. A. M. (1978). *The Principles of Quantum Mechanics*, 4th edn. Oxford University Press, London.

Duarte, F. J. (1990). Narrow-linewidth pulsed dye laser oscillators. In *Dye Laser Principles* (Duarte, F. J. and Hillman, L. W., eds.). Academic Press, New York, Chapter 4.

Duarte, F. J. (ed.). (1991). Dispersive dye lasers. In *High Power Dye Lasers*. Springer-Verlag, Berlin, Germany, Chapter 2.

Duarte, F. J. (1992). Cavity dispersion equation $\Delta\lambda \approx \Delta\theta \, (\partial\theta/\partial\lambda)^{-1}$: A note on its origin. *Appl. Opt.* **31**, 6979–6982.

Duarte, F. J. (1993). On a generalized interference equation and interferometric measurements. *Opt. Commun.* **103**, 8–14.

Duarte, F. J. (ed.). (1995a). Interferometric imaging. In *Tunable Laser Applications*, 1st edn. Marcel Dekker, New York, Chapter 5.

Duarte, F. J. (ed.). (1995b). Narrow-linewidth laser oscillators and intracavity dispersion. In *Tunable Lasers Handbook*. Academic Press, New York, Chapter 2.

Duarte, F. J. (1997). Interference, diffraction, and refraction, via Dirac's notation. *Am. J. Phys.* **65**, 637–640.

Duarte, F. J. (1998). Interference of two independent sources. *Am. J. Phys.* **66**, 662–663.

Duarte, F. J. (2003). *Tunable Laser Optics*. Elsevier Academic Press, New York.

Duarte, F. J. (2004). Comment on "reflection, refraction, and multislit interference." *Eur. J. Phys.* **25**, L57–L58.

Duarte, F. J. (2006). Multiple-prism dispersion equations for positive and negative refraction. *Appl. Phys. B* **82**, 35–38.

Duarte, F. J. (2008). Coherent electrically excited organic semiconductors: Coherent or laser emission? *Appl. Phys. B* **90**, 101–108.

Duarte, F. J. (2014). *Quantum Optics for Engineers*. CRC Press, New York.

Duarte, F. J., and Paine, D. J. (1989). Quantum mechanical description of N-slit interference phenomena. In *Proceedings of the International Conference on Lasers '88* (Sze, R. C. and Duarte, F. J., eds.). STS Press, McLean, VA, pp. 42–47.

Feynman, R. P., Leighton, R. B., and Sands, M. (1965a). *The Feynman Lectures on Physics*, Vol. III. Addison-Wesley, Reading, MA.

Feynman, R. P., Leighton, R. B., and Sands, M. (1965b). *The Feynman Lectures on Physics*, Vol. I. Addison-Wesley, Reading, MA.

Robertson, J. K. (1955). *Introduction to Optics: Geometrical and Physical*. Van Nostrand, New York.

van Kampen, N. G. (1988). Ten theorems about quantum mechanical measurements. *Phys. A* **153**, 97–113.

Wallenstein, R., and Hänsch, T. W. (1974). Linear pressure tuning of a multielement dye laser spectrometer. *Appl. Opt.* **13**, 1625–1628.

CHAPTER 3

Bass, I. L., Bonanno, R. E., Hackel, R. H., and Hammond, P. R. (1992). High-average power dye laser at Lawrence Livermore National Laboratory. *Appl. Opt.* **31**, 6993–7006.

Bennett, C. H., Bessette F., Brassard, G., Salvail, L., and Smolin, J. (1992). Experimental quantum cryptography. *J. Cryptol.* **5**, 3–28.

Dirac, P. A. M. (1978). *The Principles of Quantum Mechanics*, 4th edn. Oxford University Press, London.

Duarte, F. J. (1990). Narrow-linewidth pulsed dye laser oscillators. In *Dye Laser Principles* (Duarte, F. J. and Hillman, L. W., eds.). Academic Press, New York, Chapter 4.

Duarte, F. J. (ed.). (1991). Dispersive dye lasers. In *High Power Dye Lasers*. Springer-Verlag, Berlin, Germany.

Duarte, F. J. (1992). Cavity dispersion equation $\Delta\lambda \approx \Delta\theta\,(\partial\theta/\partial\lambda)^{-1}$: A note on its origin. *Appl. Opt.* **31**, 6979–6982.

Duarte, F. J. (1993). On a generalized interference equation and interferometric measurements. *Opt. Commun.* **103**, 8–14.

Duarte, F. J. (1997). Interference, diffraction, and refraction, via Dirac's notation. *Am. J. Phys.* **65**, 637–640.

Duarte, F. J. (1999). Multiple-prism grating solid-state dye laser oscillator: Optimized architecture. *Appl. Opt.* **38**, 6347–6349.

Duarte, F. J. (2003). *Tunable Laser Optics*. Elsevier Academic Press, New York.

Duarte, F. J. (2013). The probability amplitude for entangled polarizations: An interferometric approach. *J. Mod. Opt.* **60**, 1585–1587.

Duarte, F. J. (2014). *Quantum Optics for Engineers*. CRC Press, New York.

Duarte, F. J., and Piper, J. A. (1984). Narrow-linewidth, high-prf copper laser-pumped Dye-laser oscillators. *Appl. Opt.* **23**, 1391–1394.

Ekert, A. K. (1991). Quantum cryptography based on Bell's theorem. *Phys. Rev. Lett.* **67**, 661–663.

Everett, P. N. (1989). 300 W dye laser for field experimental site. In *Proceedings of the International Conference on Lasers '88* (Duarte, F. J., ed.). STS Press, McLean, VA, pp. 404–409.

Feng, Y., Taylor, L. R., and Calia, D. B. (2009). 25 W Raman-fiber-amplifier-based 589 nm laser for laser guide star. *Opt. Express* **17**, 19021–19026.

Feynman, R. P., Leighton, R. B., and Sands, M. (1965). *The Feynman Lectures on Physics*, Vol. III. Addison-Wesley, Reading, MA.

Haken, H. (1981). *Light*. North Holland, Amsterdam, The Netherlands.

Heisenberg, W (1927). Über den anschaulichen inhalt der quantentheoretischen kinematik und mechanic. *Z. Phys.* **43**, 172–198.

Henry, L. J., Shay, T. M., Moore, G. T., and Grosek, J. R. (2013). Seeded Raman amplifier for applications in the 1100–1500 nm spectral region. US Patent 8, 472, 486 B1.

Mandel, L., and Wolf, E. (1995). *Optical Coherence and Quantum Optics*. Cambridge University Press, Cambridge.

Mitchell, A. C. G., and Zemansky, M. W. (1971). *Resonance Radiation and Excited Atoms*. Cambridge University Press, New York.

Pique, J.-P., and Farinotti, S. (2003). Efficient modeless laser for mesospheric sodium laser guide star. *J. Opt. Soc. Am. B* **20**, 2093–2101.

Pryce, M. H. L., and Ward, J. C. (1947). Angular correlation effects with annihilation radiation. *Nature* **160**, 435.

Primmerman, C. A., Murphy, D. V., Page, D. A., Zollars, B. G., and Barclay, H. T. (1991). Compensation of atmospheric optical distortion using a synthetic beacon. *Nature* **353**, 141–143.

Ward, J. C. (1949). *Some Properties of the Elementary Particles*. Oxford University Press, Oxford.

CHAPTER 4

Born, M., and Wolf, E. (1999). *Principles of Optics*, 7th edn. Cambridge University Press, Cambridge.

Brewster, D. (1813). *A Treatise on New Philosophical Instruments for Various Purposes in the Arts and Sciences with Experiments on Light and Colours*. Murray and Blackwood, Edinburgh.

Diels, J.-C. (1990). Femtosecond dye lasers. In *Dye Laser Principles* (Duarte, F. J., ed.). Academic Press, New York, Chapter 3.

Diels, J.-C., Dietel, W., Fontaine, J. J., Rudolph, W., and Wilhelmi, B. (1985). Analysis of a mode-locked ring laser: Chirped-solitary-pulse solutions. *J. Opt. Soc. Am. B* **2**, 680–686.

Diels, J.-C., and Rudolph, W. (2006). *Ultrafast Laser Pulse Phenomena*, 2nd edn. Academic Press, New York.

Dietel, W., Fontaine, J. J., and Diels, J.-C. (1983). Intracavity pulse compression with glass: A new method of generating pulses shorter than 60 fs. *Opt. Lett.* **8**, 4–6.

Duarte, F. J. (1985). Note on achromatic multiple-prism beam expanders. *Opt. Commun.* **53**, 259–262.

Duarte, F. J. (1987a). Generalized multiple-prism dispersion theory for pulse compression in ultrafast dye lasers. *Opt. Quant. Electron.* **19**, 223–229.

Duarte, F. J. (1987b). Beam shaping with telescopes and multiple-prism beam expanders. *J. Opt. Soc. Am. A* **4**, 30.

Duarte, F. J. (1989). Transmission efficiency in achromatic nonorthogonal multiple-prism laser beam expanders. *Opt. Commun.* **71**, 1–5.

Duarte, F. J. (1990a). Narrow-linewidth pulsed dye laser oscillators. In *Dye Laser Principles* (Duarte, F. J. and Hillman, L. W., eds.). Academic Press, New York, Chapter 4.

Duarte, F. J. (1990b). Prismatic pulse compression: Beam deviations and geometrical perturbations. *Opt. Quant. Electron.* **22**, 467–471.

Duarte, F. J. (ed.). (1991). Dispersive dye lasers. In *High Power Dye Lasers.* Springer-Verlag, Berlin, Germany, Chapter 2.

Duarte, F. J. (1992). Cavity dispersion equation $\Delta\lambda \approx \Delta\theta(\partial\theta/\partial\lambda)^{-1}$: A note on its origin. *Appl. Opt.* **31**, 6979–6982.

Duarte, F. J. (1993). On a generalized interference equation and interferometric measurements. *Opt. Commun.* **103**, 8–14.

Duarte, F. J. (ed.). (1995a). Narrow-linewidth laser oscillators and intracavity dispersion. In *Tunable Lasers Handbook.* Academic Press, New York, Chapter 2.

Duarte, F. J. (ed.). (1995b). Dye lasers. In *Tunable Lasers Handbook.* Academic Press, New York, Chapter 5.

Duarte, F. J. (ed.). (1995c). Interferometric imaging. In *Tunable Laser Applications*, 1st edn. Marcel Dekker, New York, Chapter 5.

Duarte, F. J. (1999). Multiple-prism grating solid-state dye laser oscillator: Optimized architecture. *Appl. Opt.* **38**, 6347–6349.

Duarte, F. J. (2000). Multiple-prism arrays in laser optics. *Am. J. Phys.* **68**, 162–166.

Duarte, F. J. (2001a). Multiple-return-pass beam divergence and the linewidth equation. *Appl. Opt.* **40**, 3038–3041.

Duarte, F. J. (2001b). Laser sensitometer using multiple-prism beam expansion and a polarizer. US Patent 6, 236, 461.

Duarte, F. J. (2002). Secure interferometric communications in free space. *Opt. Commun.* **205**, 313–319.

Duarte, F. J. (2003). *Tunable Laser Optics*, 1st edn. Elsevier Academic Press, New York.

Duarte, F. J. (2005). Secure interferometric communications in free space: Enhanced sensitivity for propagation in the metre range. *J. Opt. A: Pure Appl. Opt.* **7**, 73–75.

Duarte, F. J. (2006). Multiple-prism dispersion equations for positive and negative refraction. *Appl. Phys. B* **82**, 35–38.

Duarte, F. J. (2009). Generalized multiple-prism dispersion theory for laser pulse compression: Higher order phase derivatives. *Appl. Phys. B* **96**, 809–814.

Duarte, F. J. (2013). Tunable laser optics: Applications to optics and quantum optics. *Prog. Quant. Electron.* **37**, 326–347.

Duarte, F. J., and Piper, J. A. (1982). Dispersion theory of multiple-prism beam expander for pulsed dye lasers. *Opt. Commun.* **43**, 303–307.

Duarte, F. J., and Piper, J. A. (1983). Generalized prism dispersion theory. *Am. J. Phys.* **51**, 1132–1134.

Duarte, F. J., and Piper, J. A. (1984). Multi-pass dispersion theory of prismatic pulsed dye lasers. *Opt. Acta* **31**, 331–335.

Duarte, F. J., Reed, B. A., and Burak, C. J. (2005). Laser sensitometer. US Patent 6, 903, 824 B2.

Duarte, F. J., Taylor, T. S., Black, A. M., Davenport, W. E., and Varmette P. G. (2011). *N*-slit interferometer for secure free-space optical communications: 527 m intra interferometric path length. *J. Opt.* **13**, 035710.

Duarte, F. J., Taylor, T. S., Clark, A. B., and Davenport, W. E. (2010). The *N*-slit interferometer: An extended configuration. *J. Opt.* **12**, 015705.

Fork, R. L., Martinez, O. M., and Gordon, J. P. (1984). Negative dispersion using pairs of prisms. *Opt. Lett.* **9**, 150–152.

Jenkins, F. A., and White, H. E. (1957). *Fundamentals of Optics*. McGraw-Hill, New York.

Lohmann, A. W., and Stork, W. (1989). Modified Brewster telescopes. *Appl. Opt.* **28**, 1318–1319.

Maker, G. T., and Ferguson, A. I. (1989). Frequency-modulation mode locking of a diode-pumped Nd:YAG laser. *Opt. Lett.* **14**, 788–790.

Marple, D. T. F. (1964). Refractive index of ZnSe, ZnTe, and CdTe. *J. Appl. Phys.* **35**, 539–542.

Newton, I. (1704). *Opticks*. Royal Society, London.

Osvay, K., Kovács, A. P., Heiner, Z., Kurdi, G., Klebniczki, J., and Csatári, M. (2004). Angular dispersion and temporal change of femtosecond pulses from misaligned pulse compressors. *IEEE J. Sel. Top. Quant. Electron.* **10**, 213–220.

Osvay, K., Kovács, A. P., Kurdi, G., Heiner, Z., Divall, M., Klebniczki, J., and Ferincz, I. E. (2005). Measurement of non-compensated angular dispersion and the subsequent temporal lengthening of femtosecond pulses in a CPA laser. *Opt. Commun.* **248**, 201–209.

Pang, L. Y., Fujimoto, J. G., and Kintzer, E. S. (1992). Ultrashort-pulse generation from high-power diode arrays by using intracavity optical nonlinearities. *Opt. Lett.* **17**, 1599–1601.

Shay, T. M., and Duarte, F. J. (2009). Tunable fiber lasers. In *Tunable Laser Applications* (Duarte, F. J., ed.). CRC Press, New York, Chapter 6.

Sirat, G. Y., Wilner, K., and Neuhauser, D. (2005). Uniaxial crystal interferometer: Principles and forecasted applications to imaging astrometry. *Opt. Express* **13**, 6310–6322.

Wyatt, R. (1978). Comment on "On the dispersion of a prism used as a beam expander in a nitrogen laser." *Opt. Commun.* **26**, 9–11.

Wynne, C. G. (1997). Atmospheric dispersion in very large telescopes with adaptive optics. *Mon. Not. R. Astron. Soc.* **285**, 130–134.

CHAPTER 5

Bennett, J. M., and Bennett, H. E. (1978). Polarization. In *Handbook of Optics* (Driscoll, W. G. and Vaughan, W., eds.). McGraw-Hill, New York.

Born, M., and Wolf, E. (1999). *Principles of Optics*, 7th edn. Cambridge University Press, New York.

Duarte, F. J. (1989). Optical device for rotating the polarization of a light beam. US Patent 4, 822, 150.

Duarte, F. J. (1990). Technology of pulsed dye lasers. In *Dye Laser Principles* (Duarte, F. J. and Hillman, L. W., eds.). Academic Press, New York, Chapter 6.

Duarte, F. J. (1992). Beam transmission characteristics of a collinear polarization rotator. *Appl. Opt.* **31**, 3377–3378.

Duarte, F. J. (1995). Solid-state dispersive dye laser oscillator: Very compact cavity. *Opt. Commun.* **117**, 480–484.

Duarte, F. J. (1999). Multiple-prism grating solid-state dye laser oscillator: Optimized architecture. *Appl. Opt.* **38**, 6347–6349.

Duarte, F. J. (2001). Laser sensitometer using a multiple-prism beam expander and a Polarizer. US Patent 6, 236, 461 B1.

Duarte, F. J. (2003). *Tunable Laser Optics*, 1st edn. Elsevier Academic Press, New York.

Duarte, F. J. (2014). *Quantum Optics for Engineers*. CRC Press, New York.

Duarte, F. J., Ehrlich, J. J., Davenport, W. E., and Taylor, T. S. (1990). Flashlamp pumped narrow-linewidth dispersive dye laser oscillators: Very low amplified spontaneous emission levels and reduction of linewidth instabilities. *Appl. Opt.* **29**, 3176–3179.

Duarte, F. J., and Piper, J. A. (1983). Generalized prism dispersion theory. *Am. J. Phys.* **51**, 1132–1134.

Dyson, F. J. (1990). Feynman's proof of Maxwell equations. *Am. J. Phys.* **58**, 209–211.

Feynman, R. P., Leighton, R. B., and Sands, M. (1965). *The Feynman Lectures on Physics*, Vol. II. Addison-Wesley, Reading, MA.

Jenkins, F. A., and White, H. E. (1957). *Fundamentals of Optics*. McGraw-Hill, New York.

Jones, R. C. (1947). A new calculus for the treatment of optical systems. *J. Opt. Soc. Am.* **37**, 107–110.

Lorrain, P., and Corson, D. (1970). *Electromagnetic Fields and Waves*. Freeman, San Francisco, CA.

Olivares, I. E., Cuadra, J. A., Aguilar, F. A., Aguirre Gomez, J. G., and Duarte, F. J. (2009). Optical method using rotating Glan-Thompson polarizers to independently vary the power of the excitation and repumping lasers in laser cooling experiments. *J. Mod. Opt.* **56**, 1780–1784.

Robson, B. A. (1974). *The Theory of Polarization Phenomena*. Clarendon Press, Oxford.

Saleh, B. E. A., and Teich, M. C. (1991). *Fundamentals of Photonics*. Wiley, New York.

CHAPTER 6

Brouwer, W. (1964). *Matrix Methods in Optical Design*. Benjamin, New York.

Duarte, F. J. (1988). Transmission efficiency in achromatic nonorthogonal multiple-prism laser beam expanders. *Opt. Commun.* **71**, 1–5.

Duarte, F. J. (1989). Ray transfer matrix analysis of multiple-prism dye laser oscillators. *Opt. Quant. Electron.* **21**, 47–54.

Duarte, F. J. (1990). Narrow-linewidth pulsed dye laser oscillators. In *Dye Laser Principles* (Duarte, F. J. and Hillman, L. W., eds.). Academic Press, New York, Chapter 4.

Duarte, F. J. (ed.). (1991). Dispersive dye lasers. In *High Power Dye Lasers*. Springer-Verlag, Berlin, Germany, Chapter 2.

Duarte, F. J. (1992). Multiple-prism dispersion and 4×4 ray transfer matrices. *Opt. Quant. Electron.* **24**, 49–53.

Duarte, F. J. (1999). Multiple-prism grating solid-state dye laser oscillator: Optimized architecture. *Appl. Opt.* **38**, 6347–6349.

Duarte, F. J. (2001). Multiple-return-pass beam divergence and the linewidth equation. *Appl. Opt.* **40**, 3038–3041.

Duarte, F. J. (2003). *Tunable Laser Optics*, 1st edn. Elsevier Academic Press, New York.

Duarte, F. J., Costela, A., Garcia-Moreno, I., Sastre, R., Ehrlich, J. J., and Taylor, T. S. (1997). Dispersive solid-state dye laser oscillators. *Opt. Quant. Electron.* **29**, 461–472.

Duarte, F. J., and Piper, J. A. (1982). Dispersion theory of multiple-prism beam expanders. *Opt. Commun.* **43**, 303–307.

Kogelnik, H. (1979). Propagation of laser beams. In *Applied Optics and Optics Engineering* (Shannon, R. R. and Wyant, J. C., eds.). Academic Press, New York, pp. 155–190.

Kostenbauer, A. G. (1990). Ray-pulse matrices: A rational treatment for dispersive optical systems. *IEEE J. Quant. Electron.* **26**, 1148–1157.

Siegman, A. (1986). *Lasers*. University Science Books, Mill Valley, CA.

Wollnik, H. (1987). *Optics of Charged Particles*. Academic Press, New York.

CHAPTER 7

Bass, I. L., Bonanno, R. E., Hackel, R. H., and Hammond, P. R. (1992). High-average power dye laser at Lawrence Livermore National Laboratory. *Appl. Opt.* **31**, 6993–7006.

Beiting, E. J., and Smith, K. A. (1979). An on-axis reflective beam expander for pulsed dye laser cavities. *Opt. Commun.* **28**, 355–358.

Belenov, E. M., Velichanskii, V. L., Zibrob, A. S., Nitikin, V. V., Sautenkov, V. A., and Uskov, A. V. (1983). Methods for narrowing the emission line of an injection laser. *Sov. J. Quant. Electron.* **13**, 792–798.

Bennetts, S., McDonald, G. D., Hardman, K. S., Debs, J. E., Carlos, C. N., Kuhn, C. C. N., Close, J. D., and Robins, N. P. (2014). External cavity diode lasers with 5 kHz linewidth and 200 nm tuning range at 1.55 μm and methods for linewidth measurement. *Opt. Express* **22**, 10642–10654.

Bernhardt, A. F., and Rasmussen, P. (1981). Design criteria and operating characteristics of a single-mode pulsed dye laser. *Appl. Phys. B* **26**, 141–146.

Blit, S., Ganiel, U., and Treves, D. (1977). A tunable, single mode, injection-locked, flashlamp pumped dye laser. *Appl. Phys.* **12**, 69–74.

Bor, Zs. (1979). A novel pumping arrangement for tunable single picosecond pulse generation with a N_2 laser pumped distributed feedback dye lasers. *Opt. Commun.* **29**, 103–108.

Born, M., and Wolf, E. (1999). *Principles of Optics*, 7th edn. Cambridge University Press, Cambridge.

Bos, F. (1981). Versatile high-power single-longitudinal-mode pulsed dye laser. *Appl. Opt.* **20**, 1886–1890.

Duarte, F. J. (1985a). Application of dye laser techniques to frequency selectivity in pulsed CO_2 lasers. In *Proceedings of the International Conference on Lasers '84* (Corcoran, K. M., Sullivan, M. D., and Stwalley, W. C., eds.). STS Press, McLean, VA, pp. 397–403.

Duarte, F. J. (1985b). Variable linewidth high-power TEA CO_2 laser. *Appl. Opt.* **24**, 34–37.

Duarte, F. J. (1985c). Multiple-prism Littrow and grazing-incidence pulsed CO_2 lasers. *Appl. Opt.* **24**, 1244–1245.

Duarte, F. J. (1985d). Note on achromatic multiple-prism beam expanders. *Opt. Commun.* **53**, 259–262.

Duarte, F. J. (1989). Transmission efficiency in achromatic nonorthogonal multiple-prism laser beam expanders. *Opt. Commun.* **71**, 1–5.

Duarte, F. J. (1990a). Narrow-linewidth pulsed dye laser oscillators. In *Dye Laser Principles* (Duarte, F. J. and Hillman, L. W., eds.). Academic Press, New York, Chapter 4.

Duarte, F. J. (1990b). Technology of pulsed dye lasers. In *Dye Laser Principles* (Duarte, F. J. and Hillman, L. W., eds.). Academic Press, New York, Chapter 6.

Duarte, F. J. (ed.). (1991a). Dispersive dye lasers. In *High Power Dye Lasers*. Springer-Verlag, Berlin, Germany, Chapter 2.

Duarte, F. J. (1991b). Dispersive excimer lasers. In *Proceedings of the International Conference on Lasers '90* (Harris, D. G. and Herbelin, J., eds.). STS Press, McLean, VA, pp. 277–279.

Duarte, F. J. (1993a). Multiple-prism grating designs tune diode lasers. *Laser Focus World* **29** (2), 103–109.

Duarte, F. J. (1993b). On a generalized interference equation and interferometric measurements. *Opt. Commun.* **103**, 8–14.

Duarte, F. J. (ed.). (1995a). Dispersive external-cavity semiconductor lasers. In *Tunable Laser Applications*. Marcel Dekker, New York, Chapter 3.

Duarte, F. J. (ed.). (1995b). Interferometric imaging. In *Tunable Laser Applications*. Marcel Dekker, New York, Chapter 5.

Duarte, F. J. (1997). Multiple-prism near-grazing-incidence grating solid-state dye laser oscillator. *Opt. Laser Technol.* **29**, 513–516.

Duarte, F. J. (1999). Multiple-prism grating solid-state dye laser oscillator: Optimized architecture. *Appl. Opt.* **38**, 6347–6349.

Duarte, F. J. (2001). Multiple-return-pass beam divergence and the linewidth equation. *Appl. Opt.* **40**, 3038–3041.

Duarte, F. J. (2003). *Tunable Laser Optics*, 1st edn. Elsevier Academic Press, New York.

Duarte, F. J. (ed.). (2009). Broadly tunable external-cavity semiconductor lasers. In *Tunable Laser Applications*, 2nd edn. CRC Press, New York, Chapter 5.

Duarte, F. J. (2014). *Quantum Optics for Engineers*. CRC Press, New York.

Duarte, F. J., and Conrad, R. W. (1987). Diffraction-limited single-longitudinal-mode multiple-prism flashlamp-pumped dye laser oscillator: Linewidth analysis and injection of amplifier system. *Appl. Opt.* **26**, 2567–2571.

Duarte, F. J., Costela, A., Garcia-Moreno, I., Sastre, R., Ehrlich, J. J., and Taylor, T. S. (1997). Dispersive solid-state dye lasers. *Opt. Quant. Electron.* **29**, 461–472.

Duarte, F. J., Ehrlich, J. J., Patterson, S. P., Russell, S. D., and Adams, J. E. (1988). Linewidth instabilities in narrow-linewidth flashlamp-pumped dye laser oscillators. *Appl. Opt.* **27**, 843–846.

Duarte, F. J., and Piper, J. A. (1980). A double-prism beam expander for pulsed dye lasers. *Opt. Commun.* **35**, 100–104.

Duarte, F. J., and Piper, J. A. (1981). Prism preexpanded grazing-incidence grating cavity for pulsed dye lasers. *Appl. Opt.* **20**, 2113–2116.

Duarte, F. J., and Piper, J. A. (1982a). Dispersion theory of multiple-prism beam expanders. *Opt. Commun.* **43**, 303–307.

Duarte, F. J., and Piper, J. A. (1982b). Comparison of prism-preexpanded and grazing incidence grating cavities for copper laser pumped dye lasers. *Appl. Opt.* **21**, 2782–2786.

Duarte, F. J., and Piper, J. A. (1984a). Narrow-linewidth, high prf copper laser-pumped dye laser oscillators. *Appl. Opt.* **23**, 1391–1394.

Duarte, F. J., and Piper, J. A. (1984b). Multi-pass dispersion theory of prismatic pulsed dye lasers. *Opt. Acta* **31**, 331–335.

Duarte, F. J., Taylor, T. S., Costela, A., Garcia-Moreno, I., and Sastre, R. (1998). Long-pulse narrow-linewidth dispersive solid-state dye-laser oscillator. *Appl. Opt.* **37**, 3987–3989.

Dupre, P. (1987). Quasiunimodal tunable pulsed dye laser at 440 nm: Theoretical development for using quad prism beam expander and one or two gratings in a pulsed dye laser oscillator cavity. *Appl. Opt.* **26**, 860–871.

Farkas, A. M., and Eden, J. G. (1993). Pulsed dye laser amplification and frequency doubling of single longitudinal mode semiconductor lasers. *IEEE J. Quant. Electron.* **29**, 2923–2927.

Flamant, P. H., and Maillard, D. J. M. (1984). Transient injection frequency-locking of a microsecond-pulsed dye laser for atmospheric measurements. *Opt. Quant. Electron.* **16**, 179–182.

Fleming, M. W., and Mooradian, A. (1981). Spectral characteristics of external-cavity controlled semiconductor lasers. *IEEE J. Quant. Electron.* **QE-17**, 44–59.

Fox, R. W., Hollberg, L., and Zibrov, A. S. (1997). Semiconductor diode lasers. In *Atomic, Molecular, and Optical Physics: Electromagnetic Radiation* (Dunning, F. B. and Hulet, R. G., eds.). Academic Press, New York, Chapter 4.

Hänsch, T. W. (1972). Repetitively pulsed tunable dye laser for high resolution spectroscopy. *Appl. Opt.* **11**, 895–898.

Harrison, J., and Mooradian, A. (1989). Linewidth and offset frequency locking of external cavity GaAlAs lasers. *IEEE J. Quant. Electron.* **QE-25**, 1152–1155.

Harvey, K. C., and Myatt, C. J. (1991). External-cavity diode laser using a grazing-incidence diffraction grating. *Opt. Lett.* **16**, 910–912.

Hawthorn, C. J., Weber, K. P., and Scholten, R. E. (2001). Littrow configuration tunable external cavity diode laser with fixed direction output beam. *Rev. Sci. Instrum.* **72**, 4477–4479.

Johnston, T. F., and Duarte, F. J. (2002). Lasers, dye. In *Encyclopedia of Physical Science and Technology*, 3rd edn., Vol. 8 (Meyers, R. A., ed.). Academic Press, New York, pp. 315–359.

Kangas, K. W., Lowenthal, D. D., and Muller, C. H. (1989). Single-longitudinal-mode, tunable, pulsed Ti:sapphire laser oscillator. *Opt. Lett.* **14**, 21–23.

Kasuya, T., Suzuki, T., and Shimoda. K. (1978). A prism anamorphic system for Gaussian beam expander. *Appl. Phys.* **17**, 131–136.

Klauminzer, G. K. (1978). Optical beam expander for dye laser. US Patent 4, 127, 828.

Kogelnik, H., and Shank., C. V. (1971). Stimulated emission in a periodic structure. *Appl. Phys. Lett.* **18**, 152–154.

Kogelnik, H., and Shank, C. V. (1972). Coupled-wave theory of distributed feedback Lasers. *J. Appl. Phys.* **43**, 2327–2335.

Laurila, T., Joutsenoja, T., Hernberg, R., and Kuittinen, M. (2002). Tunable external-cavity laser at 650 nm based on a transmission diffraction grating. *Appl. Opt.* **27**, 5632–5637.

Littman, M. G. (1978). Single-mode operation of grazing-incidence pulsed dye laser. *Opt. Lett.* **3**, 138–140.

Littman, M. G., and Metcalf, H. J. (1978). Spectrally narrow pulsed dye laser without beam expander. *Appl. Opt.* **17**, 2224–2227.

Liu, K., and Littman, M. G. (1981). Novel geometry for single-mode scanning of tunable lasers. *Opt. Lett.* **6**, 117–118.

Maeda, M., Uchino, O., Okada, T., and Miyazoe, Y. (1975). Powerful narrow-band dye laser forced oscillator. *J. Appl. Phys.* **14**, 1975–1980.

Meaburn, J. (1976). *Detection and Spectrometry of Faint Light.* Reidel, Boston, MA.

Nevsky, A. Yu., Bressel, U., Ernsting, I., Eisele, Ch., Okhapkin, M., Schiller, S., Gubenko, A., et al. (2008). A narrow-linewidth external cavity quantum dot laser for high-resolution spectroscopy in the near infrared and yellow spectral ranges. *Appl. Phys. B* **92**, 501–507.

Notomi, M., Mitomi, O., Yoshikuni, Y., Kano, F., and Tohmori, Y. (1990). Broad-band tunable two-section laser diode with external grating cavity. *IEEE Photon. Technol. Lett.* **2**, 85–87.

Pacala, T. J., McDermid, I. S., and Laudenslager, J. B. (1984). Single-longitudinal-mode operation of an XeCl laser. *Appl. Phys. Lett.* **45**, 507–509.

Saikan, S. (1978). Nitrogen-laser-pumped single-mode dye laser. *Appl. Phys.* **17**, 41–44.

Schäfer, F. P. (1990). Principles of dye laser operation. In *Dye Lasers* (Schäfer, F. P., ed.). Springer-Verlag, Berlin, Germany, Chapter 1.

Shank, C. V. (1975). Physics of dye lasers. *Rev. Mod. Phys.* **47**, 649–657.

Shank, C. V., Bjorkholm, J. E., and Kogelnik, H. (1971). Tunable distributed-feedback dye laser. *Appl. Phys. Lett.* **18**, 395–396.

Shay, T. M., and Duarte, F. J. (2009). Tunable fiber lasers. In *Tunable Lasers Applications* (Duarte, F. J., ed.). CRC Press, New York, Chapter 6.

Shoshan, I., Danon, N. N., and Oppenheim, U. P. (1977). Narrowband operation of a pulsed dye laser without intracavity beam expansion. *J. Appl. Phys.* **48**, 4495–4497.

Siegman, A. (1986). *Lasers.* University Science Books, Mill Valley, CA.

Strome, F. C., and Webb, J. P. (1971). Flashtube-pumped dye laser with multiple-prism tuning. *Appl. Opt.* **10**, 1348–1353.

Sugii, M., Ando, M., and Sasaki, K. (1987). Simple long-pulse XeCl laser with narrow-line output. *IEEE J. Quant. Electron.* **QE-23**, 1458–1460.

Uenishi, Y., Honna, K., and Nagaoka, S. (1996). Tunable laser diode using a nickel micromachined external mirror. *Electron. Lett.* **32**, 1207–1208.

Voumard, C. (1977). External-cavity-controlled 32-MHz narrow-band CW GaAlAs-diode lasers. *Opt. Lett.* **1**, 61–63.

Wadsworth, W. J., McKinnie, I. T., Woolhouse, A. D., and Haskell, T. G. (1999). Efficient distributed feedback solid state dye laser with dynamic grating. *Appl. Phys. B* **69**, 163–165.

Wallenstein, R., and Hänsch, T. W. (1974). Linear pressure tuning of a multielement dye laser spectrometer. *Appl. Opt.* **13**, 1625–1628.

Wieman, C. E., and Hollberg, L. (1991). Using diode lasers for atomic physics. *Rev. Sci. Instrum.* **62**, 1–20.

Wolf, T., Borchert, B., Drögemüller, K., and Amann, M.-C. (1991). Narrow-linewidth InGaAsP/InP metal-clad ridge-waveguide distributed feedback lasers. *Jpn. J. Appl. Phys.* **30**, L745–L747.

Wyatt, R. (1978). Narrow linewidth, short pulse operation of a nitrogen-laser-pumped dye laser. *Opt. Commun.* **26**, 429–431.

Zhu, X.-L., Lam, S.-K., and Lo, D. (2000). Distributed-feedback dye-doped solgel silica lasers. *Appl. Opt.* **39**, 3104–3107.

Zorabedian, P. (1992). Characteristics of a grating-external-cavity semiconductor laser containing intracavity prism beam expanders. *J. Lightw. Technol.* **10**, 330–335.

Zorabedian, P. (1995). Tunable external-cavity semiconductor lasers. In *Tunable Lasers Handbook* (Duarte, F. J., ed.). Academic Press, New York, Chapter 8.

CHAPTER 8

Akhmanov, S. A., Kovrigin, A. I., Kolosov, V. A., Piskarskas, A. S., Fadeev, V. V., and Khokhlov, R. V. (1966). Tunable parametric light generator with KDP crystal. *JETP Lett.* **3**, 241–245.

Armstrong, J. A., Bloemberger, N., Duccuing, J., and Pershan, P. S. (1962). Interactions between light waves in a nonlinear dielectric. *Phys. Rev.* **127**, 1918–1939.

Auyeung, J., Fekete, D., Pepper, D. M., and Yariv. A. (1979). A theoretical and experimental investigation of the modes of optical resonators with phase-conjugate mirrors. *IEEE J. Quant. Electron.* **QE-15**, 1180–1188.

Baldwin, G. C. (1969). *An Introduction to Nonlinear Optics.* Plenum Press, New York.

Barnes, N. P. (1995). Optical parametic oscillators. In *Tunable Lasers Handbook* (Duarte, F. J., ed.). Academic Press, New York, Chapter 7.

Barnes, N. P., and Corcoran, V. J. (1976). Parametric generation processes: Spectral bandwidth and acceptance angles. *Appl. Opt.* **15**, 696–699.

Bellini, M., and Hänsch, T. W. (2000). Phase-locked white-light continuum pulses: Toward a universal optical frequency-comb synthesizer. *Opt. Lett.* **25**, 1049–1051.

Berik, E., Davidenko, B., Mihkelsoo, V., Apanasevich, P., Grabchikov, A., and Orlovich, V. (1985). Stimulated Raman scattering of dye laser radiation in hydrogen: Improvement of spectral purity. *Opt. Commun.* **56**, 283–287.

Bloembergen, N. (1965). *Nonlinear Optics.* Benjamin, New York.

Bloembergen, N. (1967). The stimulated Raman effect. *Am. J. Phys.* **35**, 989–1023.

Born, M., and Wolf, E. (1999). *Principles of Optics*, 7th edn. Cambridge University Press, Cambridge.

Boyd, R. W. (1992). *Nonlinear Optics.* Academic Press, New York.

Brink, D. J., and Proch, D. (1982). Efficient tunable ultraviolet source based on stimulated Raman scattering of an excimer-pumped dye laser. *Opt. Lett.* **7**, 494–496.

Brosnan, S. J., and Byer, R. L. (1979). Optical parametric oscillators and linewidth studies. *IEEE J. Quant. Electron.* **QE-15**, 415–431.

Byer, R. L., Oshman, M. K., Young, J. F., and Harris, S. E. (1968). Visible CW parametric oscillators. *Appl. Phys. Lett.* **12**, 109–111.

Diddams, S. A. (2010). The evolving optical frequency comb. *J. Opt. Soc. Am. B* **27**, B51–B62.

Diddams, S. A., Jones, D. J., Ye, J., Cundiff, S. T., Hall, J. L., Ranka, J. K., Windeler, R. S., Holzwarth, R., Udem, T., and Hänsch, T. W. (2000). Direct link between microwave and optical frequencies with a 300 THz femtosecond laser comb. *Phys. Rev. Lett.* **84**, 5102–5105.

Diels, J.-C. (1990). Femtosecond dye lasers. In *Dye Laser Principles* (Duarte, F. J. and Hillman, L. W., eds.). Academic Press, New York, Chapter 3.

Diels, J.-C., and Rudolph, W. (2006). *Ultrafast Laser Pulse Phenomena.* Academic Press, New York.

Duarte, F. J. (2003). *Tunable Laser Optics*. Elsevier Academic Press, New York.

Eckstein, J. N., Ferguson, A. I., and Hänsch, T. W. (1978). High-resolution two-photon spectroscopy with picosecond light pulses. *Phys. Rev. Lett.* **40**, 847–850.

Fouche, D. G., and Chang, R. K. (1972). Observation of resonance Raman scattering below the dissociation limit in I_2 vapor. *Phys. Rev. Lett.* **29**, 536–539.

Giordmaine, J. A., and Miller, R. C. (1965). Tunable coherent parametric oscillation in $LiNbO_3$ at optical frequencies. *Phys. Rev. Lett.* **14**, 973–976.

Hanna, D. C., Pacheco, M. M. T., and Wong, K. H. (1985). High efficiency and high brightness Raman conversion of dye laser radiation. *Opt. Commun.* **55**, 188–192.

Harris, S. E. (1969). Tunable optical parametric oscillators. *Proc. IEEE* **57**, 2096–2113.

Hartig, W., and Schmidt, W. (1979). A broadly tunable IR waveguide Raman laser pumped by a dye laser. *Appl. Phys.* **18**, 235–241.

Holzwarth, R., Zimmermann, M., Udem, T., and Hänsch, T. W. (2001). Optical clockworks and the measurement of laser frequencies with a mode-locked frequency comb. *IEEE J. Quant. Electron.* **37**, 1493–1501.

Kumar, P., Shapiro, J. H., and Bondurant, R. S. (1984). Fluctuations in the phase-conjugate signal generated via degenerate four-wave mixing. *Opt. Commun.* **50**, 183–188.

Ludewigt, K., Birkmann, K., and Wellegehausen, B. (1984). Anti-Stokes Raman laser investigations on atomic TI and Sn. *Appl. Phys. B* **33**, 133–139.

Manners, J. (1983). XeCl laser generated infra-red SRS in barium vapour. *Opt. Commun.* **44**, 366–370.

Marshall, L. R., and Piper, J. A. (1990). Transient stimulated Raman scattering in lead vapor. *IEEE J. Quant. Electron.* **26**, 1098–1104.

Mills, D. L. (1991). *Nonlinear Optics*. Springer-Verlag, Berlin, Germany.

Milton, T. K., Reid, S. A., Kim, H. L., and McDonald, J. D. (1989). A scanning, single mode, LiNbO3, optical parametric oscillator. *Opt. Commun.* **69**, 289–293.

Orr, B. J., He, Y., and White, R. T. (2009). Spectroscopic applications of pulsed tunable optical parametric oscillators. In *Tunable Laser Applications*, 2nd edn. (Duarte, F. J., ed.). CRC Press, New York, Chapter 2.

Orr, B. J., Johnson, M. J., and Haub, J. G. (1995). Spectroscopic applications of pulsed tunable optical parametric oscillators. In *Tunable Laser Applications* (Duarte, F. J., ed.). Marcel Dekker, New York, Chapter 2.

Schmidt, W., and Appt, W. (1972). Tunable stimulated Raman emission generated by a dye laser. *Z. Naturforsch.* **28a**, 792–793.

Schomburg, H., Döbele, H. F., and Rückle, B. (1983). Generation of tunable narrow-bandwidth VUV radiation by anti-Stokes SRS in H_2. *Appl. Phys. B* **30**, 131–134.

Shen, Y. R. (1984). *The Principles of Nonlinear Optics*. Wiley, New York.

Toulouse, J. (2005). Optical nonlinearities in fibers: Review, recent examples, and systems applications. *J. Lightw. Technol.* **23**, 3625–3641.

Trutna, W. R., and Byer, R. L. (1980). Multiple-pass Raman gain cell. *Appl. Opt.* **19**, 301–312.

Trutna, W. R., Yong, J. R., Park, K., and Byer, R. L. (1979). The dependence of Raman gain on pump laser linewidth. *IEEE J. Quant. Electron.* **QE-15**, 648–655.

White, J. C., and Henderson, D. (1983). Tuning and saturation behavior of the anti-Stokes Raman laser. *Opt. Lett.* **8**, 15–17.

Wilke, V., and Schmidt, W. (1978). Tunable UV-radiation by stimulated Raman scattering. *Appl. Phys.* **16**, 151–154.

Wyatt, R., and Cotter, D. (1980). Tunable infrared generation using $6s$–$6d$ Raman transition in caesium vapour. *Appl. Phys.* **21**, 199–204.

Yariv, A. (1975). *Quantum Electronics*, 2nd edn. Wiley, New York.

Yariv, A. (1977). Compensation for atmospheric degradation of optical transmission by nonlinear optical mixing. *Opt. Commun.* **21**, 49–50.

Yariv, A. (1985). *Optical Electronics*, 3rd edn. HRW, New York.

CHAPTER 9

Allaria, E., Appio, R., Badano, L., Barletta, W. A., Bassanese, S., Biedron, S. G., Borga, A., et al. (2012). Highly coherent and stable pulses from the FERMI seeded free-electron laser in the extreme ultraviolet. *Nat. Photon.* **6**, 699–704.

Anliker, P., Luthi, H. R., Seelig, W., Steinger, J., Weber, H. P., Leutwyler, S., Schumacher, E., and Woste, L. (1977). 33-W CW dye laser. *IEEE J. Quant. Electron.* **QE-13**, 547–548.

Auerbach, M., Adel, P., Wandt, D., Fallnich, C., Unger, S., Jetschke, S., and Müller, H.-R. (2002). 10 W widely tunable narrow linewidth double-clad fiber ring laser. *Opt. Express* **10**, 139–144.

Baltakov, F. N., Barikhin, B. A., and Sukhanov, L. V. (1974). 400-J pulsed laser using a solution of rhodamine-6G in ethanol. *JETP Lett.* **19**, 174–175.

Barnes, N. P. (1995a). Transition metal solid-state lasers. In *Tunable Lasers Handbook* (Duarte, F. J., ed.). Academic Press, New York, Chapter 6.

Barnes, N. P. (1995b). Optical parametric oscillators. In *Tunable Lasers Handbook* (Duarte, F. J., ed.). Academic Press, New York, Chapter 7.

Barnes, N. P., and Williams-Byrd, J. A. (1995). Average power effects in parametric oscillators and amplifiers. *J. Opt. Soc. Am. B* **12**, 124–131.

Bass, I. L., Bonanno, R. E., Hackel, R. H., and Hammond, P. R. (1992). High-average power dye laser at Lawrence Livermore National Laboratory. *Appl. Opt.* **31**, 6993–7006.

Baving, H. J., Muuss, H., and Skolaut, W. (1982). CW dye laser operation at 200W pump power. *Appl. Phys. B.* **29**, 19–21.

Beck, R., Englisch, W., and Gürs, K. (1976). *Table of Laser Lines in Gases and Vapors.* Springer-Verlag, Berlin, Germany.

Benson, S. V. (1995). Tunable free-electron lasers. In *Tunable Lasers Handbook* (Duarte, F. J., ed.). Academic Press, New York, Chapter 9.

Berger, J. D., and Anthon, D. (2003). Tunable MEMS devices for optical networks. *Opt. Photon. News* **14** (3), 43–49.

Berger, J. D., Zhang, Y., Grade, J. D., Howard, L., Hrynia, S., Jerman, H., Fennema, A., Tselikov, A., and Anthon, D. (2001). External cavity diode lasers tuned with silicon MEMS. *IEEE LEOS Newslett.* **15** (5), 9–10.

Bernhardt, A. F., and Rasmussen, P. (1981). Design criteria and operating characteristics of a single-mode pulsed dye laser. *Appl. Phys. B* **26**, 141–146.

Bobrovskii, A. N., Branitskii, A. V., Zurin, M. V., Koshevnikov, A. V., Mishchenko, V. A., and Myl'nikov, G. D. (1987). Continuously tunable TEA CO_2 laser. *Sov. J. Quant. Electron.* **17**, 1157–1159.

Bos, F. (1981). Versatile high-power single-longitudinal-mode pulsed dye laser. *Appl. Opt.* **20**, 1886–1890.

Bradley, C. C., McClelland, J. J., Anderson, W. R., and Celotta, R. J. (2000). Magneto-optical trapping of chromium atoms. *Phys. Rev.* **61**, 053407.

Brandt, M., and Piper, J. A. (1981). Operating characteristics of TE copper bromide lasers. *IEEE J. Quant. Electron.* **QE-17**, 1107–1115.

Brau, C. A. (1990). *Free-Electron Lasers.* Academic Press, New York.

Braun, I., Ihlein, G., Laeri, F., Nöckel, J. U., Schulz-Ekloff, G., Schüth, F., Vietze, U., Weiß, Ö., and Wöhrle, D. (2000). Hexagonal microlasers based on organic dyes in nanoporous crystals. *Appl. Phys.* **70**, 335–343.

Broyer, M., Chevaleyre, J., Delacrétaz, G., and Wöste, L. (1984). CVL-pumped dye laser spectroscopic application. *Appl. Phys. B.* **35**, 31–36.

Buffa, R., Burlamacchi, P., Salimbeni, R., and Matera, M. (1983). Efficient spectral narrowing of a XeCl TEA laser. *J. Phys. D: Appl. Phys.* **16**, L125–L128.

Caro, R. G., Gower, M. C., and Webb, C. E. (1982). A simple tunable KrF laser system with narrow bandwidth and diffraction-limited divergence. *J. Phys. D: Appl. Phys.* **15**, 767–773.

Chen, H., Babin, F., Leblanc, M., and Schinn, G. W. (2003). Widely tunable single-frequency Erbium-doped fiber lasers. *IEEE Photon. Technol. Lett.* **15**, 185–187.

Chesler, R. B., and Geusic, J. E. (1972). Solid-state ionic lasers. In *Laser Handbook*, Vol. 1 (Arecchi, F. T. and Schulz-Dubois, E. O., eds.). North Holland, Amsterdam, The Netherlands, pp. 325–368.

Corzine, S. W., Bowers, J. E., Przybylek, G., Koren, U., Miller, B. I., and Soccolich, C. E. (1988). Actively mode-locked GaInAsP laser with subpicosecond output. *Appl. Phys. Lett.* **52**, 348–350.

Costela, A., Cerdan, L., and Garcia-Moreno, I. (2013). Solid state dye lasers with scattering feedback. *Prog. Quant. Electron.* **37**, 348–382.

Costela, A., Garcia-Moreno, I., and Sastre, R. (2009). Solid state dye lasers. In *Tunable Laser Applications*, 2nd edn. (Duarte, F. J., ed.). CRC Press, New York, Chapter 3.

Coutts, D. W., Wadsworth, W. J., and Webb, C. E. (1998). High average power blue generation from a copper vapour laser pumped titanium sapphire laser. *J. Mod. Opt.* **45**, 1185–1197.

Delfyett, P. J., Florez, L., Stoffel, N., Gmitter, T., Andreadakis, N., Alphonse, G., and Ceislik, W. (1992). 200 fs optical pulse generation and intracavity pulse evolution in a hybrid mode-locked semiconductor diode-laser/amplifier system. *Opt. Lett.* **17**, 670–672.

Demmler, S., Rothhardt, J., Heidt, A. M., Hartung, A., Rohwer, E. G., Bartelt, H., Limpert, J., and Tünnermann A. (2011). Generation of high quality, 1.3 cycle pulses by active phase control of an octave spanning supercontinuum. *Opt. Express* **19**, 20151–20158.

Diels, J.-C. (1990). Femtosecond dye lasers. In *Dye Laser Principles* (Duarte, F. J. and Hillman, L. W., eds.). Academic Press, New York, Chapter 3.

Diels, J.-C., and Rudolph, W. (2006). *Ultrashort Laser Pulse Phenomena,* 2nd edn. Academic Press, New York.

Dietel, W., Fontaine, J. J., and Diels, J.-C. (1983). Intracavity pulse compression with glass: A new method of generating pulses shorter than 60 fs. *Opt. Lett.* **8**, 4–6.

Dominic, V., MacCormack, S., Waarts, R., Sanders, S., Bicknese, S., Dohle, R., Wolak, E., Yeh, P. S., and Zucker, E. (1999). 110W fibre laser. *Electron. Lett.* **35**, 1158–1160.

Drever, R. W. P., Hall, J. L., Kowalski, F. V., Hough, J., Ford, G. M., Munley, A. J., and Ward, H. (1983). Laser phase and frequency stabilization using an optical resonator. *Appl. Phys. B.* **31**, 97–105.

Duarte, F. J. (1985a). Application of dye laser techniques to frequency selectivity in pulsed CO_2 lasers. In *Proceedings of the International Conference on Lasers '84* (Corcoran, K. M., Sullivan, M. D., and Stwalley, W. C., eds.). STS Press, McLean, VA, pp. 397–403.

Duarte, F. J. (1985b). Multiple-prism Littrow and grazing-incidence pulsed CO_2 lasers. *Appl. Opt.* **24**, 1244–1245.

Duarte, F. J. (1985c). Variable linewidth high-power TEA CO_2 laser. *Appl. Opt.* **24**, 34–37.

Duarte, F. J. (1991a). Dispersive excimer lasers. In *Proceedings of the International Conference on Lasers '90* (Harris, D. G. and Herbelin, J., eds.). STS Press, McLean, VA, pp. 277–279.

Duarte, F. J. (ed.). (1991b). Dispersive dye lasers. In *High Power Dye Lasers.* Springer-Verlag, Berlin, Germany, Chapter 2.

Duarte, F. J. (1993). Multiple-prism grating designs tune diode lasers. *Laser Focus World* **29** (2), 103–109.

Duarte, F. J. (ed.). (1995a). Introduction. In *Tunable Lasers Handbook.* Academic Press, New York, Chapter 1.

Duarte, F. J. (ed.). (1995b). Dye lasers. In *Tunable Lasers Handbook.* Academic Press, New York, Chapter 5.

Duarte, F. J. (ed.). (1995c). Dispersive external-cavity semiconductor lasers. In *Tunable Laser Applications* Marcel Dekker, New York, Chapter 3.

Duarte, F. J. (1999). Multiple-prism grating solid-state dye laser oscillator: Optimized architecture. *Appl. Opt.* **38**, 6347–6349.

Duarte, F. J. (2003). *Tunable Laser Optics*. Elsevier Academic Press, New York.

Duarte, F. J. (ed.). (2009). Broadly tunable external-cavity semiconductor lasers. In *Tunable Laser Applications*, 2nd edn. CRC Press, New York, Chapter 5.

Duarte, F. J. (2012). Tunable organic dye lasers: Physics and technology of high performance liquid and solid-state narrow-linewidth oscillators. *Prog. Quant. Electron.* **36**, 29–50.

Duarte, F. J. (2014). *Quantum Optics for Engineers*. CRC Press, New York.

Duarte, F. J., and Conrad, R. W. (1987). Diffraction-limited single-longitudinal-mode multiple-prism flashlamp-pumped dye laser oscillator: Linewidth analysis and injection of amplifier system. *Appl. Opt.* **26**, 2567–2571.

Duarte, F. J., Davenport, W. E., Ehrlich, J. J., and Taylor, T. S. (1991). Ruggedized narrow-linewidth dispersive dye laser oscillator. *Opt. Commun.* **84**, 310–316.

Duarte, F. J., Ehrlich, J. J., Davenport, W. E., Taylor, T. S., and McDonald, J. C. (1993). A new tunable dye laser oscillator: Preliminary report. In *Proceedings of the International Conference on Lasers '92* (Wang, C. P., ed.). STS Press, McLean, VA, pp. 293–296.

Duarte, F. J., and Hillman, L. W. (1990). Introduction. In *Dye Laser Principles* (Duarte, F. J. and Hillman, L. W., eds.). Academic Press, New York, Chapter 1.

Duarte, F. J., and James, R. O. (2003). Tunable solid-state lasers incorporating dye-doped polymer-nanoparticle gain media, *Opt. Lett.* **28**, 2088–2090.

Duarte, F. J., Liao, L. S., Vaeth, K. M., and Miller, A. M. (2006). Widely tunable green laser emission using the coumarin 545 tetramethyl dye as the gain medium. *J. Opt. A: Pre Appl. Opt.* **8**, 172–174.

Duarte, F. J., and Piper, J. A. (1981). Prism preexpanded grazing-incidence grating cavity for pulsed dye lasers. *Appl. Opt.* **20**, 2113–2116.

Duarte, F. J., and Piper, J. A. (1982). Comparison of prism-expander and grazing-incidence grating cavities for copper laser pumped dye lasers. *Appl. Opt.* **21**, 2782–2786.

Duarte, F. J., and Piper, J. A. (1984). Narrow-linewidth, high prf copper laser-pumped dye laser oscillators. *Appl. Opt.* **23**, 1391–1394.

Duarte, F. J., Taylor, T. S., Costela, A., Garcia-Moreno, I., and Sastre, R. (1998). Long-pulse narrow-linewidth dispersive solid-state dye-laser oscillator. *Appl. Opt.* **37**, 3987–3989.

Dupre, P. (1987). Quasiunimodal tunable pulsed dye laser at 440 nm: Theoretical development for using quad prism beam expander and one or two gratings in a pulsed dye laser oscillator cavity. *Appl. Opt.* **26**, 860–871.

Ell, R., Morgner, U., Kärtner, F. X., Fugimoto, J. G., Ippen, E. P., Scheuer, V., Angelow, G., et al. (2001). Generation of 5-fs pulses from a Ti:sapphire laser. *Opt. Lett.* **26**, 373–375.

Erbert, G., Bass, I., Hackel, R., Jenkins, S., Kanz, K., and Paisner, J. (1991). 43-W CW Ti:sapphire laser. In *Technical Digest, Conference on Lasers and Electro-Optics*. Paper CThH4. Optical Society of America, Washington, DC.

Everett, P. N. (1991). Flashlamp-excited dye lasers. In *High Power Dye Lasers* (Duarte, F. J., ed.). Springer-Verlag, Berlin, Germany, Chapter 6.

Fan, Y. X., Eckardt, R. C., Byer, R. L., Route, R. K., and Feigelson, R. S. (1984). AgGaS$_2$ infrared parametric oscillator. *Appl. Phys. Lett.* **45**, 313–315.

Farhoomand, J., and Pickett, H. M. (1988). A stable high power optically pumped far infrared laser system. In *Proceedings of the International Conference on Lasers '87* (Duarte, F. J., ed.). STS Press, McLean, VA, pp. 539–543.

Fedorova, K. A., Cataluna, M. A., Krestnikov, I., Livshits, D., and Rafailov, E. U. (2010). Broadly tunable high-power InAs/GaAs quantum dot external cavity diode lasers. *Opt. Express* **18**, 19438–19443.

Flamant, P. H., and Maillard, D. J. M. (1984). Transient injection frequency-locking of a microsecond-pulsed dye laser for atmospheric measurements. *Opt. Quant. Electron.* **16**, 179–182.

Fleming, M. W., and Mooradian, A. (1981). Spectral characteristics of external-cavity controlled semiconductor lasers. *IEEE J. Quant. Electron.* **QE-17**, 44–59.

Fork, R. L., Brito Cruz, C. H., Becker, P. C., and Shank, C. V. (1987). Compression of optical pulses to six femtoseconds by using cubic phase compression. *Opt. Lett.* **12**, 483–485.

Fork, R. L., Greene, B. I., and Shank, C. V. (1981). Generation of optical pulses shorter than 0.1 psec by colliding pulse mode locking. *Appl. Phys. Lett.* **38**, 671–672.

Fork, R. L., Martinez, O. E., and Gordon, J. P. (1984). Negative dispersion using pairs of prisms. *Opt. Lett.* **9**, 150–152.

Fort, J., and Moulin, C. (1987). High-power high-energy linear flashlamp-pumped dye Laser. *Appl. Opt.* **26**, 1246–1249.

Fox, R. W., Hollberg, L., and Zibrov, A. S. (1997). Semiconductor diode lasers. In *Atomic, Molecular, and Optical Physics.* Academic Press, New York, Chapter 4.

Freed, C. (1995). CO_2 isotope lasers and their applications in tunable laser spectroscopy. In *Tunabe Lasers Handbook* (Duarte, F. J., ed.). Academic Press, New York, Chapter 4.

German, K. R. (1981). Grazing angle tuner for CW lasers. *Appl. Opt.* **20**, 3168–3171.

Harstad, K. (1983). Interpulse kinetics in copper and copper halide lasers. *IEEE J. Quant. Electron.* **QE-19**, 88–91.

He, Y., and Orr, B. J. (2001). Tunable single-mode operation of a pulsed optical parametric oscillator pumped by a multimode laser. *Appl. Opt.* **40**, 4836–4848.

Hollberg, L. (1990). CW dye lasers. In *Dye Laser Principles* (Duarte, F. J. and Hillman, L. W., eds.). Academic Press, New York, Chapter 5.

Holzer, W., Gratz, H., Schmitt, T., Penzkofer, A., Costela, A., Garcia-Moreno, I., Sastre, R., and Duarte, F. J. (2000). Photo-physical characterization of rhodamine 6G in a 2-hydroxyethyl-methacrylate methyl-methacrylate copolymer. *Chem. Phys.* **256**, 125–136.

Hugi, A., Terazzi, R., Bonetti, Y., Wittmann, A., Fischer, M., Beck, M., Faist, J., and Gini, E. (2009). External cavity quantum cascade laser tunable from 7.6 to 11.4 μm. *Appl. Phys. Letts.* **95**, 061103.

James, B. W., Falconer, I. S., Bowden, M. D., Krug, P. A., Whitbourn, L. B., Stimson, P. A., and Macfarlane, J. C. (1988). Optically pumped submillimeter lasers and their applications to plasma diagnostics. In *Proceedings of the International Conference on Lasers '87* (Duarte, F. J., ed.). STS Press, McLean, VA, pp. 550–554.

Jensen, O. B., Skettrup, T., Petersen, O. B., and Larsen, M. B. (2002). Diode-pumped intracavity optical parametric oscillator in pulsed and continuous-wave operation. *J. Opt. A: Pure Appl. Opt.* **4**, 190–193.

Jeong, Y., Sahu, J. K., Payne, D. N., and Nilsson, J. (2004). Ytterbium-doped large core fiber laser with 1.36 kW continuous-wave output power. *Opt. Express* **12**, 6086–6092.

Johnston, T. F., and Duarte, F. J. (2002). Lasers, dye. In *Encyclopedia of Physical Science and Technology*, 3rd edn., Vol. 8 (Meyers, R. A., ed.). Academic Press, New York, pp. 315–359.

Kafka, J. D., and Baer, T. (1987). Prism-pair delay lines in optical pulse compression. *Opt. Lett.* **12**, 401–403.

Karnutsch, C. (2007). *Low Threshold Organic Thin Film Laser Devices.* Cuvillier, Göttingen, Germany.

Klimek, D. E., Aldag, H. R., and Russell, J. (1992). In *Conference on Lasers and Electro-Optics.* Optical Society of America, Washington, DC, p. 332.

Kner, P., Sun, D., Boucart, J., Floyd, P., Nabiev, R., Davis, D., Yuen, W., Jansen, M., and Chang-Hasnain, C. J. (2002). VCSELS. *Opt. Photon. News* **13** (3), 44–47.

Kubota, H., Kurokawa, K., and Nakazawa, M. (1988). 29-fs pulse generation from a linear-cavity synchronously pumped dye laser. *Opt. Lett.* **13**, 749–751.

Larrue, D., Zarzycki, J., Canva, M., Georges, P., Bentivegna, F., and Brun, A. (1994). Impregnated ORMOSIL matrices for efficient solid state optical gain media. *Opt. Commun.* **110**, 125–130.

Loree, T. R., Butterfield, K. B., and Barker, D. L. (1978). Spectral tuning of ArF and KrF discharge lasers. *Appl. Phys. Lett.* **32**, 171–173.

Lu, Q. Y., Bandyopadhyay, N., Slivken, S., Bai, Y., and Razeghi, M. (2013). High performance terahertz quantum cascade laser sources based on intracavity difference frequency generation. *Opt. Express* **21**, 968–973.

Ludewigt, K., Pfingsten, W., Mhlmann, C., and Wellegehausen, B. (1987). High-power vacuum-ultraviolet anti-Stokes Raman laser with atomic selenium, *Opt. Lett.* **12**, 39–41.

Maiman, T. H. (1960). Stimulated optical radiation in ruby. *Nature* **187**, 493–494.

Maslyukov, A., Sokolov, S., Kaivola, M., Nyholm, K., and Popov, S. (1995). Solid-state dye laser with modified poly(methyl methacrylate)-doped active elements. *Appl. Opt.* **34**, 1516–1518.

Maulini, R., Mohan, A., Giovannini, M., Faist, J., and Gini, E. (2006). External cavity quantum-cascade laser tunable from 8.2 to 10.4 μm using a gain element with a hetero-geneous cascade. *Appl. Phys. Lett.* **88**, 201113.

McAleavey, F. J., O'Gorman, J., Donegan, J. F., MacCraith, B. D., Hegarty, J., and Maze, G. (1997). Narrow linewidth, tunable Tm^{3+}-doped fluoride fiber laser for optical based hydrocarbon gas sensing. *IEEE J. Sel. Top. Quant. Electron.* **3**, 1103–1111.

McKee, T. J. (1985). Spectral-narrowing techniques for excimer lasers oscillators. *Can. J. Phys.* **63**, 214–219.

Miller, J. (1988). High power hydrogen Fluoride chemical lasers: Power scaling and beam quality. In *Proceedings of the International Conference on Lasers '87* (Duarte, F. J., ed.). STS Press, McLean, VA, pp. 190–217.

Mollenauer, L. F. (1985). Color center lasers. In *Laser Handbook*, Vol. 4 (Stitch, M. L. and Bass, M., eds.). North Holland, Amsterdam, The Netherlands, Chapter 2.

Mollenauer, L. F., and Bloom, D. M. (1979). Color-center laser generates picoseconds pulses and several watts CW over the 1.24–1.45-μm range. *Opt. Lett.* **4**, 247–249.

Moulton, P. F. (1986). Spectroscopic and laser characteristics of $Ti:Al_2O_3$. *J. Opt. Soc. Am. B* **3**, 125–132.

Nevsky, A. Yu., Bressel, U., Ernsting, I., Eisele, Ch., Okhapkin, M., Schiller, S., Gubenko, A., et al. (2008). A narrow-linewidth external cavity quantum dot laser for high-resolution spectroscopy in the near infrared and yellow spectral ranges. *Appl. Phys. B* **92**, 501–507.

Olivares, I. E., Duarte, A. E., Saravia, E. A., and Duarte, F. J. (2002). Lithium isotope separa-tion with tunable diode lasers. *Appl. Opt.* **41**, 2973–2977.

Orr, B. J., He, Y., and White, R. T. (2009). Spectroscopic applications of pulsed tunable optical parametric oscillators. In *Tunable Laser Applications*, 2nd edn. (Duarte, F. J., ed.). CRC Press, New York, Chapter 2.

Orr, B. J., Johnson, M. J., and Haub, J. G. (1995). Spectroscopic applications of pulsed tunable optical parametric oscillators. In *Tunable Laser Applications* (Duarte, F. J., ed.). Marcel Dekker, New York, Chapter 2.

Osvay, K., Kovács, A. P., Kurdi, G., Heiner, Z., Divall, M., Klebniczki, J., and Ferincz, I. E. (2005). Measurements of non-compensated angular dispersion and the subsequent tem-poral lengthening of femtosecond pulses in a CPA laser. *Opt. Commun.* **248**, 201–209.

Pacala, T. J., McDermid, I. S., and Laudenslager, J. B. (1984). Single-longitudinal-mode oper-ation of an XeCl laser. *Appl. Phys. Lett.* **45**, 507–509.

Pacheco, D. P., Aldag, H. R., Itzkan, I., and Rostler, P. S. (1988). A solid-state flashlamp-pumped dye laser employing polymer hosts. In *Proceedings of the International Conference on Lasers '87* (Duarte, F. J., ed.). STS Press, McLean, VA, pp. 330–337.

Pang, L. Y., Fujimoto, J. G., and Kintzer, E. S. (1992). Ultrashort-pulse generation from high-power diode arrays by using intracavity optical nonlinearities. *Opt. Lett.* **17**, 1599–1601.

Piper, J. A. (1974). Increased efficiency and new CW transitions in the helium-iodine laser system. *J. Phys. D: Appl. Phys.* **7**, 323–328.

Piper, J. A. (1976). Simultaneous CW laser oscillation on transitions of Cd^+ and I^+ in a hollow-cathode $He-CdI_2$ discharge. *Opt. Commun.* **19**, 189–192.

Piper, J. A. (1978). A transversely excited copper halide laser with large active volume. *IEEE J. Quant. Electron.* **QE-14**, 405–407.

Piper, J. A., and Gill, P. (1975). Output characteristics of the He-Zn laser. *J. Phys. D: Appl. Phys.* **8**, 127–134.

Powers, P. E., Ellington, R. J., Pelouch, W. S., and Tang, C. L. (1993). Recent advences in the Ti:sapphire-pumped high-repetition-rate femtosecond optical parametric oscillator. *J. Opt. Soc. Am. B* **10**, 2163–2167.

Rhodes, C. K. (ed.). (1979). *Excimer Lasers.* Springer-Verlag, Berlin, Germany.

Rifani, M., Yin, Y.-Y., Elliott, D. S., Jay, M. J., Jang, S.-H., Kelley, M. P., Bastin, L., and Kahr, B. (1995). Solid-state dye laser from stereoscopic host-guest interactions. *J. Am. Chem. Soc.* **117**, 7572–7573.

Ruddock, I. S., and Bradley, D. J. (1976). Bandwidth-limited subpicosecond generation in mode-locked CW dye laser. *Appl. Phys. Lett.* **29**, 296–297.

Salvatore, R. A., Schrans, T., and Yariv, A. (1993). Wavelength tunable source of subpicosecond pulses from CW passively mode-locked two-section multiple-quantum-well laser. *IEEE Photon. Technol. Lett.* **5**, 756–758.

Samuel, I. D. W., and Turnbull, G. A. (2007). Organic semiconductor lasers. *Chem. Rev.* **107**, 1272–1295.

Schäfer, F. P. (ed.). (1990). *Dye Lasers.* Springer-Verlag, Berlin, Germany.

Scheps, R. (1993). Low-threshold dye laser pumped by visible laser diodes. *IEEE Photon. Lett.* **5**, 1156–1158.

Schneider, T. R., and Cox, J. D. (1988). The nuclear pumping of lasers—Revisited. In *Proceedings of the International Conference on Lasers '87* (Duarte, F. J., ed.). STS Press, McLean, VA, pp. 234–240.

Schröder, T., Boller, K.-J., Fix, A., and Wallenstein R. (1994). Spectral properties and numerical modelling of a critically phase-matched nanosecond LiB_3O_5 optical parametric oscillator. *Appl. Phys. B.* **58**, 425–438.

Shan, X., Siddiqui, A. S., Simeonidou, D., and Ferreira, M. (1991). Rebroadening of spectral linewidth with shorter wavelength detuning away from the gain curve peak in external cavity semiconductor lasers sources. In *Conference on Lasers and Electro-Optics*, Optical Society of America, Washington, DC, pp. 258–259.

Shand, M. L., and Walling, J. C. (1982). A tunable emerald laser. *IEEE J. Quant. Electron.* **QE-18**, 1829–1830.

Shay, T. M., and Duarte, F. J. (2009). Tunable fiber lasers. In *Tunable Laser Applications*, 2nd edn. (Duarte, F. J., ed.). CRC Press, New York, Chapter 6.

Shay, T., Hanson, F., Gookin, D., and Schimitschek, E. J. (1981). Line narrowing and enhanced efficiency of an HgBr laser by injection locking. *Appl. Phys. Lett.* **39**, 783–785.

Singh, S., Dasgupta, K., Sasi, K., Manohar, K. G., Nair, L. G., and Chatterjee, U. K. (1994). High-power high-repetition-rate copper-vapor-pumped dye laser. *Opt. Eng.* **33**, 1894–1904.

Smith, R. S., and DiMauro, L. F. (1987). Efficiency and linewidth improvements in a grazing-incidence dye laser using an intracavity lens and spherical end mirror. *Appl. Opt.* **26**, 855–859.

Smith, R. G., Geusic, J. E., Levinstein, H. J., Rubin, J. J., Singh, S., and Van Uitert, L. G. (1968). Continuous optical parametric oscillation in $Ba_2NaNb_5O_{15}$. *Appl. Phys. Lett.* **12**, 308–309.

Srinivasan, B., Tafoya, J., and Jain, R. K. (1999). High power watt-level CW operation of diode-pumped 2.7 mm fiber lasers using efficient cross-relaxation and energy transfer mechanisms. *Opt. Express* **4**, 490–495.

Sugii, M., Ando, M., and Sasaki, K. (1987). Simple long-pulse XeCl laser with narrow-line output. *IEEE J. Quant. Electron.* **QE-23**, 1458–1460.

Sugiyama, A., Nakayama, T., Kato, M., and Maruyama, Y. (1996). Characteristics of a dye laser amplifier transversely pumped by copper vapor lasers with a two-dimmensional calculation model. *Appl. Opt.* **36**, 5849–5854.

Sze, R. C., and Harris, D. G. (1995). Tunable excimer lasers. In *Tunable Lasers Handbook* (Duarte, F. J., ed.). Academic Press, New York, Chapter 3.

Sze, R. C., Kurnit, N. A., Watkins, D. E., and Bigio, I. J. (1986). Narrow band tuning with small long-pulse excimer lasers. In *Proceedings of the International Conference on Lasers '85* (Wang, C. P., ed.). STS Press, McLean, VA, pp. 133–144.

Tallman, C., and Tennant, R. (1991). Large-scale excimer-laser-pumped dye lasers. In *High Power Dye Lasers* (Duarte, F. J., ed.). Springer-Verlag, Berlin, Germany, Chapter 4.

Tang, K. Y., O'Keefe, T., Treacy, B., Rottler, L., and White, C. (1987). Kilojoule output XeCl dye laser: Optimization and analysis. In *Proceedings: Dye Laser/Laser Dye Technical Exchange Meeting, 1987* (Bentley, J. H., ed.). U. S. Army Missile Command, Redstone Arsenal, Al, pp. 490–502.

Tavella, F., Willner, A., Rothhardt, J., Hädrich, S., Seise, E., Düsterer, S., Tschentscher, T., et al. (2010). Fiber-amplifier pumped high average power few-cycle pulse non-collinear OPCPA. *Opt. Express* **18**, 4689–4694.

Tenenbaum, J., Smilanski, I., Gabay, S., Levin, L. A., Erez, G., and Lavi, S. (1980). Structure of 510.6 and 578.2 nm copper laser lines. *Opt. Commun.* **32**, 473–477.

Tittel, F. K., Marowski, G., Nighan, W. L., Zhu, Y., Sauerbrey, R. A., and Wilson, W. L. (1986). Injection-controlled tuning of an electron-beam excited $XeF(C \rightarrow A)$ laser. *IEEE J. Quant. Electron.* **QE-22**, 2168–2173.

Tratt, D. M., Kar, A. K., and Harrison, R. G. (1985). Spectral control of gain-switched lasers by injection-seeding: Application to TEA CO_2 lasers. *Prog. Quant. Opt.* **10**, 229–266.

Varangis, P. M., Li, H., Liu, G. T., Newell, T. C., Stintz, A., Fuchs, B., Malloy, K. J., and Lester, L. F. (2000). Low-threshold quantum dot lasers with 201 nm tuning range. *Electron. Lett.* **36**, 1544–1545.

Walling, J. C., and Peterson, O. G. (1980). High gain laser performance in alexandrite. *IEEE J. Quant. Electron.* **QE-16**, 119–120.

Walling, J. C., Peterson, O. G., Jensen, H. P., Morris, R. C., and O'Dell, E. W. (1980a). Tunable alexandrite lasers. *IEEE J. Quant. Electron.* **QE-16**, 1302–1315.

Walling, J. C., Peterson, O. G., and Morris, R. C. (1980b). Tunable CW alexandrite laser. *IEEE J. Quant. Electron.* **QE-16**, 120–121.

Wang, Y., Larotonda, M. A., Luther, B. M., Alessi, D., Berrill, M., Shlyatpsev, V. N., and Rocca, J. J. (2005). Demonstration of high-repetition-rate tabletop soft-X-ray lasers with saturated output at wavelengths down to 13.9 nm and gain down to 10.9 nm. *Phys. Rev. A.* **72**, 053807.

Webb, C. E. (1991). High-power dye lasers pumped by copper vapor lasers. In *High Power Dye Lasers* (Duarte, F. J., ed.). Springer-Verlag, Berlin, Germany, Chapter 5.

Willett, C. S. (1974). *An Introduction to Gas Lasers: Population Inversion Mechanisms.* Pergamon Press, New York.

Woodward, B. W., Ehlers, V. J., and Lineberger, W. C. (1973). A reliable, repetitively pulsed, high-power nitrogen laser. *Rev. Sci. Instrum.* **44**, 882–887.

Yang, T. T., Burde, D. H., Merry, G. A., Harris, D. G., Pugh, L. A. Tillotson, J. H., Turner, C. E., and Copeland, D. A. (1988). Spectra of electron-beam pumped XeF lasers. *Appl. Opt.* **27**, 49–57.

Yodh, A. G., Bai, Y., Golub, J. E., and Mossberg, T. W. (1984). Grazing-incidence dye lasers with and without intracavity lenses: A comparative study. *App. Opt.* **23**, 2040–2042.

Yoshida, S., Fujii, H., Amano, S., Endho, M., Sawano, T., and Fujioka, T. (1988). Highly efficient chemically pumped oxygen iodine laser. In *Proceedings of the International Conference on Lasers '87* (Duarte, F. J., ed.). STS Press, McLean, VA, pp. 223–229.

Zhang, D., Zhao, J., Yang, O., Liu, W., Fu, Y., Li, C., Luo, M., Hu, S. Q., and Wang L. (2012). Compact MEMS external cavity tunable laser with ultra-narrow linewidth for coherent detection. *Opt. Express* **20**, 19670–19682.

Zorabedian, P. (1992). Characteristics of a grating-external-cavity semiconductor laser containing intracavity prism beam expanders. *J. Lightw. Technol.* **10**, 330–335.

Zorabedian, P. (1995). Tunable external-cavity semiconductor lasers. In *Tunable Lasers Handbook* (Duarte, F. J., ed.). Academic Press, New York, Chapter 8.

CHAPTER 10

Altman, J. H. (1977). Sensitometry of black- and white- materials. In *The Theory of the Photographic Process* (James, T. H., ed.). Eastman Kodak Company, Rochester, NY, pp. 481–516.

Bennett, C. H., and Brassard, G. (1984). Quantum cryptography: Public key distribution and coin tossing. In *Proceedings of the IEEE International Conference on Computers Systems and Signal Processing*. Bangalore, India.

Boffi, P., Piccinin, D., Mottarella, D., and Martinelli, M. (2000). All optical free-space processing for optical communications signals. *Opt. Commun.* **181**, 79–88.

Dainty, J. C., and Shaw, R. (1974). *Image Science*. Academic Press, New York.

Deutsch, D. (1992). Quantum computation. *Phys. World* **5** (6), 57–61.

Dirac, P. A. M. (1978). *The Principles of Quantum Mechanics*, 4th edn. Oxford University Press, London.

Duarte, F. J. (1985). Note on achromatic multiple-prism beam expanders. *Opt. Commun.* **53**, 259–262.

Duarte, F. J. (1987). Beam shaping with telescopes and multiple-prism beam expanders. *J. Opt. Soc. Am. A* **4**, 30.

Duarte, F. J. (ed.). (1991). Dispersive dye lasers. In *High Power Dye Lasers*. Springer-Verlag, Berlin, Germany, pp. 7–43.

Duarte, F. J. (1993a). Electro-optical interferometric microdensitometer system. US Patent 5, 255, 069.

Duarte, F. J. (1993b). On a generalized interference equation and interferometric measurements. *Opt. Commun.* **103**, 8–14.

Duarte, F. J. (ed.). (1995). Interferometric imaging. In *Tunable Laser Applications*. Marcel Dekker, New York, pp. 153–178.

Duarte, F. J. (1996). Generalized interference equation and optical processing. In *Proceedings of the International Conference on Lasers '95* (Corcoran, V. J. and Goldman, T. A., eds.). STS Press, McLean, VA, pp. 615–617.

Duarte, F. J. (2001). Laser sensitometer using multiple-prism beam expansion and a polarizer. US Patent 6, 236, 461.

Duarte, F. J. (2002). Secure interferometric communications in free space. *Opt. Commun.* **205**, 313–319.

Duarte, F. J. (2005). Secure interferometric communications in free space: Enhanced sensitivity for propagation in the metre range. *J. Opt. A: Pure Appl. Opt.* **7**, 73–75.

Duarte, F. J. (2013). The probability amplitude for entangled polarizations: An interferometric approach. *J. Mod. Opt.* **60**, 1585–1587.

Duarte, F. J. (2014). *Quantum Optics for Engineers*. CRC Press, New York.

Duarte, F. J., and Paine, D. J. (1989). Quantum mechanical description of N-slit interference phenomena. In *Proceedings of the International Conference on Lasers '88* (Sze, R. C. and Duarte, F. J., eds.). STS Press, McLean, VA, pp. 42–27.

Duarte, F. J., and Piper, J. A. (1982). Dispersion theory of multiple-prism beam expanders for pulsed dye lasers. *Opt. Commun.* **43**, 303–307.

Duarte, F. J., Reed, B. A., and Burak, C. J. (2005). Laser sensitometer. US Patent 6, 903, 824 B2.

Duarte, F. J., Taylor, T. S., Black, A. M., Davenport, W. E., and Varmette, P. G. (2011). N-slit interferometer for secure free-space optical communications: 527 m intra interferometric path length. *J. Opt.* **13**, 035710.

Duarte, F. J., Taylor, T. S., Black, A. M., and Olivares, I. E. (2013). Diffractive patterns super-imposed over propagating *N*-slit interferograms. *J. Mod. Opt.* **60**, 136–140.

Duarte, F. J., Taylor, T. S., Clark, A. B., and Davenport, W. E. (2010). The *N*-slit interferometer: An extended configuration. *J. Opt.* **12**, 015705.

Olivares, I. E., Cuadra, J. A., Aguilar, F. A., Aguirre-Gomez, J. G., and Duarte, F. J. (2009). *J. Mod. Opt.* **56**, 1780–1784.

Pryce, M. H. L., and Ward, J. C. (1947). Angular correlation effects with annihilation radiation. *Nature* **160**, 435.

Turunen, J. (1986). Astigmatism in laser beam optical systems. *Appl. Opt.* **25**, 2908–2911.

Ward, J. C. (1949). *Some Properties of the Elementary Particles*. DPhil thesis, Oxford University Press, Oxford.

Willebrand, H. A., and Ghuman, B. S. (2001). Fiber optics without fiber. *IEEE Spectr.* **38** (8), 40–45.

Yu, S. T. S., and Gregory, D. A. (1996). Optical pattern recognition: Architectures and techniques. *Proc. IEEE* **84**, 733–775.

CHAPTER 11

Born, M., and Wolf, E. (1999). *Principles of Optics*, 7th edn. Cambridge University Press, Cambridge.

Byer, R. L., Paul, J., and Duncan, M. D. (1977). In *Laser Spectroscopy III* (Hall, J. L. and Carlsten, J. L., eds.). Springer-Verlag, Berlin, Germany, pp. 414–416.

Demtröder, W. (2008). *Laser Spectroscopy*, 4th edn., Vol. 1. Springer-Verlag, Berlin, Germany.

Duarte, F. J. (ed.). (1991). Dispersive dye lasers. In *High Power Dye Lasers*. Springer-Verlag, Berlin, Germany, Chapter 2.

Duarte, F. J. (1993). On a generalized interference equation and interferometric measurements. *Opt. Commun.* **103**, 8–14.

Duarte, F. J. (1995). Solid-sate dispersive dye laser oscillator: Very compact cavity. *Opt. Commun.* **117**, 480–484.

Duarte, F. J. (2003). *Tunable Laser Optics*, 1st edn. Elsevier Academic Press, New York.

Duarte, F. J. (2004). Comment on "Reflection, refraction, and multislit interference." *Eur. J. Phys.* **25**, L57–L58.

Duarte, F. J. (2007). Coherent electrically-excited organic semiconductors: Visibility of inter-ferograms and emission linewidth. *Opt. Lett.* **32**, 412–414.

Duarte, F. J. (2008). Coherent electrically excited organic semiconductors: Coherent or laser emission? *Appl. Phys. B* **90**, 101–108.

Duarte, F. J. (ed.). (2010). Electrically-pumped organic-semiconductor coherent emission: A review. In *Coherence and Ultrashort Pulsed Laser Emission*. Intech, Rijeka, Croatia.

Duarte, F. J. (2014). *Quantum Optics for Engineers*. CRC Press, New York.

Duarte, F. J., Liao, L. S., and Vaeth, K. M. (2005). Coherence characteristics of electrically excited tandem organic light-emitting diodes. *Opt. Lett.* **30**, 3072–3074.

Feynman, R. P. (1965). *The Feynman Lectures on Physics: Exercises*. Addison-Wesley, Reading, MA.

Feynman, R. P., Leighton, R. B., and Sand, M. (1965). *The Feynman Lectures on Physics*, Vol. III. Addison-Wesley, Reading, MA.

Fischer, A., Kullmer, R., and Demtröder, W. (1981). Computer controlled Fabry-Perot wave-meter. *Opt. Commun.* **39**, 277–282.

Gardner, J. L. (1985). Compact Fizeau wavemeter. *Appl. Opt.* **24**, 3570–3573.

Hanbury Brown, R., and Twiss, R. Q. (1956). A test of a new type of stellar interferometer on Sirius. *Nature* **178**, 1046–1048.

Konishi, N., Suzuki, T., Taira, Y., Kato, H., and Kasuya, T. (1981). High precision wavelength meter with Fabry-Perot optics. *Appl. Phys.* **25**, 311–316.

Meaburn, J. (1976). *Detection and Spectrometry of Faint Light*. Reidel, Boston, MA.

Michelson, A. A. (1927). *Studies in Optics*. University of Chicago, Chicago, IL.

Snyder, J. J. (1977). Fizeau wavelength meter. In *Laser Spectroscopy III* (Hall, J. L. and Carlsten, J. L., eds.). Springer-Verlag, Berlin, Germany, pp. 419–420.

Steel, W. H. (1967). *Interferometry*. Cambridge University Press, Cambridge.

Wallenstein, R., and Hänsch, T. W. (1974). Linear pressure tuning of a multielement dye laser spectrometer. *Appl. Opt.* **13**, 1625–1628.

CHAPTER 12

Born, M., and Wolf, E. (1999). *Principles of Optics*, 7th edn. Cambridge University Press, Cambridge.

Duarte, F. J. (1981). *Investigation of the Optically-Pumped Iodine Dimer Laser in the Visible*. Macquarie University, Sydney, Australia.

Duarte, F. J. (1983). Prism-grating system for laser wavelength measurements. *J. Phys. E: Sci. Instrum.* **16**, 599–601.

Duarte, F. J. (1985). Note on achromatic multiple-prism beam expanders. *Opt. Commun.* **53**, 259–262.

Duarte, F. J. (1990). Narrow-linewidth pulsed dye laser oscillators. In *Dye Laser Principles* (Duarte, F. J. and Hillman, L. W., eds.). Academic Press, New York, Chapter 4.

Duarte, F. J., and Piper, J. A. (1982). Dispersion theory of multiple-prism beam expander for pulsed dye lasers. *Opt. Commun.* **43**, 303–307.

Duarte, F. J., and Piper, J. A. (1983). Generalized prism dispersion theory. *Am. J. Phys.* **51**, 1132–1134.

Duarte, F. J., and Piper, J. A. (1984). Multi-pass dispersion theory of prismatic pulsed dye lasers. *Opt. Acta* **31**, 331–335.

Gerstenkorn, S., and Luc, P. (1978). *Atlas du Spectre d'Absorption de la Molecule d'Iode*. CNRS, Paris, France.

Maystre, D. (1980). Integral methods. In *Electromagnetic Theory of Gratings* (Petit, R., ed.). Springer-Verlag, Berlin, Germany, Chapter 3.

Meaburn, J. (1976). *Detection and Spectrometry of Faint Light*. Reidel, Boston, MA.

Morris, M. B., and McIlrath, T. J. (1979). Portable high-resolution laser monochromator-interferometer with multichannel electronic readout. *Appl. Opt.* **24**, 4145–4151.

Newton, I. (1704). *Opticks*. The Royal Society, London.

CHAPTER 13

Barnes, N. P. (1995a). Transition metal solid-state lasers. In *Tunable Lasers Handbook* (Duarte, F. J., ed.). Academic Press, New York, Chapter 6.

Barnes, N. P. (1995b). Optical parametric oscillators. In *Tunable Lasers Handbook* (Duarte, F. J., ed.). Academic Press, New York, Chapter 7.

Duarte, F. J. (2003). *Tunable Laser Optics*. Elsevier Academic Press, New York.

Duarte, F. J. (2009). Appendix on optical quantities and conversion units. In *Tunable Laser Applications* (Duarte, F. J., ed.), 2nd edn. CRC Press, New York, Chapter 15.

Duarte, F. J., Costela, A., Garcia-Moreno, I., and Sastre, R. (2000). Measurements of $\partial n/\partial T$ in solid-state dye-laser gain media. *Appl. Opt.* **39**, 6522–6523.

Duarte, F. J., and James, R. O. (2003). Tunable solid-state lasers incorporating dye-doped polymer-nanoparticle gain media. *Opt. Lett.* **28**, 2088–2090.

Herzberg, G. (1950). *Spectra of Diatomic Molecules*. Van Nostrand, New York.

Marple, D. T. F. (1964). Refractive index of ZnSe, ZnTe, and CdTe. *J. Appl. Phys.* **35**, 539–542.

Schott North America. (2001). *Optical Glass Data Sheets*. Schott, Duryea, PA.

Wolfe, W. L. (1978). Properties of optical materials. In *Handbook of Optics* (Driscoll, W. G. and Vaughan, W., eds.). McGraw-Hill, New York.

Index